GIS and RS: Practical Machine Learning Tools and Techniques

GIS and RS: Practical Machine Learning Tools and Techniques

Editor: Dilan Thomas

MURPHY & MOORE
www.murphy-moorepublishing.com

www.murphy-moorepublishing.com

⊛ MURPHY & MOORE

Cataloging-in-Publication Data

GIS and RS : practical machine learning tools and techniques / edited by Dilan Thomas.
     p. cm.
Includes bibliographical references and index.
ISBN 978-1-63987-745-4
1. Geographic information systems. 2. Remote sensing. 3. Geographic information systems--Equipment and supplies.
4. Remote sensing--Equipment and supplies. 5. Information storage and retrieval systems--Geography. 6. Space optics.
I. Thomas, Dilan.
G70.212 .G57 2023
910.285--dc23

Murphy & Moore Publishing
1 Rockefeller Plaza,
New York City,
NY 10020, USA

ISBN 978-1-63987-745-4

Contents

Permissions

List of Contributors

Index

Preface

Machine learning (ML) refers to an artificial intelligence (AI) technique that teaches computers to learn from experiences. The algorithms of ML utilize computational techniques to learn information directly from data rather than using a preconceived equation as a model. ML is divided into two main categories, which include supervised learning and unsupervised learning. Each of them has diverse uses in geographic information system (GIS) and remote sensing (RS). ML is a key component of spatial analysis in GIS. It is extremely helpful for analyzing data in a variety of domains, including processing of satellite images. ML tools are primarily used in the processing of remote sensing data for interpretation, filtering and prediction. This book unravels the recent studies on machine learning tools and techniques for GIS and RS. As machine learning is emerging at a rapid pace, its contents will help the readers understand the modern concepts and applications of the subject. The book will serve as a valuable source of reference for graduate and postgraduate students.

Significant researches are present in this book. Intensive efforts have been employed by authors to make this book an outstanding discourse. This book contains the enlightening chapters which have been written on the basis of significant researches done by the experts.

Finally, I would also like to thank all the members involved in this book for being a team and meeting all the deadlines for the submission of their respective works. I would also like to thank my friends and family for being supportive in my efforts.

Editor

Integrating Remote Sensing and Landscape Characteristics to Estimate Soil Salinity using Machine Learning Methods

Nan Wang [1], Jie Xue [1], Jie Peng [2], Asim Biswas [3], Yong He [4] and Zhou Shi [1,5,*]

[1] Institute of Agricultural Remote Sensing and Information Technology Application, College of Environmental and Resource Sciences, Zhejiang University, Hangzhou 310058, China; wangnanfree@zju.edu.cn (N.W.); xj2019@zju.edu.cn (J.X.)

[2] College of Plant Science, Tarim University, Alar 843300, China; 11414049@zju.edu.cn

[3] School of Environmental Sciences, University of Guelph, Guelph, ON N1G2W1, Canada; biswas@uoguelph.ca

[4] College of Biosystems Engineering and Food Science, Zhejiang University, Hangzhou 310058, China; yhe@zju.edu.cn

[5] Key Laboratory of Spectroscopy Sensing, Ministry of Agriculture, Hangzhou 310058, China

* Correspondence: shizhou@zju.edu.cn

Abstract: Soil salinization, one of the most severe global land degradation problems, leads to the loss of arable land and declines in crop yields. Monitoring the distribution of salinized soil and degree of salinization is critical for management, remediation, and utilization of salinized soil; however, there is a lack of thorough assessment of various data sources including remote sensing and landscape characteristics for estimating soil salinity in arid and semi-arid areas. The overall goal of this study was to develop a framework for estimating soil salinity in diverse landscapes by fusing information from satellite images, landscape characteristics, and appropriate machine learning models. To explore the spatial distribution of soil salinity in southern Xinjiang, China, as a case study, we obtained 151 soil samples in a field campaign, which were analyzed in laboratory for soil electrical conductivity. A total of 35 indices including remote sensing classifiers (11), terrain attributes (3), vegetation spectral indices (8), and salinity spectral indices (13) were calculated or derived and correlated with soil salinity. Nine were used to model and estimate soil salinity using four predictive modelling approaches: partial least squares regression (PLSR), convolutional neural network (CNN), support vector machine (SVM) learning, and random forest (RF). Testing datasets were divided into vegetation-covered and bare soil samples and were used for accuracy assessment. The RF model was the best regression model in this study, with $R^2 = 0.75$, and was most effective in revealing the spatial characteristics of salt distribution. Importance analysis and path modeling of independent variables indicated that environmental factors and soil salinity indices including digital elevation model (DEM), B10, and green atmospherically resistant vegetation index (GARI) showed the strongest contribution in soil salinity estimation. This showed a great promise in the measurement and monitoring of soil salinity in arid and semi-arid areas from the integration of remote sensing, landscape characteristics, and using machine learning model.

Keywords: soil salinity; remote sensing; machine learning; predictive mapping

1. Introduction

Soil salinization is a severe environmental hazard posing a considerable threat to global land degradation [1,2]. Soil salinity exerts negative impacts on ecosystem health, soil quality, and crop breeding and harvest, affecting approximately 20% of irrigated farmlands and agricultural ecosystems worldwide [3,4]. Soil salinity results from a complicated progression related to climate, groundwater, topography, and human activities [5,6]. For example, salt-induced degradation is more pronounced in semi-arid and arid regions. Human activities such as tillage in a natural environment characterized by low precipitation, high soil evaporation, and high groundwater level [7,8] make cultivated soils more vulnerable to serious salinization problems. The demand for natural resources and food from the increasing population require more land to be used for farming, including marginal, vulnerable, and already degraded lands such deserts and lands affected by salinity [9,10]. Therefore, careful monitoring, quantitative assessment and analysis, and mapping to reveal the temporal and spatial distribution of soil salinity have become pressing concerns for land management and reclamation of salinized soil [11,12].

Under complex human activities and geological cycles, soil salinization exhibits high spatial and temporal variability [2], monitoring of which can provide information for land management [12]. Traditional methods to detect and assess soil salinity require intensive regular field work and laboratory analyses, which are time- and cost-consuming [13]. Sampling in a large area is almost impractical for frequent experimentation [14,15]. To solve the problem, in situ measurements using proximal sensing electromagnetic (EM) induction instruments have been used to estimate soil apparent electrical conductivity (EC_a) and simplify field sampling work [16]. The measurement of soil salinity at individual points or at a field-scale rather than large spatial scales often limit the use of these proximal soil sensors [12,17,18]; although the estimation accuracy has been improved, it still requires a large amount of manpower. As an alternative, digital soil mapping (DSM) can be used to predict soil properties, with limited soil samples and environmental covariates derived from remote sensing and ancillary data. DSM helps in estimating the spatial distribution of soil salinity at a large scale, from either sparse or discrete samples [19,20]. Satellite remote sensing technology can provide substantial soil information, a salient advantage of broad spatial coverage and periodic measurements over traditional field surveys or even surveys with proximal soil sensing [12,21]. Salt in soil exhibits particular absorption features of the soil surface within the electromagnetic spectrum range, while non-saline soils show higher reflectance in visible and near-infrared wavelengths. This provides theoretical support for using multispectral sensors and hyperspectral data to estimate soil salinity [22,23]. Data collected from various multispectral sensors, such as the moderate resolution imaging spectroradiometer (MODIS), the Landsat series, IKONOS, the HuanJing series, and the GaoFen series have been applied to extract salinization information across the world [1].

As the process of salinization is affected by topography, vegetation cover, and soil moisture, the use of remote sensing images alone cannot indicate the distribution of salinity efficiently [24]. Surface salt can be sufficiently monitored from satellite imagery only if the soils are bare and dry. The topography and vegetation cover largely affect the migration and deposition of salt, and humid weather brings moisture to the soils and turns the surface colors darker in the imagery, which sharply decreases the accuracy of soil salinity detection [25]. Different covariates contribute differently to the formation of soil salt. Choosing the independent variables with strong explanatory properties of soil salinity to estimate soil salinity helps to improve the accuracy of the estimation and the speed of calculation. Correlation coefficients are used to measure the correlation between covariates and soil salinity, so as to evaluate whether the covariates can be used for modeling. In the process of modeling training datasets, the RF regression model can be used to obtain variable importance. The obtained RF variable importance is accuracy-based importance (MDA, mean decrease accuracy) [1]. In order to eliminate the influence of multicollinearity on interpretation, the partial least squares path modeling (PLS-PM) is used to evaluate the interaction of independent variables [26]. Evaluating the correlation between covariates to soil salinity and each other is beneficial to reveal the key indicators of soil salts formation or estimation. Various studies have successfully employed satellite remote sensing data and auxiliary data to reveal the distribution of soil salinity on the basis of the correlations between several indices derived from information on soil properties, environmental factors, spectral bands, and soil reflectance spectra at different spatial scales [16,23,27,28]. As radar has a robust capability for penetrating land cover in all weather, a new opportunity for estimating soil density may be presented by use of radar imagery, including Sentinel-1, Seasat SAR, JERS-1 SAR, and ERS-1/2 SAR [29].

Construction of models between covariates (e.g., salinity controlling factors) and dependent variables (e.g., soil salinity) can accurately estimate the distribution and extent of soil salinity. The models for estimating soil salinity can be divided into linear regression models and nonlinear regression models. Most linear models such as multiple factor regression (MRF), inverse density weighted (IDW) regression, and partial least squares regression (PLSR) have been applied to determine soil salinity [30]. However, most linear models show poor estimation accuracy in regions with high spatial variability in salinity [28]. Recently, strong momentum has been gained in soil salinity mapping using machine learning models such as support vector machine (SVM), cubist, and random forest (RF) [31,32]. The neural network (NN) models such as back propagation neural network (BPNN), multi-layer perceptron neural network (MLP-NN), and convolutional neural network (CNN) are gradually being used for salt estimation [33]. Different independent variables are used to model soil salinity, and appear in different estimation accuracies [34]. The formation of soil salinity is a complex process controlled by multiple factors. Thus, simple linear sum of effects of multiple factors may not reveal the actual situation, while nonlinear models can better fit the contributions of various factors of soil salinity.

The challenge of efficiently estimating soil salinity in regions with high spatial variability and diverse landscapes by machine learning models has recently gained attention of researchers [28,35]. A comprehensive assessment of contributions from multiple factors of soil salinization using machine learning models remains lacking. A thorough assessment of environmental covariates that are highly correlated with soil salinity should be done to quantify spatial and temporal variability of soil salinity in conjunction with adaptable machine learning models and appropriate remote sensing imageries. The overall goal of this research was to use machine learning methods to develop a framework in estimating soil salinity in diverse landscapes by fusing information from satellite images and landscape characteristics. Specifically, this research aimed to (a) explore suitable covariates to reduce interference by soil moisture, vegetation, and other factors on the remote sensing estimation of soil salinization; (b) estimate soil salinity by employing partial least squares regression (PLSR), convolutional neural network (CNN), support vector machine (SVM), and random forest (RF) models using factors derived from remote sensing imagery and landscape characteristics; (c) quantitatively estimate and map the soil salinization in diverse landscapes of southern Xinjiang, China, as a case study area with high accuracy; and (d) analyze and identify the important factors. For areas with diverse and inaccessible landscapes, these will provide a framework to improve soil salinity monitoring and mapping, which could improve land management and planning and assessment of reclamation activities.

2. Materials and Methods

2.1. Study Area

The study area is in the center of Aksu district, southern Xinjiang Province, northwestern China (40°41'–41°20' N, 80°42'–81°20' E) (Figure 1). It extends from east to west for 70 km and from north to south for 92 km. There is a north–south provincial highway (S215) running through the entire study area. The study area is a typical alluvial fan from northwest to southeast and experiences long daylight hours with sufficient solar thermal resources. The climate is a typical continental warm temperate arid climate, with a low average annual rainfall of 60 mm and a high average annual evaporation of 2000 mm [16]. The complexity of landform, arid climatic conditions, intense evapotranspiration, and high level of underground water contribute to the accumulation of soil salinity [36]. The principal type of land use is desert in the center, and recently cultivated land in the northwest and south of the study area. Despite the soil salinization in desert areas, there are still some halophytes growing in the northwest of the central part, including *Tamarix*, *Halocnemum strobilaceum*, *Halostachys caspica*, reeds, and *Alhagi sparsifolia* Shap. [28]. The phenomenon of salt accumulation in the surface layer of the soil is particularly severe in the dry season.

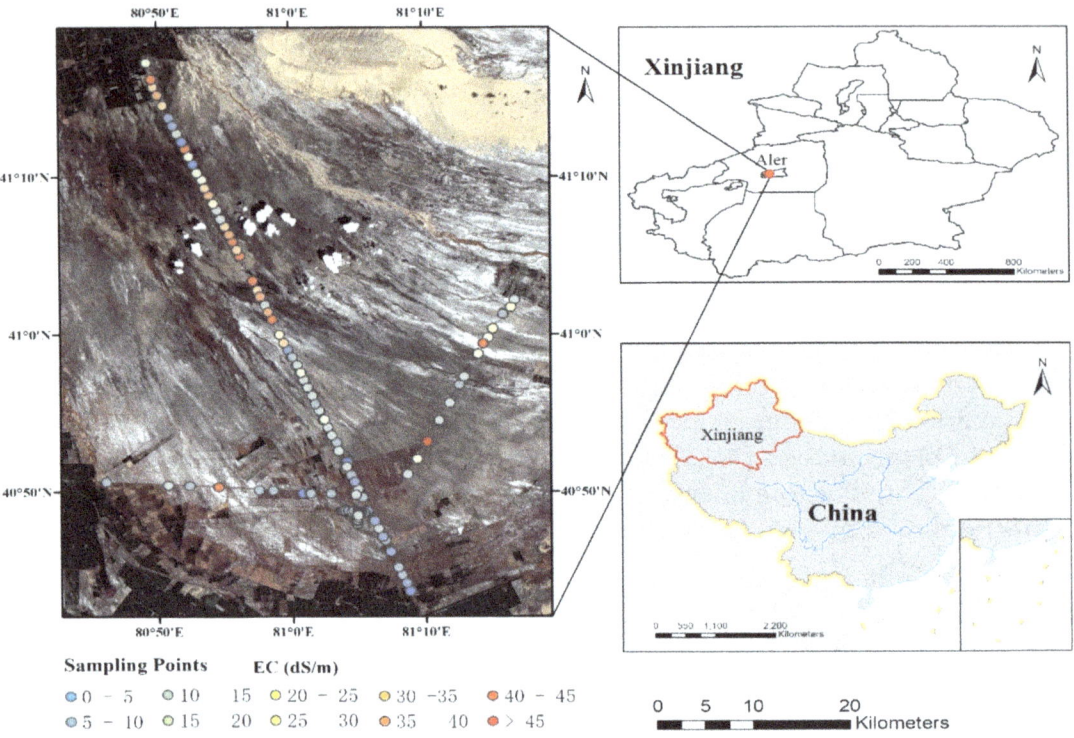

Figure 1. Geographical extent of the study area and the locations of sampling points. A total of 151 points were measured for soil electrical conductivity (EC) in two directions.

2.2. Soil Sampling and Laboratory Analysis

Soil samples were collected on 27 July 2017 along the S215 provincial highway (Figure 1). Along the sides of the road, sample plots of uniform landform of 30 × 30 m were chosen. The topographic features of the study area were uniform and flat, and the Quincunx sampling method [37] was used to collect 5 topsoil samples (0–20 cm). The Trimble Juno SB handheld GPS was used to record the latitude and longitude of each sampling point. Trimble Juno SB handheld GPS used differential positioning to characterize the geographic location of the point with accuracy of 2–5 m, which can be used for the scale of the study area. To prevent moisture evaporation, we placed samples in sealed plastic bags. Geolocations of the sampling points were imported and overlaid on the synchronized remote sensing image in ArcGIS to check the distribution. A total of 151 sampling points were sampled for further analysis. Soil samples from the same pixel were fully mixed and homogenized, and then they were subsampled to retain 300 g of soil for further laboratory analysis. Visually observable stones and leaves were first removed and then soil samples were air-dried, ground, and sieved to 2 mm size. Considering that the soil conductivity in the study area is relatively high, the soil leachate was obtained with a 1:5 soil/water ratio using a LeiCi DDS-307 (ShengKe, Shanghai, China) conductivity meter to measure the electrical conductivity.

2.3. Satellite Imagery and Preprocessing

Satellite imagery from Landsat-8 and Sentinel-1 were downloaded for the same date of field sampling. Landsat-8 was launched by the Landsat program on 11 February 2013 and comprises an operational land imager (OLI) and a thermal infrared sensor (TIRS). These two sensors monitor 11 bands

(Table 1) [38]. The OLI provides 8 bands at a spatial resolution of 30 m and a panchromatic band at 15 m, and the TIRS provides 2 thermal infrared bands at a spatial resolution of 100 m. The Sentinel-1 satellite is an earth observation satellite from the European Space Agency's Copernicus Project (GMES), and it was launched on 3 April 2014. It carries a C-band synthetic aperture radar, which can provide all-day images in various weather conditions. Sentinel-1A data was acquired in the interferometric wide (IW) mode C-band, with dual polarization vertical transmit vertical receive (VV)/vertical transmit horizontal receive (VH) at a spatial resolution of 20 m [39,40]. The orbit number of Sentinel-1A image covering the study area was 17635 (Table 2) [41,42].

There were a few clouds in the middle of the Landsat-8 image, and the "Fmask" tool was used to remove clouds [36]. To characterize landforms, we applied radiation correction and atmospheric correction using Fast Line-of-sight Atmospheric Analysis of Spectral Hypercubes (FLAASH). After applying orbital correction, thermal noise removal, radiation correction, Lee filtering, and range Doppler terrain correction to Sentinel-1 images, we converted radar backscattering coefficients of the two VV and VH bands from digital number (DN) to a decibel (dB) [32]. In addition, both Landsat-8 data and Sentinel-1 data were resampled to the same spatial resolution of 30 m using resample function in ArcGIS. After preprocessing, single bands were used as independent variables, or bandmath function was used to calculate vegetation spectral indices, salinity spectral indices, and backscattering coefficients from original bands.

Table 1. Landsat-8 specifications for bands. Bands 1 to 9 are operational land imager (OLI) spectral bands and bands 10 and 11 are thermal infrared sensor (TIRS) spectral bands.

Sensor	Spectral Band	Wavelength (μm)	Resolution (m)
OLI	Band 1—Coastal/Aerosol	0.433–0.453	30
	Band 2—Blue	0.450–0.515	30
	Band 3—Green	0.525–0.600	30
	Band 4—Red	0.630–0.680	30
	Band 5—Near Infrared (NIR)	0.845–0.885	30
	Band 6—Short Wavelength Infrared (SWIR)	1.560–1.600	30
	Band 7—Short Wavelength Infrared (SWIR)	2.100–2.300	30
	Band 8—Panchromatic	0.500–0.680	15
	Band 9—Cirrus	1.360–1.390	30
TRIS	Band 10—Long Wavelength Infrared	10.30–11.30	100
	Band 11—Long Wavelength Infrared	11.50–12.50	100

Table 2. Characteristics of Sentinel-1A data.

Parameter	Value
Pass direction	Ascending
Mode	Interferometric wide (IW)
Polarization	VH, VV
Temporal resolution	12 days
Spatial resolution	20 m
Wavelength	5.6 cm
Radiometric stability	0.5 dB (3σ)
Radiometric accuracy	1 dB (3σ)
Phase error	5°
Orbit number	17,635
Swath	250 km

2.4. Auxiliary Data

In most cases, single-band data and auxiliary data such as vegetation cover and landform characteristics or topographic factors were coupled to characterize the distribution of soil salinity [24,43]. Auxiliary data included remote sensing-based auxiliary data, radar-based auxiliary data, and DEM-based auxiliary data (Table 3), and was used for estimation. Remote sensing-based auxiliary data included vegetation spectral indices and salinity spectral indices. Halophytes can grow in saline soil in arid areas; thus, vegetation indices can also be used to characterize soil salinity. Previous studies used vegetation indices including normalized difference vegetation index (NDVI), generalized difference vegetation index (GDVI), green atmospherically resistant vegetation index (GARI), extended NDVI (ENDVI), and enhanced vegetation index (EVI) to indicate soil salinity by monitoring the halophytic properties of plants, and salinity indices including salinity indexes can be used to directly indicate the content of soil salts [28,44]. Radar-based auxiliary data consist of backscattering coefficients after removing the influence of vegetation [32,45]. To reveal the impact of vegetation on the accuracy of estimation, we calculated vegetation coverage (VFC) for the study area. VFC was used to identify vegetation covered (VFC > 45%) and bare soil pixels (VFC < 45%). A water cloud model was applied to remove the effect of vegetation water content (VWC) on backscattering coefficients [12,33]. A digital elevation model (DEM) at a spatial resolution of 30 m with terrain attributes including elevation, slope, and aspect derived from the DEM in ArcGIS were regarded as DEM-based auxiliary data. DEM data were obtained from the Shuttle Radar Topography Mission (SRTM; https://www2.jpl.nasa.gov/srtm/). In the study area, the DEM of each pixel was available, and the Landsat-8, sentinel-1A, and DEM data were all resampled to a spatial resolution of 30 m.

Table 3. Thirty-five indices including auxiliary data (remote sensing data, terrain attributes, vegetation spectral indices, salinity spectral indices, and radar data) used to estimate soil EC in this study along with their abbreviations, calculation formulae, and the reference of calculation.

Auxiliary Data	Land Surface Parameters	Abbreviation	Formulations	References
Remote sensing data	Bands 1–7	B1–B7		
	Band 10	B10		Landsat-8
	Band 11	B11		
Salinity spectral indices	Brightness index	BI	$[(B4)^2 + (B5)^2]^{0.5}$	[46]
	Salinity index	SI	$(B4 \times B2)^{0.5}$	[46]
	Salinity index 1	SI1	$(B4 \times B3)^{0.5}$	[33]
	Salinity index 3	SI3	$[(B4)^2 + (B3)^2]^{0.5}$	[33]
	Salinity index I	S1	$B2/B4$	[47]
	Salinity index II	S2	$(B2 - B4)/(B2 + B4)$	[47]
	Salinity index III	S3	$B3 \times B4/B2$	[48]
	Salinity index IV	S4	$B2 \times B4/B3$	[48]
	Salinity index V	S5	$B4 \times B5/B3$	[48]
	Salinity index VI	S6	$B6/B7$	[49]
	Salinity index VII	S7	$(B6 - B7)/(B6 + B7)$	[49]
	Salinity index VII	S8	$(B3 + B4)/2$	[47]
	Salinity index IX	S9	$(B3 + B4 + B5)/2$	[47]
	Soil salinity index	SI_T	$B4/B5 \times 100$	[50]
Vegetation spectral indices	Normalized difference vegetation index	NDVI	$(B5 - B4)/(B5 + B4)$	[51]
	Green atmospherically resistant vegetation index	GARI	$\{B5 - [B3 + \gamma \times (B2 - B4)]\}/\{B5 + [B3 + \gamma \times (B2 - B4)]\}$	[51]
	Extended NDVI	ENDVI	$(B5 + B7 - B4)/(B5 + B7 + B4)$	[36]
	Generalized difference vegetation index	GDVI	$(B5^2 - B4^2)/(B5^2 + B4^2)$	[52]
	Non-linear vegetation index	NLI	$(B5^2 - B4)/(B5^2 + B4)$	[53]
	Canopy response salinity index	CRSI	$(B5 \times B4 - B3 \times B2)/(B5 \times B4 + B3 \times B2)^{0.5}$	[52]
	Enhanced vegetation index	EVI	$g \times (B5 - B4)/(B5 + C1 \times B4 - C2 \times B2 + L)$	[54]
Terrain attributes	Elevation	DEM		[44]
	Slope	S		SAGA GIS
	Aspect	A		SAGA GIS
Radar data	Backscattering coefficients of VH band	VH	$(\sigma^0 - \sigma_{veg\_VH}^0)/L$	[46]
	Backscattering coefficients of VV band	VV	$(\sigma^0 - \sigma_{veg\_VV}^0)/L$	[46]

2.5. Modeling Strategies

In this study, the total soil samples were divided into 2 parts: 103 samples for training and 48 samples for testing. The soil samples were sorted from low to high according to the measured EC values, and 1 in every 3 samples were randomly selected to comprise the testing datasets. Training datasets and testing datasets were independent from each other. To choose the variables sharing significant influence on soil EC and to improve the efficiency of estimation, we applied factor correlation and significance analysis to filter independent variables. Suitable covariates were then employed to quantitatively reveal the spatial distribution of soil salinity. The PLSR, CNN, SVM, and RF models were used in this study.

PLSR is a typical linear regression model that integrates the advantages of principal component analysis into regression. It shows superiority in a situation in which the number of variables is very large with strong collinearity and noise [28]. Among previous investigations, PLSR has proven to be the most widely used linear regression model for estimating soil attributes [33,55]. The partial least squares regression model firstly extracts a principal component from the independent variable matrix and the dependent variable matrix. The principal components need to contain the variation information in their respective matrices as much as possible and maximize the correlation between the dependent variable components and the independent variable components. After the first component extraction, the regression of the 2 components to their respective source matrices was established, and the estimation accuracy was used to evaluate the results. If not, the residuals of the two variable matrices after the previous regression were used for another component extraction and regression until a satisfied accuracy was obtained [19]. Assuming a total of k rounds of component extraction and regression process, we finally obtained a regression model composed of k independent variable components [13,55]. In the modeling, the number of component extraction and regression processes ranged from 1 to 10, and the regression model with 6 extractions showed the best performance.

Support vector machine regression is a supervised machine learning method developed on the basis of statistical learning theory and can avoid overfitting [23,34]. SVM shows high estimation accuracy when modeling variables without collinearity. The principle of modeling is to find hyperplanes in high-dimensional space. It uses a nonlinear mapping algorithm to convert linearly inseparable samples into high-dimensional features using the principle of structural risk minimization (SRM) based on the Vapnik–Chervonenkis (VC) dimension in order to make them linearly separable [34,56]. For nonlinear models, the support vector machine regression model enables high-dimensional feature spaces to use linear algorithms to perform linear analysis on the nonlinear features of samples [23]. In SVM modeling, radial basis function was used as the kernel function. The cost range was set from 0.0001 to 0.1, and the gamma range was set from 1 to 1000. The best result was obtained when the cost was 10 and the gamma was 0.01.

RF is a flexible ensemble-learning algorithm based on decision trees. The basic idea of RF is to generate multiple independent decision trees using random samples and finally to yield one single estimation determined by each decision tree in the forest [16]. RF has an advantage of reducing the risk of overfitting by taking the average of decision trees. In addition, RF is relatively stable when faced with extreme values because they only affect one decision tree and are unlikely to affect the result. The parameters of the model were the number of trees (ntree) and the number of variables selected when each node is split (mtry). In order to obtain the best model, we selected the root mean square error of the corresponding model of different mtry, and then selected the mtry of the model that can obtain the smallest error value as the optimal number of variables. The parameters of the model were the number of trees (ntree) and the number of variables selected when splitting each node (mtry) [22,33]. In RF modeling, the number of trees is 500. In order to obtain the best model, we used mtry ranging from 1 to 20 to loop with the minimum root mean squared error used as the criterion. The best result was obtained when the number of variables at each split was 9, the node size was 12, and 10 features were randomly extracted at each node of each tree.

CNN is an efficient deep learning model widely used in computer vision and image classification [57]. CNN is developed on the basis of a neural network (NN) algorithm and features feedforward performance with deep structure. In recent research, NN has been efficiently applied to estimate soil salinity using satellite images, and CNN is also used for estimation [23]. It simulates the process of biological visual perception mechanism and can perform supervised learning and unsupervised learning. A CNN regression model typically consists of an input layer, a set of successive hidden layers including a convolutional layer, a pooling layer, and an output layer [58]. During the processing of activation function and the convolution kernel, the feature vectors of the input layer are calculated several times in the hidden layer, which is used to fit the regression relationship and display the estimation in the output layer. The procedure performs as a fully connected multilayer perceptron. The CNN model was framed with the tensorflow module. In the CNN modeling procedure, the one-dimensional data of 9 attributes was calculated into a 3×3 two-dimensional matrix; firstly, there were 4 hidden layers, and a 2×2 convolution kernel with ReLU as an activation function was used in the convolution layer. In each convolution, through the calculation of the 2×2 convolution kernel, each pixel was calculated to twice the original. Considering that there were only 9 soil properties used, the input matrix size was not large, the dimension of the input data was 9, and the dimension of the output data was 1. After 4 calculations, each pixel thickened to 256 pixels, and $3 \times 3 \times 256$ neurons were obtained, which produced three-dimensional data. Then, two fully connected layers were used to reduce the dimensionality of the data to one dimension to achieve regression. To prevent overfitting, the learning rate was initiated at 0.000005 and reduced to 0.98 of the former figure every 20,000 times. The probability parameter of the drop-out layer was 0.75, and a total of 200,000 iterations were performed. In addition, the GradientDescentOptimizer function was used in Tensorflow to achieve gradient descent. Momentum was not used, but the attenuation of the learning rate can also help improve training efficiency and accelerate to convergence.

2.6. Partial Least Squares Path Modeling

In the interpretation of the characteristics of the independent variables, multicollinearity of them will affect the importance of modeling, and thus partial least squares path modeling (PLS-PM) can be used to assess the interaction of independent variables [26]. A path analysis model can decompose the influence of independent variables on dependent variables into direct and indirect impacts, making the causality of variables more specific. The model of partial least squares (PLS) path modeling (PM) is a variance-based structural equation modeling (SEM) technique that is widely used to explain the connections between variables [59,60]. It can model causal paths between latent variables (LV) to explain an inner model, and the measured variables (MV) to explain the outer model [61]. In the outer model, the influence between LV and MV are quantified by weights (λ). In the inner model, the links between LV are quantified by path coefficients (β) to explain the connection coefficient (r) of independent variables and dependent variables [59].

2.7. Accuracy Assessment and Uncertainty Assessment

To evaluate the accuracy of statistical models for soil salt estimation, we adopted several accuracy indicators including R^2 (Equation (1)) and root mean squared error (RMSE) (Equation (2)) were adopted [21,62].

$$R^2 = 1 - \frac{\sum_{i=1}^{n} \left(Y_{testing,i} - Y_{model,i} \right)^2}{\sum_{i=1}^{n} \left(Y_{testing,i} - \overline{Y}_{testing} \right)^2} \tag{1}$$

$$RMSE = \sqrt{\frac{\sum_{i=1}^{n} \left(Y_{testing,i} - Y_{model,i} \right)^2}{n}} \tag{2}$$

where n is the number of samples, $Y_{testing,i}$ is the ith measured EC, and $Y_{model,i}$ is the ith simulated EC by each model.

To reduce the specificity of the particular training set selection, we used 50× non-parametric bootstrapping to measure the uncertainty associated with the estimation. Bootstrapping was performed with the bagging method to randomly extract data from the training dataset [13]. After each sample was collected, the sample was replaced, while un-extracted data were not used in the modeling set and were considered out-of-bag (OOB) [63]. In the training process of modeling, we used a fivefold cross-validation method to obtain the best parameters of the model with the minimum of root mean squared error (RMSE). For each testing process and each grid point, we constructed 50× PLSR, CNN, SVM, and RF models to calculate the estimated EC. Then, the average of 50 soil EC values from each regression model was obtained as the final result.

We used 90% confidence intervals (CIs) (Equation (3)) to indicate that the true value of soil EC value has 90% possibility within the interval between upper and lower CIs limits [64]. Moreover, the uncertainty (Equation (4)) was used to estimate the prediction.

$$CIs = \overline{Y} \pm a \times SD \tag{3}$$

$$Uncertainty = \frac{CI_{upper} - CI_{lower}}{\overline{Y}} \tag{4}$$

where $\overline{Y}$ is the average soil EC value of the 50× estimations. The number of a was 1.676 when the number of repetitions is 50, and the confidence interval is 90%. SD is the standard deviation of estimations. CI_{upper} and CI_{lower} are the lower and upper bounds of CIs.

3. Results

3.1. Descriptive Statistics for Estimated EC

Figure 2 shows the spatial distribution and violin plot of training and testing datasets. Randomly selected testing data were distributed at various locations of the sampling points in the study area. Table 4 presents descriptive statistics of EC for training datasets; testing datasets; and all samples under bare soil, vegetation cover, and the entire sampling group. The descriptive statistics of measured EC value of training datasets and the testing datasets were as close as possible so that the modeling could be applied to as much testing data as possible. There was a high variation in EC from 2.09 dS m^{-1} to 46.70 dS m^{-1}, both extreme values found in bare soil. The average EC of all samples was 19.45 dS m^{-1}, and the median was 16.67 dS m^{-1}. The standard deviation was 9.98, the kurtosis was 0.39, the skewness was 0.94, and the CV was 51%, indicating variability in measured EC that showed near normal distribution. Vegetation mostly covered sampling points in the northwest, and the ratio of vegetation-covered soil to bare soil across the sampling points was approximately 1:2. Overall, the measured EC values were used as a grading standard of soil salinity according to the natural soil salinization classification standard: non-saline soil, EC ≤ 7.5 dS m^{-1}; lightly salinized soil, 7.5 dS m^{-1} < EC ≤ 15 dS m^{-1}; moderately salinized soil, 15 dS m^{-1} < EC ≤ 30 dS m^{-1}; severe salinized soil, 30 dS m^{-1} < EC ≤ 60 dS m^{-1}; and saline soil, EC > 60 dS m^{-1} [65,66]. The EC values of soil samples were mainly in the moderate and severe salinity categories. However, the coefficient of variation of 51% indicated a high variation in the measured soil EC. Spatially, severe salinity was mainly observed in the middle of the study area, mostly in desert and a few areas covered with vegetation.

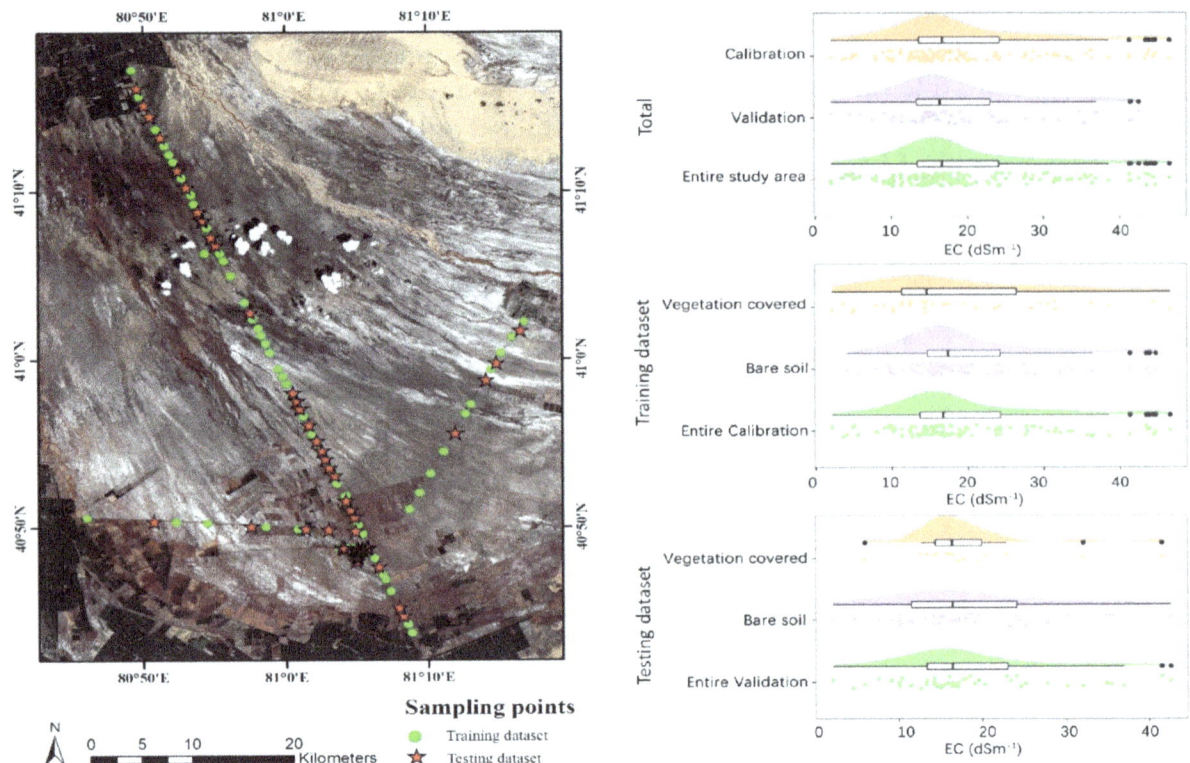

Figure 2. Spatial distribution and violin plot of training and testing datasets.

Table 4. Descriptive statistics of EC (dSm^{-1}).

	Datasets	Amount	Min	Max	Mean	Median	SD	Kutosis	Skewness	CV
Training	Bare soil	68	4.10	44.80	20.47	17.40	9.76	0.56	1.09	48%
	Vegetation covered	35	2.09	46.70	18.77	14.58	11.48	0.04	0.85	61%
	Entire datasets	103	2.09	46.70	19.89	16.73	10.35	0.27	0.93	52%
Testing	Bare soil	33	2.09	42.60	18.45	16.43	9.57	0.27	0.79	52%
	Vegetation covered	15	5.90	41.50	18.61	16.37	8.51	3.28	1.58	46%
	Entire datasets	48	2.09	42.60	18.50	16.40	9.16	0.68	0.94	50%
Total	Bare soil	101	2.09	44.80	19.81	16.73	9.70	0.46	0.98	49%
	Vegetation covered	50	2.09	46.70	18.72	15.71	10.59	0.45	0.96	57%
	Entire study area	151	2.09	46.70	19.45	16.67	9.98	0.39	0.94	51%

3.2. Selection of Independent Variables

Including all the remote sensing groups (11), terrain attributes (3), vegetation spectral indices (8), and salinity spectral indices (13), we assessed a total of 35 indices for significance and relationships to the measured EC values using Pearson correlation analysis [67]. Among them, relationships of 9 indices (B2, B5, SI, SI1, SI3, S8, S9, CRSI, S) were significant at $p < 0.05$, and 12 (B3, B6, B7, B10, B11, S1, S2, S6, GDVI, NLI, EVI, DEM) were significant at $p < 0.01$ probability level (Table 5). A total of 13 indices were not significantly related to EC. There were nine indices (S5, S6, GARI, SI_T, NDVI, GDVI, CRSI, A, VH) negatively correlated with EC; GDVI exhibited the strongest negative relationship with a correlation coefficient of −0.76. DEM exhibited the strongest positive relationship with a correlation coefficient of 0.41. Overall, B6, B7, B10, B11, S6, GARI, GDVI, EVI, and DEM were selected as independent variables to estimate soil salinity on the basis of R > 0.23. For chosen factors, there were four remote sensing indices, three vegetation spectral indices, one terrain attribute, and one salinity spectral index.

Table 5. Correlation coefficients between 35 indices and measured EC values.

Category	Factor	R	Category	Factor	R
	B1	0.15		S6	−0.44 **
	B2	0.19 *		S7	0.03
	B3	0.21 **	Salinity spectral indices	S8	0.19 *
	B4	0.14		S9	0.19 *
Remote sensing data	B5	0.18 *		SI_T	−0.01
	B6	0.26 **		NDVI	−0.01
	B7	0.23 **		GARI	−0.23 *
	B10	0.40 **		ENDVI	0.09
	B11	0.39 **	Vegetation spectral indices	GDVI	−0.76 **
	BI	0.14		NLI	0.22 **
	SI	0.18 *		CRSI	−0.21 *
	SI1	0.19 *		EVI	0.23 **
	SI3	0.18 *		DEM	0.41 **
	S1	0.22 **	Terrain attributes	S	0.17 *
Salinity spectral indices	S2	0.22 **		A	−0.06
	S3	0.14		VH	−0.03
	S4	0.14	Radar data	VV	0.01
	S5	−0.04			

** Significant at the 0.01 probability level. * Significant at the 0.05 probability level.

3.3. Evaluation of the Accuracy of Estimations

After selecting independent variables, we developed predictive models using PLSR, CNN, SVM, and RF to estimate soil salinity. The bootstrap sampling method was used 50 times to calculate the uncertainty, and a fivefold cross-validation method was used to obtain the best parameters of the model with the minimum of RMSE. During the testing process, the simulated soil EC was evaluated against the measured EC (Table 6). After parameter optimization and bootstrapping on the training datasets for each model, we found the average estimation to be $R^2 = 0.52$ for the PLSR model, $R^2 = 0.53$ for the CNN model, $R^2 = 0.68$ for the SVM model, and $R^2 = 0.75$ for the RF model. The accuracy of bootstrap, vegetation-covered dataset, and bare soil dataset is shown in Table 6. Comparing the accuracy of estimation among vegetation-covered, bare soil, and total samples, we found the results from vegetation-covered areas to show higher accuracy but greater dispersion than the testing datasets of total samples for PLSR, SVM, and RF models. The accuracy of bootstrap for each model was similar to the accuracy of the testing datasets, while the RF model was lower. Figure 3 shows the uncertainty value of each testing point in order to assess the uncertainty of modeling. The mean uncertainty value of CNN model was 0.03, which was the lowest when compared to PLSR (0.06), RF (0.06), and SVM (0.08). Figure 4 shows the scatterplots of measured and estimated EC values of the testing data. For PLSR, SVM, and RF models, the estimated EC values were close to the measured EC values when the soil salinity was lower than 25 dS m^{-1}. However, the estimated EC values were underestimated for high measured EC values. For CNN models, the estimated and measured values were distributed on both sides of the 1:1 line of the scatterplots with higher dispersion. Overall, modeling with RF resulted in the highest R^2 and the lowest MAE, while the PLSR model yielded the lowest R^2 and highest MAE. Thus, by comparing four models above, we concluded that the RF model is the best regression model.

Table 6. Accuracy comparison of partial least squares regression (PLSR), convolutional neural network (CNN), support vector machine (SVM), and random forest (RF) models.

Models	Total		Bootstrap		Vegetation Covered		Bare Soil	
	R^2	RMSE	R^2	RMSE	R^2	RMSE	R^2	RMSE
PLSR	0.52	7.32	0.42	8.28	0.58	6.77	0.51	7.60
CNN	0.53	6.44	0.54	7.98	0.51	6.13	0.54	6.58
SVM	0.68	7.53	0.52	7.88	0.73	6.86	0.67	7.85
RF	0.75	7.33	0.56	7.84	0.76	6.82	0.75	7.59

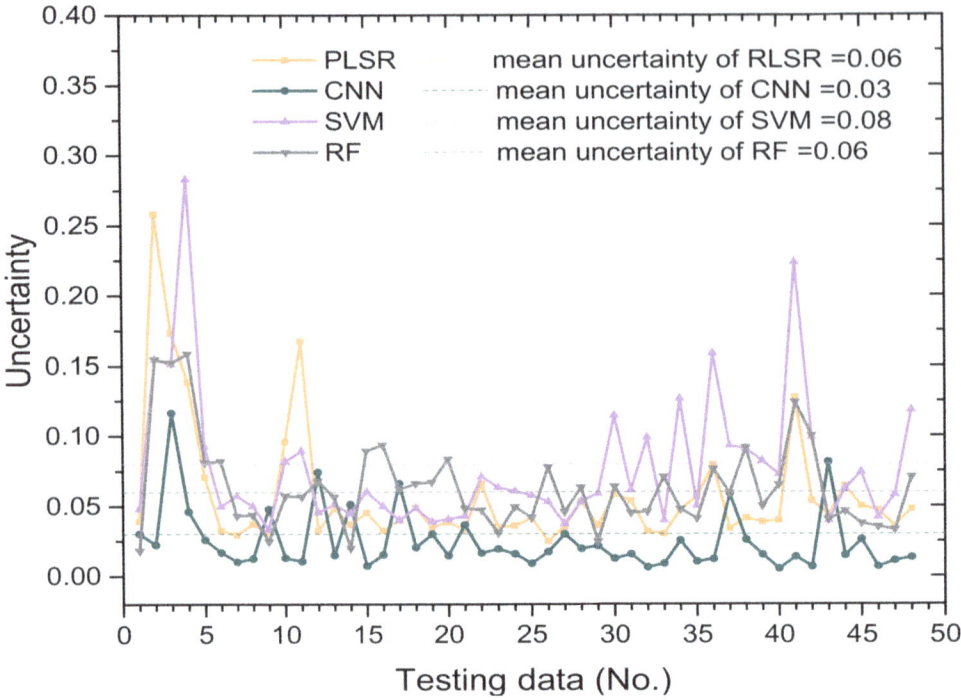

Figure 3. Uncertainty value of 48 testing points.

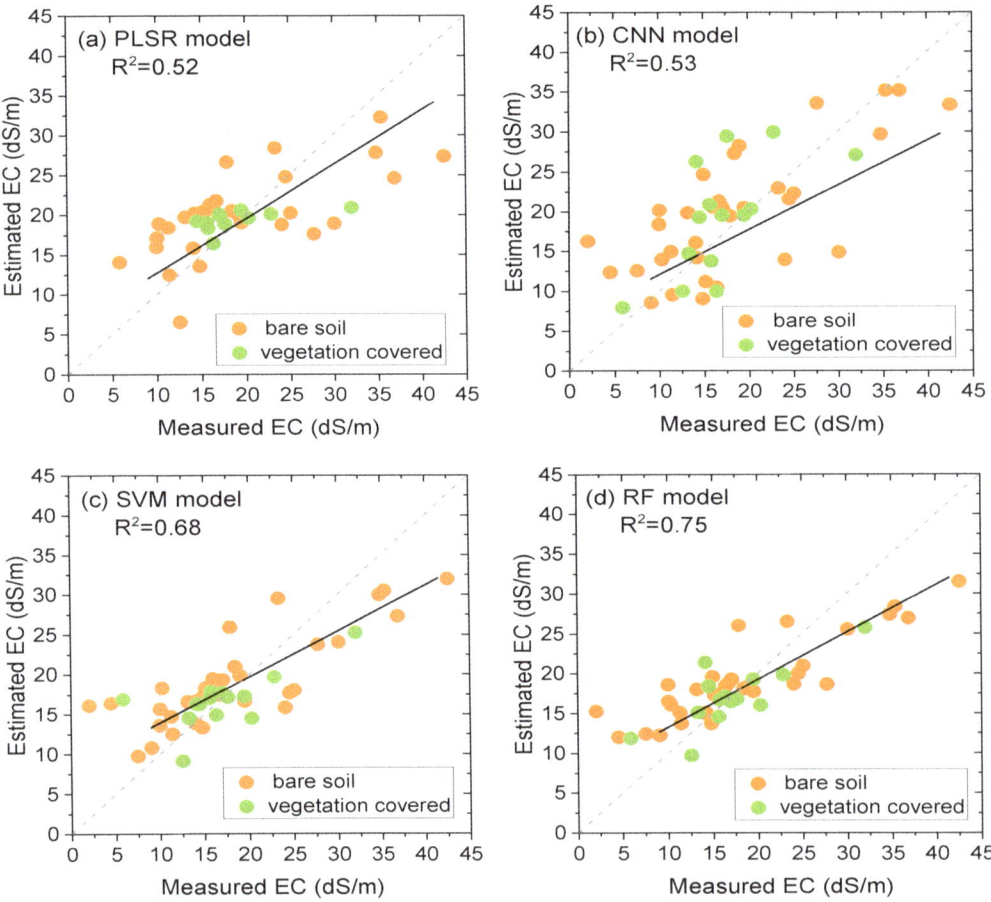

Figure 4. Performance of EC estimation by (**a**) PLSR model, (**b**) CNN model, (**c**) SVM model, and (**d**) RF model by soil samples in bare soil (orange) and vegetation covered (green).

3.4. Mapping of EC and Its Uncertainty

Finally, the RF model was employed to construct an EC map of the study area; after applying bootstrap on each pixel, we obtained 50 maps of soil EC. The average of the 50 maps was shown as the final map of the study area (Figure 5a). The estimated EC values varied from 7.19 to 38.3 dS m^{-1}. The northwest and the south of the study area exhibited relatively low soil salinity content. Generally, farmlands are located in these areas, and due to the farmland improvement policies and measurements, the EC values were low to make it suitable for growing crops. Figure 1 shows that the measured EC values of the soil samples in the southern part of the study area were lower than 15 dS m^{-1}; thus, the soil EC values calculated by the RF model were similar to the measured result. However, in the south of the study area, the EC values of some farmlands were within the moderately salinized soil class (15 dS m^{-1} < EC ≤ 30 dS m^{-1}). A high amount of salt was mainly concentrated in the northwest to the middle of the study area covering desert areas. Further, the EC values decreased from these areas to the south of the study area. From northwest to southeast of the study area, the EC values changed from low to high and then low again. Regardless of the existence of crops in the farmland, and comparing EC values to the distribution of halophytes, the highest estimated EC value appeared in the area where halophytes grew. In addition, the EC values of some vegetation-covered areas were in the moderately and severely salinized soil categories (30 dS m^{-1} < EC ≤ 60 dS m^{-1}), while in general, the EC values of the vegetation-covered areas were lower.

The spatial distribution of the uncertainty value of the estimation is shown in Figure 5b. Among all pixels in the study area, the uncertainty value ranged from 0.01 to 0.27. The uncertainty values were low (< 0.15) in the northeast to the middle of the study area because most of the sampling points were distributed there, while they were relatively higher (> 0.15) in northwest and south of the study area because of the low density of sampling points. As Figures 3 and 4 show, for the data with lower (< 10 dS m^{-1}) and higher (> 40 dS m^{-1}) soil EC value, the deviation between the estimated value and the measured value was larger, and the uncertainty was stronger. Therefore, as shown in Figure 5, the estimated value of soil salinity in the farmland in the northwest and south of the study area was lower, and the uncertainty value was higher.

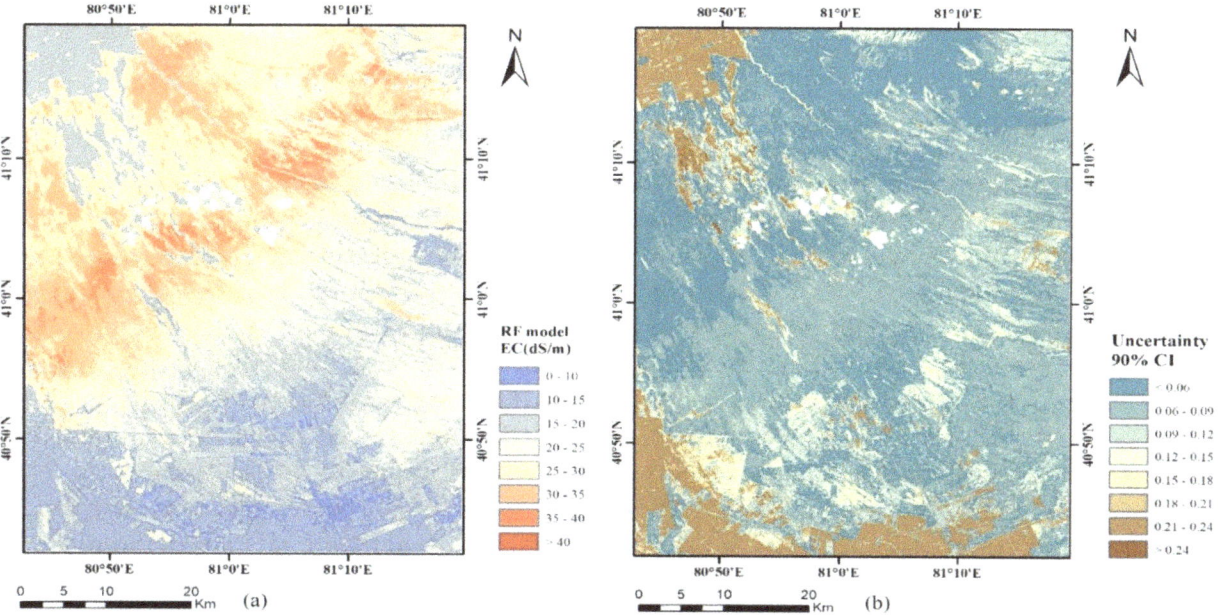

Figure 5. EC map derived from random forest model (**a**) and uncertainty map of soil EC values (**b**).

4. Discussion

4.1. Estimation Capabilities of Different Models

For uncertainty assessment (Figure 3), four models showed relatively high uncertainty in data where EC value was lower than 10 dS m^{-1} or higher than 40 dS m^{-1}. Most soil EC values of data were in the range of 10 dS m^{-1} to 25 dS m^{-1}, and thus the estimation in this interval was more accurate when modeling. The plot of mean uncertainty showed that the linear model (PLSR) had the highest uncertainty, followed by classical machine learning model (SVM) and tree-based machine learning model (RF), while the neural network model (CNN) had the lowest uncertainty. According to the estimation accuracy of the four models on the testing datasets (Table 6), the estimation capabilities were ranked as follows: RF regression > SVM regression > CNN regression > PLSR.

In general, the estimation accuracy of the nonlinear model was better than that of the linear model. In regions with high spatial variability in salinity, we found that social management, environmental factors, and geological conditions synergistically affect the accumulation of salt on the soil surface. Linear regression cannot accurately simulate this complex process, and thus the estimation ability of linear models was poor [28]. However, in the model building process, nonlinear regression showed the algorithm's ability to incorporate a large number of disparate predictors [14]. For the CNN model, it needed a large number of samples for training in the modeling process to obtain the best simulation. Due to the lack of training data, the training times cannot be set too much to prevent overfitting. Therefore, the CNN model did not show its advantages in the training of small samples [34]. Similar to other neural network models (e.g., MLP-NN) used in the latest research, NN methods were less flexible in terms of the parameterization and computational efficiency compared to RF model [34].

Due to the classification framework of decision trees, the RF model performed with the best estimation capabilities among the four models. In areas with large soil salt heterogeneity, each regression tree in the random forest was classified and regressed, which fully reflected the difference in salt distribution. For the RF model, the input of all independent variables and automatic screening by the machine was beneficial to regression, and all the independent variables (38) were put into the model for test. However, the results showed the best accuracy of R^2 was 0.46, which was much lower than the accuracy after filtering the independent variables. Thus, the RF model was regarded as the best model with nine indicators including B6, B7, B10, B11, S6, GARI, GDVI, EVI, and DEM. In building tree-structured regression models, different observations of variable selection biases towards variates with splits can seriously affect the results [17,22].

4.2. Sensitivity of Water and Vegetation Coverage

It can be seen from the modeling procedure that the contributions of different independent variables on estimating soil salinity were different. For example, the movement and accumulation of salts on the soil surface is controlled by ecological, geomorphic, climatic, and hydrologic factors. These factors affect the soil–water balance [25,68]. The water resources in the study area came from precipitation, irrigation, and rivers, and the sampling time in July was the wet season in Xinjiang. There are three weather stations around the study area: Aksu station in the northwest, Avati station in the southwest, and Alar station in the southeast. Intermittent precipitation was recorded from 22 July to 24 July in accordance with the hourly observation data of these weather stations. A precipitation of 0.1–1.6 mm per hour, which is considered light rain, was recorded within this time. Moreover, precipitation as high as 10.4 mm per hour occurred in Aksu station on 25 July. Due to the precipitation right before the sampling work, the surface salt might have leached into the deep soil by the rainwater and reduced salt content at the surface layer in comparison to the previous study [28]. The soil backscattering coefficient is affected by vegetation, surface roughness, and radar system parameters, and the vegetation scattering theory only shows its superiority in areas with homogeneous vegetation [69]. For arid and semiarid areas where vegetation cover is only partial, the contribution of soil to backscatter is generally smaller than that of the vegetation [70,71]. Due to the precipitation and the growth of halophytes, the soil moisture in the study area was generally higher than usual in the short term, and the variability was not significant in these places. The soil backscattering coefficient of Sentinel-1 was thus not highly relevant for modeling (Table 5), and the main factor affecting the soil–water balance was the presence of vegetation in this study [23,44].

For the testing datasets, the accuracy was intermediate between the vegetation-covered and the bare soil datasets for PLSR, SVM, and RF models (Table 6). Vegetation showed a strong negative correlation with soil salinity. In areas with high vegetation coverage, the use of vegetation spectral indices as independent variables improved estimation accuracy efficiently. For bare soil areas and areas with sparse vegetation coverage, the growth of vegetation and soil moisture had little effect on the formation of salts, resulting in poor estimation accuracy [16,28]. For the CNN model, repeated convolution on the dataset was performed with small samples, weakening the effect of a single independent variable on the dependent variable. Therefore, the sensitivity of EC to moisture and vegetation was not significant, and the dispersion of the estimated values was uniform (Figure 4b). As a result, in the dry season, the spectral indices, vegetation spectral indices, and soil moisture data can be used to characterize and distinguish saline and non-saline soils. In the wet season, however, precipitation affects the water–salt balance and causes the salt on the soil surface to penetrate into the deep soil in a short period of time. This leads to abnormal effects of soil moisture and vegetation on salinity estimation. Therefore, the dry season is more suitable for estimation of soil salinity than the wet season [11,31,50,72].

4.3. Interaction of the Independent Variables

Studying the contribution of independent variables to dependent variables and the interrelationship of independent variables can obtain the strongest factors affecting soil salinization. The correlation can provide a theoretical knowledge for subsequent research and improvement of the indices that characterizes soil salinity, which helps in soil salinity estimation, saline soil management, and farmland reclamation. To quantify the importance of the nine independent variables in RF modeling process, we used importance analysis and path modeling. We calculated 50× RF modeling, with the averaged importance of indicators being shown in Figure 6. DEM exhibited the highest importance of 0.45 for modeling, followed by B10 with modeling importance of 0.14 and GARI with modeling importance of 0.14. Independent variables were grouped according to their attributes. Terrain attributes exhibited the highest importance of 0.45. The remote sensing data had the second largest contribution with importance of 0.32, followed by vegetation spectral indices with importance of 0.21. The salinity spectral indices exhibited the least modeling importance of 0.089.

In order to eliminate the influence of multiple collinearity among the nine independent variables on the evaluation of the independent variable interaction, we grouped the independent variables into partial least squares regression path modeling [26]. The framework of path analysis was set first by determination of remote sensing band data on auxiliary data and then the inhibition by terrain attributes and salinity indices of plant growth [46]. The PLS-PM was then used to explain the interactions between groups of independent variables (Figure 7). It can evaluate the fitness and the predictive ability of the model [73]. The vegetation spectral indices and salinity spectral indices were calculated from the raw remotely sensed data, and thus the remote sensing data directly contributed to these two independent variable groups [74]. Terrain attributes with $r = 0.44$ contributed the most to estimated EC among the four independent variable groups (Figure 6) [75]. Terrain attributes were related to vegetation with $r = 0.70$ and with salinity spectral indices with $r = 0.32$. In the estimation of soil salinity, only the vegetation spectral indices had a negative influence. The roots of plants can absorb salt, and thus the salt content of the soil where vegetation grows was relatively low, with saline soil also inhibiting the growth of vegetation. The interaction of the three groups showed that DEM greatly affected the distribution of vegetation and the accumulation of salt, being a key factor in the estimation of salt in this study. In southern Xinjiang, DEM and vegetation coverage can be evaluated as important covariates to improve the equation that characterize soil salinity [76]. The relevance ranking of the four independent variable groups matched the modeling importance of the RF model, indicating that the RF model exhibited accurate modeling of independent variables.

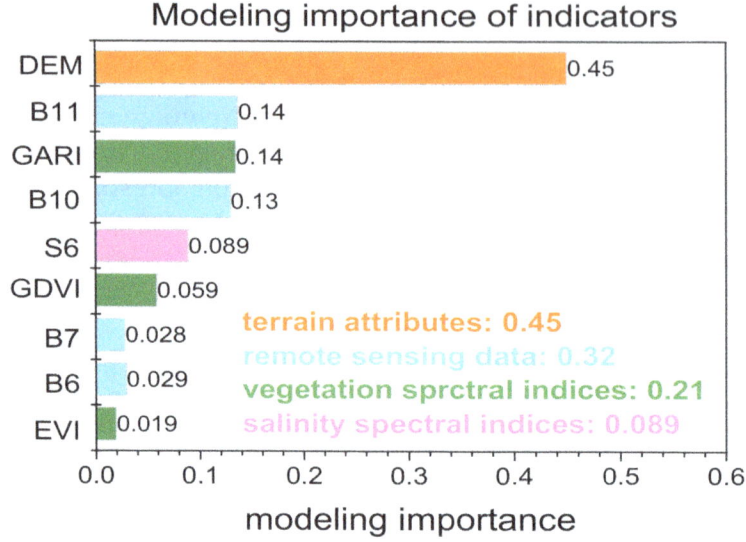

Figure 6. Importance of independent variables when modeling RF.

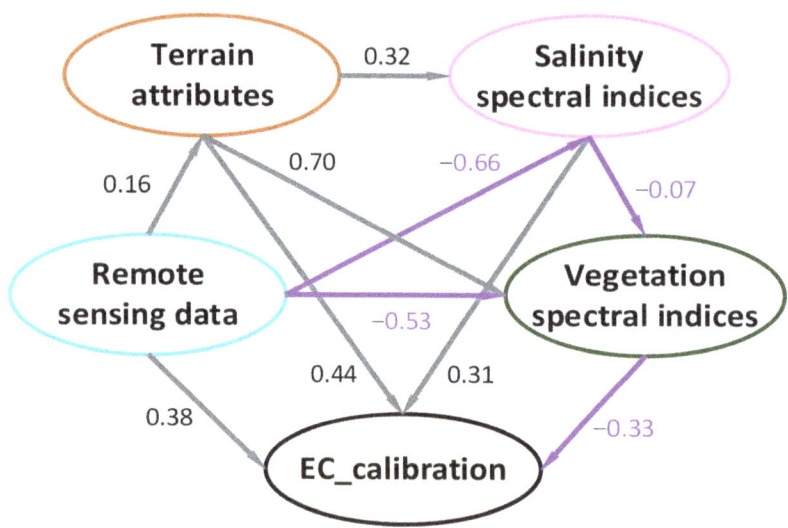

Figure 7. Path analysis of training dataset.

4.4. Effects of Human Activities on Soil Salinization

Figures 6 and 7 indicate that DEM was the factor with the greatest positive correlation to estimate soil salinity. In previous studies, population, cultivated area, and number of livestock were the main factors for differences in salinization in a basin [77]. In general, humans prefer to settle in low-lying areas near water. For example, a town is distributed along the Tarim River in the southern part of the study area. The farmlands in this part of the study area were reclaimed to make it more suitable for human life, leading to higher salt content in the soil outside the oasis regions [78]. The topography of the alluvial fan in oasis regions affected the distribution of soil salinity, and the water and soil nutrients accumulated in the middle and lower reaches. In addition, there was a large amount of artificially reclaimed farmland in the south of the study area. Therefore, vegetation grew in these areas and the soil salt content was reduced. During precipitation and irrigation, water accumulates in areas with lower elevation, and the groundwater level rises and brings salt to the surface of the soil. Thus, if there were no interference from human activities, the soil EC should show a low value in the northwest of the study area, the upper part of the alluvial fan [28]. Due to the reclamation in the southern portion of the study area, the soil salinity content decreased significantly compared with levels before the reclamation, which changed the original spatial distribution of soil salinity [16,28]. Soil EC showed lower values in the lower terrain, and there was a strong positive correlation between elevation and soil salinity (Table 5 and Figure 6). The farmland in the northwest was well cultivated, and the soil EC was relatively low compared to the soil around this area. The estimated soil EC value of the farmland in the northwest was found to be under 15 dS m^{-1}, while the estimated soil EC value of the surroundings was above 30 dS m^{-1}. Most of the farmland in the south of the study area was newly cultivated since 2010 but had not been completely improved, and thus the salt content was still higher than in the farmland in the northwest. This result shows farmland reclamation can partially de-salinize the soil with high salt content; therefore, the salinity of the top soil will drop sharply, making it suitable for crop planting.

5. Conclusions

In this study, nine independent variables involving four categories (remote sensing data, terrain attributes, salinity spectral indices, and vegetation spectral indices) and four models (PLSR, CNN, SVM, and RF) were selected to estimate soil salinity in Xinjiang. The main results are as follows:

1. Overall, among the 35 factors considered, 9 indices (B6, B7, B10, B11, S6, GARI, GDVI, EVI, and DEM) were significant and contributed to the estimation of soil salinity. The random forest regression model was the best model in this study, with better model performance and accuracy measures ($R^2 = 0.75$, RMSE = 7.33 dS m^{-1}).

2. According to the EC map derived from the random forest model, the northwest and the south of the study area showed relatively low soil salinity. Salt was mainly concentrated in the northwest to the middle of the study area covering the desert.

3. Farmland reclamation affected the distribution of soil salinity, resulting in a strong positive correlation between elevation and soil EC. Moreover, the variability of soil salinity was sensitive to moisture and vegetation. Dry season was more conducive for the estimation of soil salinity.

Author Contributions: All authors had substantial contributions to this article. Conceptualization, N.W., J.P., and Z.S.; data curation, J.X., J.P., and Y.H.; formal analysis, N.W., J.X., and A.B.; funding acquisition, J.P. and Z.S.; methodology, N.W., A.B., and Y.H.; software, N.W., J.X., and J.P.; supervision, Z.S.; visualization, N.W., J.X., and J.P.; writing—original draft, N.W., J.P., and A.B. All authors have read and agreed to the published version of the manuscript.

References

1. Wang, J.; Ding, J.; Yu, D.; Ma, X.; Zhang, Z.; Ge, X.; Teng, D.; Li, X.; Liang, J.; Lizaga, I.; et al. Capability of Sentinel-2 MSI data for monitoring and mapping of soil salinity in dry and wet seasons in the Ebinur Lake region, Xinjiang, China. *Geoderma* **2019**, *353*, 172–187. [CrossRef]

2. Ren, D.; Wei, B.; Xu, X.; Engel, B.; Li, G.; Huang, Q.; Xiong, Y.; Huang, G. Analyzing spatiotemporal characteristics of soil salinity in arid irrigated agro-ecosystems using integrated approaches. *Geoderma* **2019**, *356*, 113935. [CrossRef]

3. Gorji, T.; Sertel, E.; Tanik, A. Monitoring soil salinity via remote sensing technology under data scarce conditions: A case study from Turkey. *Ecol. Indic.* **2017**, *74*, 384–391. [CrossRef]

4. Wicke, B.; Smeets, E.; Dornburg, V.; Vashev, B.; Gaiser, T.; Turkenburg, W.; Faaij, A. The global technical and economic potential of bioenergy from salt-affected soils. *Energy Environ. Sci.* **2011**, *4*, 2669–2681. [CrossRef]

5. Ivushkin, K.; Bartholomeus, H.; Bregt, A.K.; Pulatov, A.; Kempen, B.; Sousa, L. Global mapping of soil salinity change. *Remote Sens. Environ.* **2019**, *231*, 111260. [CrossRef]

6. Gorji, T.; Yildirim, A.; Hamzehpour, N.; Tanik, A.; Sertel, E. Soil salinity analysis of Urmia Lake Basin using

Landsat-8 OLI and Sentinel-2A based spectral indices and electrical conductivity measurements. *Ecol. Indic.* **2020**, *112*, 106173. [CrossRef]

7.   Jiang, Q.; Peng, J.; Biswas, A.; Hu, J.; Zhao, R.; He, K.; Shi, Z. Characterising dryland salinity in three dimensions. *Sci. Total Environ.* **2019**, *682*, 190–199. [CrossRef]

8.   Ma, Z.; Zhou, L.; Yu, W.; Teng, H.; Shi, Z. Improving TMPA 3B43 V7 datasets using land surface characteristics and ground observations on the Qinghai-Tibet Plateau. *Int. J. IEEE Geosci. Remote. Sens. Lett.* **2018**, *15*, 178–182. [CrossRef]

9.   Shi, Z.; Cheng, J.; Huang, M.; Zhou, L. Assessing Reclamation Levels of Coastal Saline Lands with Integrated Stepwise Discriminant Analysis and Laboratory Hyperspectral Data. *Pedosphere* **2006**, *16*, 154–160. [CrossRef]

10. Peng, J.; Ji, W.; Ma, Z.; Li, S.; Chen, S.; Zhou, L.; Shi, Z. Predicting total dissolved salts and soluble ion concentrations in agricultural soils using portable visible near-infrared and mid-infrared spectrometers. *Biosyst. Eng.* **2016**, *152*, 94–103. [CrossRef]

11. Wu, W.; Muhaimeed, A.S.; Al-Shafie, W.M.; Al-Quraishi, A.M.F. Using L-band radar data for soil salinity mapping—a case study in Central Iraq. *Environ. Res. Commun.* **2019**, *1*, 081004. [CrossRef]

12. Davis, E.; Wang, C.; Dow, K. Comparing Sentinel-2 MSI and Landsat 8 OLI in soil salinity detection: A case study of agricultural lands in coastal North Carolina. *Int. J. Remote Sens.* **2019**, *40*, 6134–6153. [CrossRef]

13. Wang, J.; Ding, J.; Abulimiti, A.; Cai, L. Quantitative estimation of soil salinity by means of different modeling methods and visible-near infrared (VIS-NIR) spectroscopy, Ebinur Lake Wetland, Northwest China. *PeerJ* **2018**, *6*, e4703. [CrossRef] [PubMed]

14. Vermeulen, D.; Van Niekerk, A. Machine learning performance for predicting soil salinity using different combinations of geomorphometric covariates. *Geoderma* **2017**, *299*, 1–12. [CrossRef]

15. Dakak, H.; Huang, J.; Zouahri, A.; Douaik, A.; Triantafilis, J. Mapping soil salinity in 3-dimensions using an EM38 and EM4Soil inversion modelling at the reconnaissance scale in central Morocco. *Soil Use Manag.* **2017**, *33*, 553–567. [CrossRef]

16. Hu, J.; Peng, J.; Zhou, Y.; Xu, D.; Zhao, R.; Jiang, Q.; Fu, T.; Shi, Z. Quantitative Estimation of Soil Salinity Using UAV-Borne Hyperspectral and Satellite Multispectral Images. *Remote Sens.* **2019**, *11*, 736. [CrossRef]

17. Wang, X.; Zhang, F.; Ding, J.; Kung, H.; Latif, A.; Johnson, V.C. Estimation of soil salt content (SSC) in the Ebinur Lake Wetland National Nature Reserve (ELWNNR), Northwest China, based on a Bootstrap-BP neural network model and optimal spectral indices. *Sci. Total Environ.* **2018**, *615*, 918–930. [CrossRef]

18. Mulder, V.L.; de Bruin, S.; Schaepman, M.E.; Mayr, T.R. The use of remote sensing in soil and terrain mapping—A review. *Geoderma* **2011**, *162*, 1–19. [CrossRef]

19. McBratney, A.B.; Santos, M.L.M.; Minasny, B. On digital soil mapping. *Geoderma* **2003**, *117*, 3–52. [CrossRef]

20. Jenny, H. *Factors of Soil Formation*; McGraw-Hill: New York, NY, USA, 1941.

21. Bai, L.; Wang, C.; Zang, S.; Zang, S.; Wu, C.; Luo, J.; Wu, Y. Mapping Soil Alkalinity and Salinity in Northern Songnen Plain, China with the HJ-1 Hyperspectral Imager Data and Partial Least Squares Regression. *Sensors* **2018**, *18*, 3855. [CrossRef]

22. Sultanov, M.; Ibrakhimov, M.; Akramkhanov, A.; Bauer, C.; Conrad, C. Modelling End-of-Season Soil Salinity in Irrigated Agriculture Through Multi-temporal Optical Remote Sensing, Environmental Parameters, and In Situ Information. *PFG-J. Photogramm. Remote Sens. Geoinf. Sci.* **2019**, *86*, 221–233. [CrossRef]

23. Hoa, P.V.; Giang, N.V.; Binh, N.A.; Hai, L.V.H.; Phan, T.; Hasanlou, M.; Bui, D.T. Soil Salinity Mapping Using SAR Sentinel-1 Data and Advanced Machine Learning Algorithms: A Case Study at Ben Tre Province of the Mekong River Delta (Vietnam). *Remote Sens.* **2019**, *11*, 128. [CrossRef]

24. Fathizad, H.; Ali, H.A.M.; Sodaiezadeh, H.; Kerry, R.; Taghizadeh-Mehrjardi, R. Investigation of the spatial and temporal variation of soil salinity using random forests in the central desert of Iran. *Geoderma* **2020**, *365*, 114233. [CrossRef]

25. Masoud, A.A.; Koike, K.; Atwia, M.G.; El-Horiny, M.M.; Gemail, K.S. Mapping soil salinity using spectral mixture analysis of landsat 8 OLI images to identify factors influencing salinization in an arid region. *Int. J. Appl. Earth Obs.* **2019**, *83*, 101944. [CrossRef]

26. Sanches, F.L.F.; Fernandes, A.C.P.; Ferreira, A.R.L.; Cortes, R.M.V.; Pacheco, F.A.L. A partial least squares - Path modeling analysis for the understanding of biodiversity loss in rural and urban watersheds in Portugal. *Sci. Total Environ.* **2018**, *626*, 1069–1085. [CrossRef]

27. Ma, L.; Ma, F.; Li, J.; Ge, J.; Yang, S.; Wu, D.; Feng, J.; Ding, J. Characterizing and modeling regional-scale variations in soil salinity in the arid oasis of Tarim Basin, China. *Geoderma* **2017** *305*, 1–11. [CrossRef]

28. Peng, J.; Biswas, A.; Jiang, Q.; Zhao, R.; Hu, J.; Hu, B.; Shi, Z. Estimating soil salinity from remote sensing and terrain data in southern Xinjiang Province, China. *Geoderma* **2019**, *337*, 1309–1319. [CrossRef]

29. Yang, R.M.; Guo, W.W. Using Sentinel-1 Imagery for Soil Salinity Prediction Under the Condition of Coastal Restoration. *IEEE J. Sel. Top. Appl. Earth Obs. Remote Sens.* **2019**, *12*, 1482–1488. [CrossRef]

30. Chi, Y.; Sun, J.; Liu, W.; Wang, J.; Zhao, M. Mapping coastal wetland soil salinity in different seasons using an improved comprehensive land surface factor system. *Ecol. Indic.* **2017**, *74*, 384–391. [CrossRef]

31. Yu, H.; Liu, M.; Du, B.; Wang, Z.; Hu, L.; Zhang, B. Mapping Soil Salinity/Sodicity by using Landsat OLI Imagery and PLSR Algorithm over Semiarid West Jilin Province, China. *Sensors* **2019**, *107*, 105517. [CrossRef]

32. Wu, W.; Zucca, C.; Muhaimeed, A.S.; Ziadat, F.; Nangia, V.; Payne, W.B. Soil salinity prediction and mapping by machine learning regression in Central Mesopotamia, Iraq. *Land Degrad. Dev.* **2018**, *29*, 4005–4014. [CrossRef]

33. Liu, Y.; Zhang, F.; Wang, C.; Wu, S.; Liu, J.; Xu, A.; Pan, K.; Pan, X. Estimating the soil salinity over partially vegetated surfaces from multispectral remote sensing image using non-negative matrix factorization. *Geoderma* **2019**, *354*, 113887. [CrossRef]

34. Wang, F.; Shi, Z.; Biswas, A.; Yang, S.; Ding, J. Multi-algorithm comparison for predicting soil salinity. *Geoderma* **2020**, *365*, 114211. [CrossRef]

35. Nouri, H.; Borujeni, S.C.; Alaghmand, S.; Anderson, S.J.; Sutton, P.C.; Parvazian, S.; Beecham, S. Soil Salinity Mapping of Urban Greenery Using Remote Sensing and Proximal Sensing Techniques; The Case of Veale Gardens within the Adelaide Parklands. *Sustainability* **2018**, *10*, 2826. [CrossRef]

36. Wang, J.; Ding, J.; Yu, D.; Teng, D.; He, B.; Chen, X.; Ge, X.; Zhang, Z.; Wang, Y.; Yang, X.; et al. Machine learning-based detection of soil salinity in an arid desert region, Northwest China: A comparison between Landsat-8 OLI and Sentinel-2 MSI. *Sci. Total Environ.* **2020**, *707*, 136092. [CrossRef]

37. Cheng, J.L.; Shi, Z.; Zhu, Y.W. Assessment and mapping of environmental quality in agricultural soils of Zhejiang Province, China. *J. Environ. Sci.* **2007**, *19*, 50–54. [CrossRef]

38. Sabri, E.M.; Boukdir, A.; Karaoui, I.; Arioua, A.; Messlouhi, R.; Idrissi, A.E.A. Modelling soil salinity in Oued El Abid watershed, Morocco. *E3S Web Conf.* **2018**, *37*, 04002. [CrossRef]

39. Taghadosi, M.M.; Hasanlou, M.; Eftekhari, K. Soil salinity mapping using dual-polarized SAR Sentinel-1 imagery. *Int. J. Remote Sens.* **2018**, *40*, 237–252. [CrossRef]

40. Zhu, L.; Walker, J.P.; Tsang, L.; Huang, H.; Ye, N.; Rudiger, C. Soil moisture retrieval from time series multi-angular radar data using a dry down constraint. *Remote Sens. Environ.* **2019**, *231*, 111237. [CrossRef]

41. Alifu, H.; Vuillaume, J.F.; Johnson, B.A.; Hirabayashi, Y. Machine-learning classification of debris-covered glaciers using a combination of Sentinel-1/-2 (SAR/optical), Landsat 8 (thermal) and digital elevation data. *Geomorphology* **2020**, *369*, 1–18. [CrossRef]

42. Qiu, S.; Zhu, Z.; He, B. Fmask 4.0: Improved cloud and cloud shadow detection in Landsats 4–8 and Sentinel-2 imagery. *Remote Sens. Environ.* **2019**, *231*, 111205. [CrossRef]

43. Muller, E.; DeÂcamps, H. Modeling soil moisture-reflectance. *Remote Sens. Environ.* **2000**, *76*, 173–180. [CrossRef]

44. Taghizadeh-Mehrjardi, R.; Minasny, B.; Sarmadian, F.; Malone, B.P. Digital mapping of soil salinity in Ardakan region, central Iran. *Geoderma* **2014**, *213*, 15–28. [CrossRef]

45. Huete, A.; Didan, K.; Miura, T.; Rodriguez, E.P.; Gao, X.; Ferreira, L.G. Overview of the radiometric and biophysical performance of the MODIS vegetation indices. *Remote Sens. Environ.* **2002**, *83*, 195–213. [CrossRef]

46. Kumar, K.; Hari, P.K.S.; Arora, M.K. Estimation of water cloud model vegetation parameters using a genetic algorithm. *Hydrol. Sci. J.* **2012**, *57*, 776–789. [CrossRef]

47. Douaoui, A.E.K.; Nicolas, H.; Walter, C. Detecting salinity hazards within a semiarid context by means of combining soil and remote-sensing data. *Geoderma* **2006**, *134*, 217–230. [CrossRef]

48. Abbas, A.; Khan, S. Using Remote Sensing Techniques for Appraisal of Irrigated Soil Salinity. In *Proceedings of International Congress on Modelling and Simulation*; Oxley, L., Kulasiri, D., Eds.; Modelling & Simulation Soc Australia & New Zealand Inc.: Christchurch, New Zealand, 2007; pp. 2632–2638.

49. Bannari, A.; Guedon, A.M.; El-Harti, A.; Cherkaoui, F.Z.; El-Chmari, A. Characterization of Slightly and Moderately Saline and Sodic Soils in Irrigated Agricultural Land using Simulated Data of Advanced Land Imaging (EO-1) Sensor. *Commun. Soil Sci. Plant Anal.* **2008**, *39*, 2795–2811. [CrossRef]

50. Alexakis, D.D.; Daliakopoulos, I.N.; Panagea, I.S.; Tsanis, I.K. Assessing soil salinity using WorldView-2 multispectral images in Timpaki, Crete, Greece. *Geocarto Int.* **2016**, *33*, 321–338. [CrossRef]

51. Gitelson, A.; Kaufman, Y.; Merzlyak, M. Use of a Green Channel in Remote Sensing of Global Vegetation from EOS-MODIS. *Remote Sens. Environ.* **1996**, *58*, 289–298. [CrossRef]

52. Wu, W.; Al-Shafie, W.M.; Mhaimeed, A.S.; Ziadat, F.; Nangia, V.; Payne, W.B. Soil Salinity Mapping by Multiscale Remote Sensing in Mesopotamia, Iraq. *IEEE J. Sel. Top. Appl. Earth Obs. Remote Sens.* **2014**, *7*, 4442–4452. [CrossRef]

53. Goel, N.S.; Qin, W. Influences of canopy architecture on relationships between various vegetation indices and LAI and Fpar: A computer simulation. *Remote Sens. Rev.* **1994**, *10*, 309–347. [CrossRef]

54. Scudiero, E.; Skaggs, T.H.; Corwin, D.L. Regional scale soil salinity evaluation using Landsat 7, western San Joaquin Valley, California, USA. *Geoderma Reg.* **2014**, *2–3*, 82–90. [CrossRef]

55. Zhang, X.; Huang, B. Prediction of soil salinity with soil-reflected spectra: A comparison of two regression methods. *Sci. Rep.* **2019**, *9*, 5067–5075. [CrossRef] [PubMed]

56. Nurmemet, I.; Sagan, V.; Ding, J.; Halik, U.; Abliz, A.; Yakup, Z. A WFS-SVM Model for Soil Salinity Mapping in Keriya Oasis, Northwestern China Using Polarimetric Decomposition and Fully PolSAR Data. *Remote Sens.* **2018**, *10*, 598. [CrossRef]

57. Ma, X.; Dai, Z.; He, Z.; Ma, J.; Wang, Y.; Wang, Y. Learning Traffic as Images: A Deep Convolutional Neural Network for Large-Scale Transportation Network Speed Prediction. *Sensors* **2017**, *17*, 818. [CrossRef] [PubMed]

58. Alonso-Monsalve, S.; Suárez-Cetrulo, A.L.; Cervantes, A.; Quintana, D. Convolution on neural networks for high-frequency trend prediction of cryptocurrency exchange rates using technical indicators. *Expert Syst. Appl.* **2020**, *149*, 113250–113264. [CrossRef]

59. Henseler, J.; Hubona, G.; Ray, P.A. Using PLS path modeling in new technology research: Updated guidelines. *Ind. Manag. Data Syst.* **2016**, *116*, 2–20. [CrossRef]

60. Danks, N.; Sharma, P.; Sarstedt, M. Model selection uncertainty and multimodel inference in partial least squares structural equation modeling (PLS-SEM). *J. Bus. Res.* **2020**, *113*, 13–24. [CrossRef]

61. Oliveira, C.F.; do Valle Junior, R.F.; Valera, C.A.; Rodrigues, V.S.; Fernandes, L.F.S.; Pacheco, F.A.L. The modeling of pasture conservation and of its impact on stream water quality using Partial Least Squares-Path Modeling. *Sci. Total Environ.* **2019**, *697*, 134081. [CrossRef]

62. Huang, J.; Wu, J.; Fang, Y.; Wu, J.; Huang, J. Comparison of partial least square regression, support vector machine, and deep-learning techniques for estimating soil salinity from hyperspectral data. *J. Appl. Remote Sens.* **2018**, *12*, 022204.

63. Pan, L.; Politis, D.N. Bootstrap prediction intervals for linear, nonlinear and nonparametric autoregressions. *J. Stat. Plan. Inference* **2016**, *177*, 1–27. [CrossRef]

64. Zhou, Y.; Xue, J.; Chen, S.; Zhou, Y.; Liang, Z.; Wang, N.; Shi, Z. Fine-Resolution Mapping of Soil Total Nitrogen across China Based on Weighted Model Averaging. *Remote Sens.* **2020**, *12*, 85. [CrossRef]

65. Fernández-Buces, N.; Siebe, C.; Cram, S.; Palacio, J.L. Mapping soil salinity using a combined spectral response index for bare soil and vegetation: A case study in the former lake Texcoco, Mexico. *J. Arid Environ.* **2006**, *65*, 644–667. [CrossRef]

66. Zhang, T.; Qi, J.; Gao, Y.; Ouyang, Z.; Zeng, S.; Zhao, B. Detecting soil salinity with MODIS time series VI data. *Ecol. Indic.* **2015**, *52*, 480–489. [CrossRef]

67. Shahabi, M.; Jafarzadeh, A.A.; Neyshabouri, M.R.; Ghorbani, M.A.; Kamran, K.V. Spatial modeling of soil salinity using multiple linear regression, ordinary kriging and artificial neural network methods. *Arch. Agron. Soil Sci.* **2016**, *63*, 151–160. [CrossRef]

68. Ma, L.; Yang, S.; Simayi, Z.; Gu, Q.; Li, J.; Yang, X.; Ding, J. Modeling variations in soil salinity in the oasis of Junggar Basin, China. *Land Degrad. Dev.* **2018**, *29*, 551–562. [CrossRef]

69. Patel, N.R.; Mukund, A.; Parida, B.R. Satellite-derived vegetation temperature condition index to infer root zone soil moisture in semi-arid province of Rajasthan, India. *Geocarto Int.* **2019**, 1–17. [CrossRef]

70. Hajj, M.E.; Baghdadi, N.; Zribi, M.; Belaud, G.; Cheviron, B.; Courault, D.; Charron, F. Soil moisture retrieval over irrigated grassland using X-band SAR data. *Remote Sens. Environ.* **2016**, *176*, 202–218. [CrossRef]

71. Zhang, L.; Meng, Q.; Yao, S.; Wang, Q.; Zeng, J.; Zhao, S.; Ma, J. Soil Moisture Retrieval from the Chinese GF-3 Satellite and Optical Data over Agricultural Fields. *Sensors* **2018**, *18*, 2675. [CrossRef]

72. Huang, J.; Koganti, T.; Santos, F.A.M.; Triantafilis, J. Mapping soil salinity and a fresh-water intrusion in three-dimensions using a quasi-3d joint-inversion of DUALEM-421S and EM34 data. *Sci. Total Environ.* **2017**, *577*, 395–404. [CrossRef]

73. Schuberth, F.; Rademaker, M.E.; Henseler, J. Estimating and assessing second-order constructs using PLS-PM: The case of composites of composites. *Ind. Manag. Data Syst.* **2020**, *120*, 2211–2241. [CrossRef]

74. Racetin, I.; Krtalic, A.; Srzic, V.; Zovko, M. Characterization of short-term salinity fluctuations in the Neretva River Delta situated in the southern Adriatic Croatia using Landsat-5 TM. *Ecol. Indic.* **2020**, *110*, 105924. [CrossRef]

75. Thiam, S.; Villamor, G.B.; Kyei-Baffour, N.; Matty, F. Soil salinity assessment and coping strategies in the coastal agricultural landscape in Djilor district, Senegal. *Land Use Policy* **2019**, *88*, 104191. [CrossRef]

76. Elia, S.; Todd, H.S.; Dennis, L.C. Regional-scale soil salinity assessment using Landsat ETM+ canopy reflectance. *Remote Sens. Environ.* **2015**, *169*, 335–343.

77. Xu, H.; Chen, C.; Zheng, H.; Luo, G.; Yang, L.; Wang, W.; Wu, S.; Ding, J. AGA-SVR-based selection of feature subsets and optimization of parameter in regional soil salinization monitoring. *Int. J. Remote Sens.* **2020**, *41*, 4470–4495. [CrossRef]

78. Ding, J.; Yu, D. Monitoring and evaluating spatial variability of soil salinity in dry and wet seasons in the Werigan–Kuqa Oasis, China, using remote sensing and electromagnetic induction instruments. *Geoderma* **2014**, *235–236*, 316–322. [CrossRef]

A Comparison of Two Machine Learning Classification Methods for Remote Sensing Predictive Modeling of the Forest Fire in the North-Eastern Siberia

Piotr Janiec [1,2] and Sébastien Gadal [1,2,*]

[1]   Aix Marseille Univ, Université Côte d'Azur, Avignon Université, CNRS, ESPACE, UMR 7300, Avignon, 13545 Aix-en-Provence CEDEX 04, France; piotr.janiec@etu.univ-amu.fr

[2]   Department of Ecology and Geography, Institute of Natural Sciences, North-Eastern Federal University, 67000 Yakutsk, Russia

*   Correspondence: sebastien.gadal@univ-amu.fr

Abstract: The problem of forest fires in Yakutia is not as well studied as in other countries. Two methods of machine learning classifications were implemented to determine the risk of fire: MaxENT and random forest. The initial materials to define fire risk factors were satellite images and their products of various spatial and spectral resolution (Landsat TM, Modis TERRA, GMTED2010, VIIRS), vector data (OSM), and bioclimatic variables (WORLDCLIM). The results of the research showed a strong human influence on the risk in this region, despite the low population density. Anthropogenic factors showed a high correlation with the occurrence of wildfires, more than climatic or topographical factors. Other factors affect the risk of fires at the macroscale and microscale, which should be considered when modeling. The random forest method showed better results in the macroscale, however, the maximum entropy model was better in the microscale. The exclusion of variables that do not show a high correlation, does not always improve the modeling results. The random forest presence prediction model is a more accurate method and significantly reduces the risk territory. The reverse is the method of maximum entropy, which is not as accurate and classifies very large areas as endangered. Further study of this topic requires a clearer and conceptually developed approach to the application of remote sensing data. Therefore, this work makes sense to lay the foundations of the future, which is a completely automated fire risk assessment application in the Republic of Sakha. The results can be used in fire prophylactics and planning fire prevention. In the future, to determine the risk well, it is necessary to combine the obtained maps with the seasonal risk determined using indices (for example, the Nesterov index 1949) and the periodic dynamics of forest fires, which Isaev and Utkin studied in 1963. Such actions can help to build an application, with which it will be possible to determine the risk of wildfire and the spread of fire during extreme events.

Keywords: wildfires; MaxENT; random forest; risk modeling; GIS; multi-scale analysis; Yakutia; Artic; Siberia

1. Introduction

The disturbance regime in the boreal forest is extremely variable. Every year in Siberia, millions of hectares of forest are burned. Forest fires are one of the main factors causing not only long-term, harmful changes in plant ecosystems, but also contribute to the deterioration of living conditions in society, especially in the event of wildfires. Taiga fire is a natural phenomenon. Fires determine the normal, ecological functioning of the forest in this region. Forest fires are an inseparable part of the natural cycle. After the fire, there are favorable conditions for the young generation of trees. Wildfires are important for the indigenous peoples of Siberia. The territory after the wildfire turns into a pasture and people harvest berries for winter stocks. Fires have deep consequences and issues to the transformation of the landscape in the specific context of permafrost. Fire in forest areas where permafrost is spread is considered as an important factor in modeling the surface and influencing geomorphological processes.

Boreal forest plays a big role in the circulation of carbon dioxide in the regional and global scale. Changes in fire regime and climate in this region have already started, and they have an impact on the carbon releasing dynamic not only during the fire, but also years after the fire when the permafrost is thawing [1,2]. Forest fires in Russia, in general, and in the Republic of Sakha (Yakutia) are one of the main and most common natural threats. Currently, the area covered by forest fires has increased significantly. According to the results of scientific research, not the least role in the spread of fires is played by the human factor. Almost half of all forest fires that have occurred in Yakutia are caused by humans [2]. The occurrence and spread of forest fires are influenced by a complex of different factors that mutually reinforce each other and create conditions that contribute to the forest fire.

Geoinformation systems and modern remote sensing methods are a popular and effective means of identifying the most important factors affecting fires. GIS technologies are effective for building the necessary long-term fire safety models in countries such as Canada, the USA, and most of the European countries, where sustainable forest management is wide. GIS forest fire risk models are commonly used, but most of them do not include the human factor [3]. In Russia, the Nesterov index, from the 1940s is still in use [4]. There is large interest in the development of an integrated fire hazard assessment system integrated with GIS, remote sensing, meteorological, and government data. A system like that, available for forestry, firefighters, and local authorities, can improve the fight against fires and reduce the damage, thanks to faster detection of fire ignition and focusing on the most endangered places. The purpose of this work was geoinformation modeling the long-term wildfire risk in the Republic of Sakha (Yakutia) on a macro- and microscale. To achieve this goal, the following tasks were defined: (1) study structure and regime of the wildfires in the boreal forest; (2) define the most important factors that cause forest fires in the region of the studies; (3) develop and implement statistical methods to determine the relationships between the wildfires and its factors; (4) develop spatial and statistical analysis methods for building a model of wildfire risk in macro- and microscale; and (5) validation of the built model and create wildfire risk maps at the macro-and microscale.

The methodology of the study of wildfire risk is based on forest management and geographical developments. The field of studies is built on Canadian, U.S., and European approaches to deal with natural hazards as in the works of San-Miguel-Ayanz [3], Tedim [5], Oliveira [6], Cardille [7], Martinez [8], and others. Knowledge on the boreal forests in Yakutia was taken from the publications of such authors as A.P. Isaev [9], E.I. Troeva [2], M.M. Cherosov [10], and others.

Work was divided into two main stages. The first stage presents the methodology validating the impact of individual fire factors on fire occurrence. The result of the analysis is a determination of dependencies between factors and exclusion from the further classification of factors that do not affect fires. In the second step, based on the literature analysis [11–13] two methods of machine learning classifications were implemented to determine the risk of fire: MaxENT and random forest. The initial materials to define fire risk factors were satellite images and their products of various spatial and spectral resolution (Landsat TM, Modis TERRA, GMTED2010, VIIRS), vector data (OSM), and bioclimatic variables (WORLDCLIM).

The results in the form of long-term fire risk maps can be used for fire prevention and planning fire-fighting measures in the territory of the republic. The results can help create an application that can be used to determine the risk of fire and the spread of fire during a disaster. This work has the potential to lay the foundations for the future of a fully automated application of fire risk assessment in the Sakha Republic.

2. Material and Methods

2.1. Yakutia

To correctly understand fire regimes in Yakutia, it is important to know the features of their origin. The geographical, geological, climatological, and ecological position in the landscape make it possible to explain the complexity of the fire phenomenon. Yakutia is a very specific region, with occurrence extremely high and low temperatures, a thick layer of permafrost, specific geological structures, and the occurrence of light taiga forests dependent on fire regimes. Due to the high complexity of the territory, it was decided to study wildfires on two scales to examine the dependencies between fires and their factors on the regional and global scale. At the macroscale, the territory of research was the territory of the Sakha Republic; on a microscale, one of the regions of the republic was chosen, which was the Nyurbinskii district (Figure 1). This region was chosen due to the high number of fire incidents in recent years.

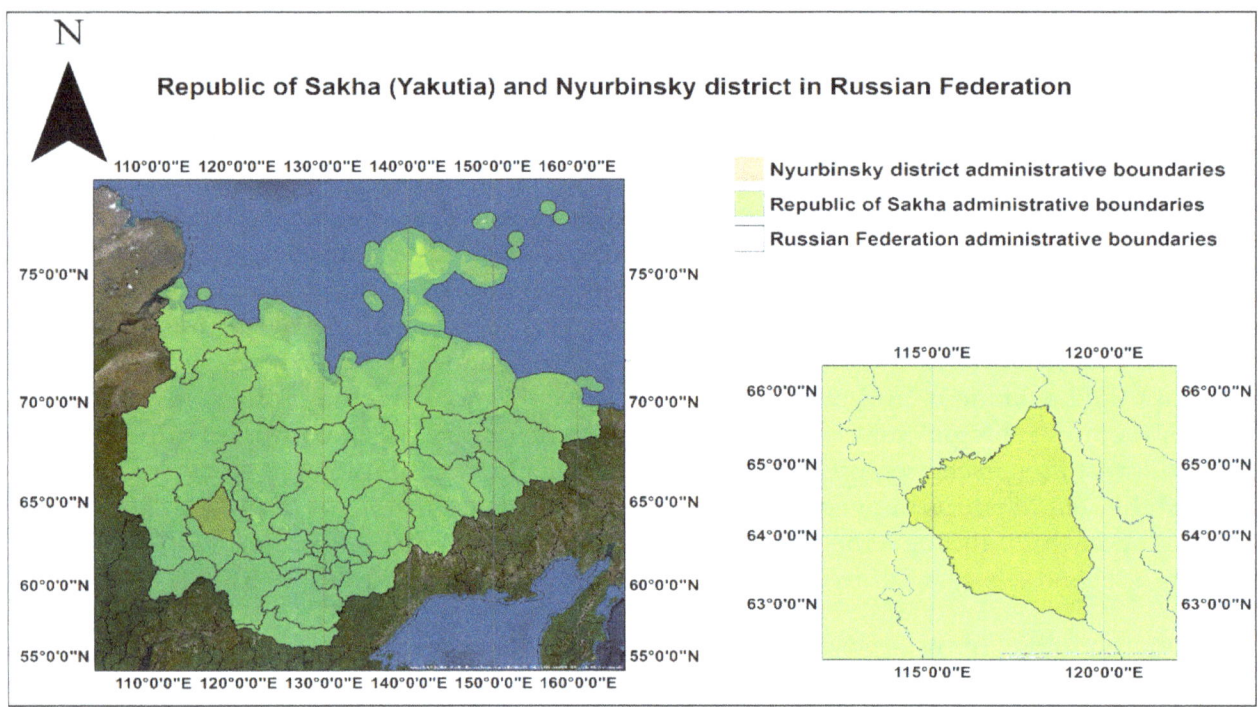

Figure 1. The geographical extent of the study area (source: Earthstar Geographics, low resolution 15 m imagery).

The Republic of Sakha (Yakutia) is in northeastern Siberia and occupies 1/5 of the territory of the Russian Federation. The territory of the republic from the east and south is closed by mountain ranges; in the north, it has access to the Arctic Ocean. The relief and geological structure are distinguished by a complex and diverse structure. The orography of the territory determines the characteristics of a sharply continental climate, permafrost, soil, vegetation, wildlife, grasslands, and grazing land, and influences the nature of human economic activity [14]. A characteristic feature of the climate of Yakutia is a sharp continentality, which is manifested in large annual fluctuations in temperature and a relatively small amount of precipitation. The main factor of this state is the Siberian anticyclone. In the study area, the average air temperature during the winter months is −30 °C, and the average temperature of the coldest month (January) is −35 °C. The average annual precipitation is 200–400 mm [15]. The permafrost thickness in the territory is sometimes more than 100 m. Seasonal thawing varies between 0.8 and 3.3 m, depending on the landscape and the type of soil. Permafrost in the conditions of Yakutia has an impact on all soil processes [16]. Sakha can be divided into three great vegetation belts. About 40% of Yakutia is located above the Arctic Circle, and all of this is covered with permafrost, which greatly affects the region's ecology and limits the forests in the southern region. The arctic and subarctic tundra define a middle region where lichens and mosses grow like big green carpets and are favorite pastures for deer. In the southern part of the tundra belt along the rivers, scattered dwarf Siberian pines and larches grow. Below the tundra is a vast region of the taiga zone [2].

According to the state report "On the state and environmental protection of the Republic of Sakha (Yakutia) in 2018, the forested area occupies 82% of the territory of Yakutia, but the forest is 54% of the territory. The area of the forest zone is 252,820.0 million hectares. The forest cover is very diverse, from 11.5% in the Verkhoyansk district to 91.7% in South Yakutia. In the Nyurbinsky district, the forest area is 4416.3 thousand hectares and is more than 84% of the total area. The timber reserved is about 9.2 milliard m^3 and 96% of it is coniferous. The average timber reserve per one hectare is 58 m^3. The biggest average timber reserves are of *Pinus Sibirica* stands (188 m^3/ha) and *Picea* spp. (130 m^3/ha). The average timber reserves of *Pinus sylvestris* is 104 m^3/ha, *Larix* spp. is 62 m^3/ha, and *Betula Pendula* 41 m^3/ha.

A total of 95.6% of forest species in Yakutia are coniferous. The major species are *Larix* (4 main types: *Larix cajanderi, L. gmelinii, L. sibirica,* and *L. czekanowskii (L. sibirica x L. gmelinii)*), which represent 77.5% of the total forest resources. The second main species is *Pinus sylvestris* (6.5% of the total area) and *Pinus obovata* (0.24% of the total area). In southwest Yakutia, *Pinus sibirica* and *Abies sibirica* occur. *Picea ajanensis* is characteristic of the south mountainous regions. Other mountainous regions are occupied by *Pinus pumilo* (4.6% of the total forest area). The main deciduous tree species are *Betula pendula, B. pubescens, B. ermannii, Populus staveolens, P. tremula, Chosenia arbutifolia,* and *Salix* spp. Deciduous trees occupy only two million hectares, which are 1.24% of the total forest area.

Tree stands are well adapted to growing under extremely hard conditions of a dry climate and occurrence of permafrost. Forests are also very well adapted to the recovery process after fires. The process is determined by several factors. The most important factors are high seed production, good seed germination, and highly adaptive potential. *Larix cajanderi* is highly adapted to the existence with frequent fires. This species is pouring seeds in late summer in the ripening year. Seeds are being covered by needle litter and snow, where it creates the right conditions for sprouting and uses the spring moisture. The best conditions for forest growth are during the first years after a fire. Litter is destroyed, and the soil is enriched with ashy elements. The upper soil horizon's moisture is increased due to inflow from lower horizons [10].

Nyurbinsky ulus (region) is in the middle of the Vilyi River and occupies the territory adjacent to both Vilyi and its main tributary Marha. The region has an area of approximately 52.4 thousand sq km. Nyurbinsky region is part of the West Yakutia natural zone. It is characterized by plains and plateaus and the key elements are soil and vegetation on which relief has a big influence. According to Kurzhuev, the region is located on the Viluy Plateau, which is part of the Central Yakut Plateau. There is a characteristic occurrence of cryolithozone relief forms such as alases, yuryakhs, and bulgunyakhs. The number of days with snow cover is 210–225 days. The average annual temperature in the region is negative and in Markha it is −11.1 °C. The coldest month is January with an absolute minimum of –60 °C, and the warmest is July (the absolute maximum is +37 °C). The average amount of precipitation per year is 260–280 mm [14]. The maximum is in July. In the geobotanical regionalization of Yakutia, the Nyurbinsky region is situated in the boreal region in the taiga zone—a subzone of the middle taiga forests—sub-province, Central Yakutian middle taiga. The area of the region is dominated by Larix forests and Pinus Sylvestris forests. The river's valleys are characterized by rich meadows as well as common steppe and forest-steppe landscapes. Dry belts of alas vegetation are represented by the *Carex duriuscula* steppes [14].

2.2. Spatial Monitoring Factors for Fire Forest Monitoring in Yakutia

2.2.1. Fires Data

In the Republic of Sakha (Yakutia), open data on forest fires are not available. There is a database of forest fires at the Ministry of Environmental Protection (https://minpriroda.sakha.gov.ru), but these data are tabular, and without georeferencing, it is impossible to create a GIS database necessary for this type of research. These data are understated due to the addition to the database of only those fires in which there was action taken to extinguish the fire. Fires that are far from human activity often do not extinguish, so are not classified in the database. The most popular global data source on fire data is the Fire Information for Resource Management System (FIRMS). Data available in the service are collected from the VIIRS 375 m, 750 m, and MODIS Collection 6 Active Fire Product. The data are collected from 2002 until now. For the studies, we chose data from the MODIS Collection 6 sensor because of their longer availability and enough spatial resolution. We used data between 2001 and 2018 from the FIRMS fire archive. The shapefiles were in the Geographic WGS84 projection. The confidence values ranged from 0% to 100% and ranged in one of three fire classes (low-confidence fire, nominal-confidence fire, or high-confidence fire) [17].

2.2.2. Factors Data

Forest fire regimes are extremely diverse and vary due to the spread of fires and climate changes [8]. In difficult to manage forest areas like in Yakutia, it is necessary to properly characterize the factors that are causing forest fires. Researchers have used several different variables to assess wildfire risk [18–21]. Due to the lack of data, it was decided to investigate only long-term fire factors data like constant, long-term factors such as slope, aspect, fuel, climate, NDVI, etc. Constant factors are those factors that do not change rapidly, but gradually, in the long perspective. Constant factors can be calculated in medium- or long-term periods before the fire season [3]. In Yakutia, there is a very big data shortage. If some data are already collected, it is very difficult to access them. For these reasons, mostly global datasets were used.

Variables were divided into four groups: (1) meteorological (precipitations, temperature, maximum summer temperature, radiation); (2) NDVI; (3) landform (elevation, slope, slope direction); and (4) human activity (distance from roads, distance from settlements, distance from rivers). All data types were in other formats and had different spatial resolutions (Table 1).

Table 1. Technical description of the data.

Factor	Source	Format	Resolution	Preprocessing
Radiation Precipitations Temperature Max temperature	WorldClim	GeoTiff	30 s	ROI
NDVI	MODIS/LANDSAT	Hdf/GeoTiff	250 m/30 m	Mosaic, resampling, atmospheric correction ROI
Elevation Slope Slope direction	GMTED2010	GeoTiff	7.5 s	Mosaic, resampling ROI
Distance from settelments Distance from roads Distance from water lines	OSM	Shape		Eulicdean distance/ROI

To collect bioclimatic variables, the WorldClim dataset of global climate layers was used. The WorldClim dataset has a spatial resolution of about 1 km^2 [22]. WorldClim is a set of global climate layers (gridded climate data), specifically developed for ecological modeling on GIS. Currently, WorldClim provides several datasets for different temporal scenarios (past, current, and future conditions). In this work, data for the current condition scenario (1970–2000) were used. WordClim bioclimatic variables are analysis-ready data so the preprocessing was not necessary.

NDVI is greatly used in the evaluation of the phenology and productivity of the vegetation. Onigemo et al. claims that the values of NDVI obtained through images at the peak of drought were related to the content moisture and fresh phytomass, showing its potential to estimate fire risk [23]. Illaera et al. found a good correlation between the NDVI values and the location of wildfires [24]. To assess NDVI in macro-and microscale, it was decided to use NDVI from two sources: MODIS and LANDSAT images. The Landsat data were used for the determination of NDVI in the Nyurbinsky region. As the region of interest, the following paths and rows were defined: 129-15; 129-16; 130-14; 130-15; 130-16; 131-15. There were selected available images from the year with the smallest cloud cover. Images without cloud cover at the peak of the growing season (July) were available only for Landsat 5 in 2009. Before creating the mosaic, atmospheric correction was necessary. It allowed for improving images from level 1 to level 2. For NDVI calculations, bands 3 and 4 were used by the formula described by J.W. Rouse in 1973 [25]. To determine the average NDVI in the vegetation season for Yakutia between 2001 and 2018 with MODIS, dataset MOD13A2 Version 6 was used. MOD13A2 provides NDVI values with a resolution of 1 km. The product is derived with a monthly interval. There were selected images at the peak of the growing season (July) from each year. At the territory of Yakutia, there are the following tiles, which were defined as region of interests: h21v01, h22v01, h23v01, h21v02, h22v02, h23v02, h24v02, h23v03, h24v03, h25v03. After the selection of the data yearly, an NDVI mosaic was created and the mean NDVI calculated between 2001 and 2015.

To model landform factor data, it was decided to use one of the most used in these types of works, digital elevation models: the Global Multi-resolution Terrain Elevation Data 2010 (GMTED2010). It incorporates the current best available global elevation data. GMTED2010 is commonly used for radiometric and geometric correction, cover mapping, and extraction of drainage features for hydrologic modeling [26]. There was a necessity for mosaicking. For the GMTED, the following entities were used: GMTED2010N50E120, GMTED2010N50E150, GMTED2010N70E120, and GMTED2010N70E150. From the DEM, the information about elevation, slope, and slope direction was derived.

To investigate the relationship between the fire and the presence of a human, it was necessary to obtain information on the distance of fires from roads, buildings, and rivers. The data in shape format were downloaded from Open Street Map (OSM). According to previous works on this subject [27,28], to achieve this goal, the Euclidean distance was used. Euclidean distance is a straight line between two points in Euclidean space. The locations were converted to raster. The resolution of the raster was defined by the shortest of the width or height of the extent of the input feature, in the input spatial reference, divided by 250.

After the preprocessing, two regions of interest (ROI) were defined for each obtained raster. Each of them was clipped using the forest cover of the Nyurbinsky region and Republic of Sakha (Yakutia) forest cover. The forest cover was made available by the Institute of Biological Problems of Cryolithozone in Yakutia. The same ROI was made for fire points between 2001–2015 from the FIRMS dataset.

2.3. Modeling the Risk Assessment of Forest Fires

2.3.1. Fire Factors Analysis

The first step after the preprocessing (Figure 2) was the classification of the values in each raster, for ordering data and eliminating individual pixels strongly deviating from the average pixels were probably deviating due to measurement error or instrument inaccuracy. Due to a large amount of data, the surface to cover (3,084,000 km^2), and computing capabilities, it was necessary to simplify the data to some extent but retain their characteristics. Jenks's optimization method was used. The Jenks classification is designed to identify data clusters as well as maintain a representation of all data in the set. The optimization lasts as long as the limits of the intervals are obtained, respecting the principle of the smallest possible differentiation of the observations contained in them, with the greatest distance between the intervals at the same time. This method aims to reduce the variance within classes and maximize the variance between classes. This is done by striving to minimize the average deviations of each class from the middle class while maximizing the deviations of each class from the means of other groups [29]. New values in rasters have been extracted to each fire point from the FIRMS dataset of over 17 years in the territory of Yakutia and the Nyurbinsky region.

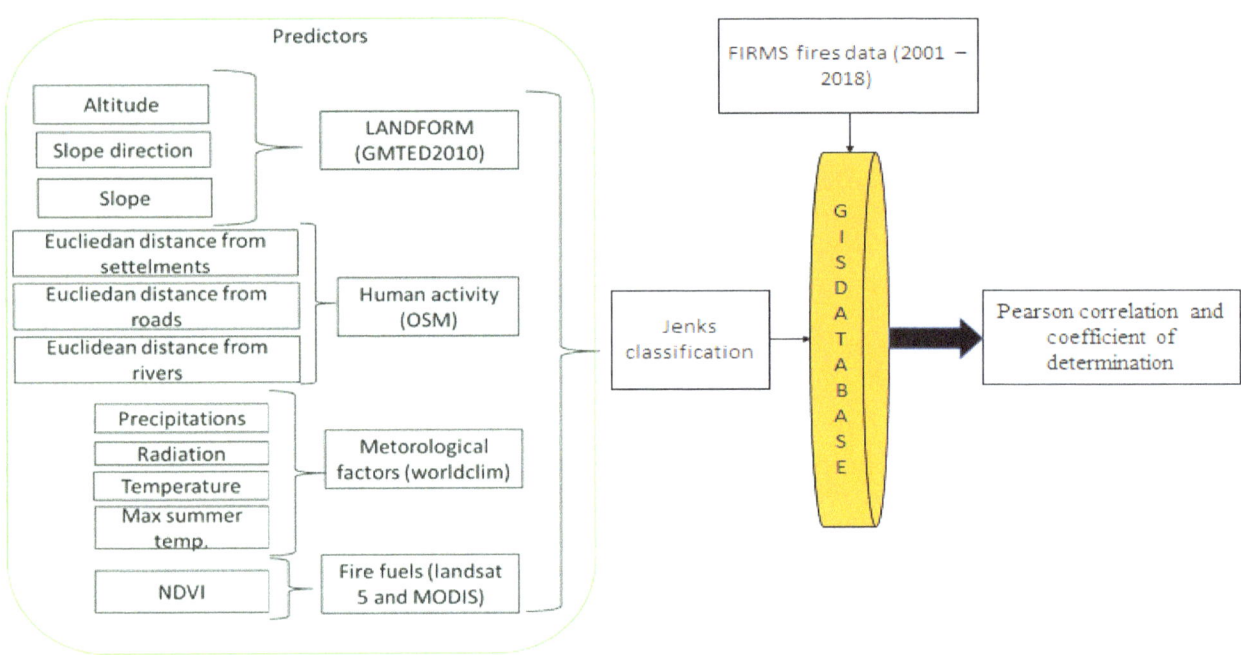

Figure 2. Schema of the analysis of the relationship between the fires and their factors.

The next step was to calculate the area of each class in each factor. Then, all fires between 2001–2018 were summed, separately for each category in each obtained earlier categorized raster. With these data, it was possible to determine the regression, and correlation between the dependent variables and explanatory variables. The same calculation scheme was repeated for each variable. In the last step, Pearson correlation and coefficient of determination between the fire data and factors data were calculated for each model.

2.3.2. Long-Term Fire Risk Modeling

The next step was to build risk models using two machine learning methods (Figure 3). The training dataset was built from the FIRMS fire data between 2001–2015. All registered fire points at the territory in the Republic of Sakha Yakutia and Nyurbinsky region in this period were used. The variables were divided, according to the previously obtained correlation coefficient. As the predictors, two types of datasets were examined. A dataset with a coefficient of correlation equal or higher than satisfactory (higher than 0.6) and a dataset with all 11 previously selected fire factors. All predictor rasters have been resampled and converted to grid format. Such a set of training data and predictor data were created, separately for the territory of Yakutia and the territory of the Nyurbinsky region.

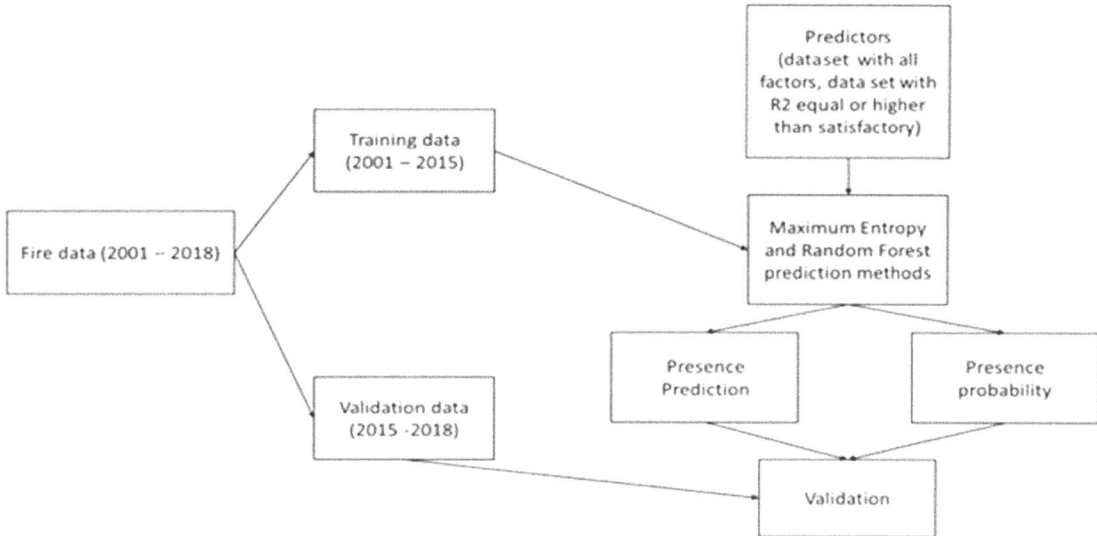

Figure 3. Long-term fire risk modeling methodology.

In the simulation, two types of prediction methods were selected, referring to the works of authors such as Peters [11], Parisien [30] (maximum entropy prediction model) and Oliveira [6] (random forest prediction model). In their studies, the authors showed a good correlation between the risk models obtained when using machine learning and forest fires. The authors demonstrated the superiority of methods using machine learning over traditional ones.

The maximum entropy prediction model is a widely used and accepted statistical method to produce predicting probability distributions. The model is adapted in diverse topics such as thermodynamics, economics, forensics, imaging technologies, and recently, ecology. Maximum entropy can provide accurate predictions of patterns in macroecology and help identify the mechanisms that matter most [31]. The algorithm is widely used for mapping species distributions [32] and conservation planning [31]. In recent years, the model of maximum entropy began to be used in the ecology of fires. Maximum entropy is a density estimation method based on a probability distribution. It is a presence-only machine learning algorithm that iteratively contrasts environmental predictor values at occurrence locations with those of a large background sample taken across the study area [33]. Maximum entropy has proved to be an enormously powerful tool for reconstructing images from many types of data [34].

The biggest advantage of a random forest is that it is a very flexible method, and it can be used in different types of problems. The random forest algorithm is a fully nonparametric machine learning method for data analysis. Classification and regression random forest is competitive with the best available methods and superior to most methods in common use [35]. The application of random forest can be an effective methodology to predict fire occurrence in different sites. The random forest algorithm is a technique developed by Breiman (2001) [35]. It combines a large set of decision trees, which is the biggest advantage compared to a simple decision tree algorithm. Each tree is trained by a set of variables, which are randomly selected from the training dataset.

All classifications were carried out using SAGA—6.4.0 and its modules "Maximum Entropy Presence Prediction", "Random Forest Presence Prediction (ViGrA) Classification". After the processing, two types of models were obtained: presence prediction and presence probability maps. Presence prediction maps are only meant to determine if there is a possibility of a fire. All raster pixels are classified into two classes:

- Absence—there is the possibility of a forest fire.
- Presence—there is no possibility of the forest fire.

Presence probability maps are meant to determine how high the possibility is of the fire. All raster pixels are classified into six classes:

- Very low—very low possibility of the fire.
- Low—low possibility of fire.
- Moderate—moderate possibility of fire.
- High—high possibility of fire.
- Very high—very high possibility of fire.
- Extreme—extreme possibility of fire.

To obtain only six possibility classes, it was necessary to reclassify the models. For reclassification, the natural breaks method was used.

For the validation of the results, raw fire points data between 2015 and 2018 in the Republic of Sakha (Yakutia) and the Nyurbinsky region were used. To carry out statistical analysis pixel count in each probability class in the presence probability maps and each class at the presence prediction maps were summed. Fire points in each class were summed and divided by the pixel sum in each class. This process was carried out to consider the surface of each class when verifying models. Next, the percentage share of fires in each class was calculated. The last step was regression and correlation analysis.

3. Results

3.1. Fire Factors

In the first part of the studies, attempts were made to uncover the main factors affecting the possibility of a wildfire. Of the 11 preselected factors, not all of them showed a good correlation with fire points (Table 2).

Table 2. Coefficient of correlation for each variable in the Republic of Sakha (Yakutia) and the Nyurbinsky region.

Factor	Yakutia	Nyurbinksy
	R	R
Radiation	0.97	0.39
Precipitations	−0.33	−0.99
Temperature	−0.41	0.20
Max temperature	0.93	0.47
NDVI	0.91	0.97
DEM	−0.76	0.28
Slope	−0.97	−0.73
Slope direction	−0.43	0.15
Settlements	0.97	0.86
Roads	−0.80	−0.82
Water lines	0.91	−0.54

In Yakutia (macroscale), the correlation above 70% was shown by factors such as solar radiation, maximum summer temperature, NDVI, elevation, slope, distance from roads, distance from settlements, and distance from rivers.

The problem was studied in the scale of the region using the example of the Nyurbinsky. In this example, only five factors out of 11 showed a good correlation. A very strong relationship between fires and precipitation was demonstrated. As the amount of precipitation falls, the number of points of ignition increases, which was not observed on a macroscale. The distribution of NDVI in microscale is almost the same as for the Yakutia. With an increase of the slope, fewer fires are observed, as in the macroscale.

3.2. Fire Risk

The next step was to create fire hazard models using two types of modeling methods. Fire risk models in the territory of Yakutia are shown in Figures 4 and 5. Results of the modeling differed, but both methods gave satisfying results. For the territory of the Republic of Sakha (Yakutia), high coefficients of correlation for each prediction method can be observed (Tables 3 and 4). Coefficients were slightly higher for the random forest prediction method. In this method, the model with 11 variables did not give considerably better results than the model with nine variables.

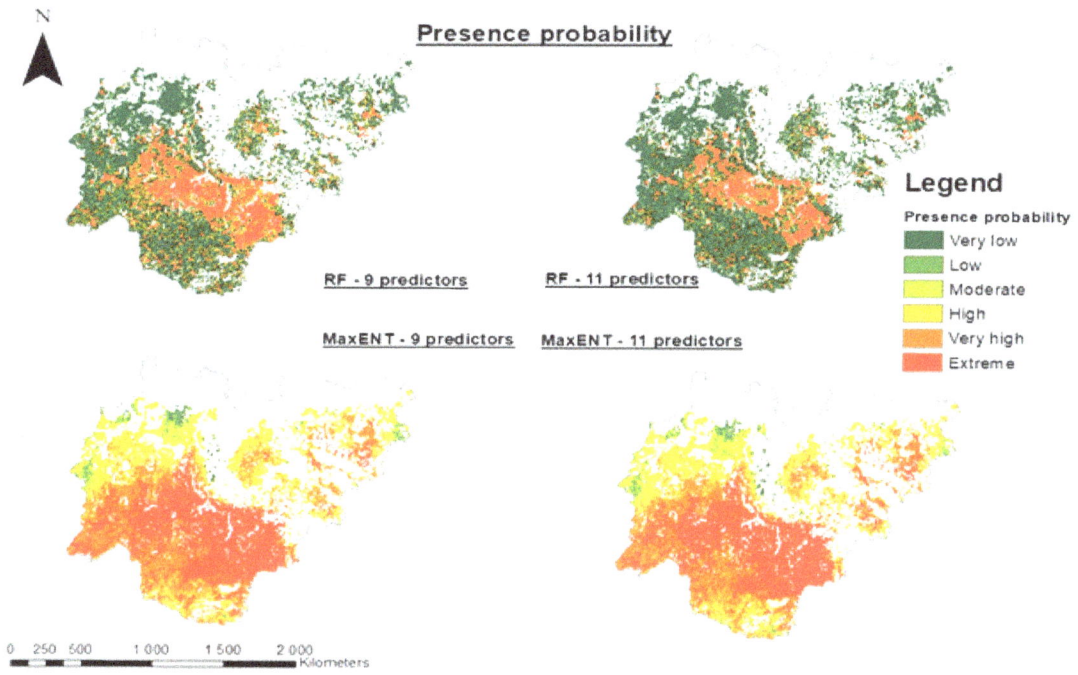

Figure 4. Results of the long-term wildfire presence probability in Yakutia (macroscale).

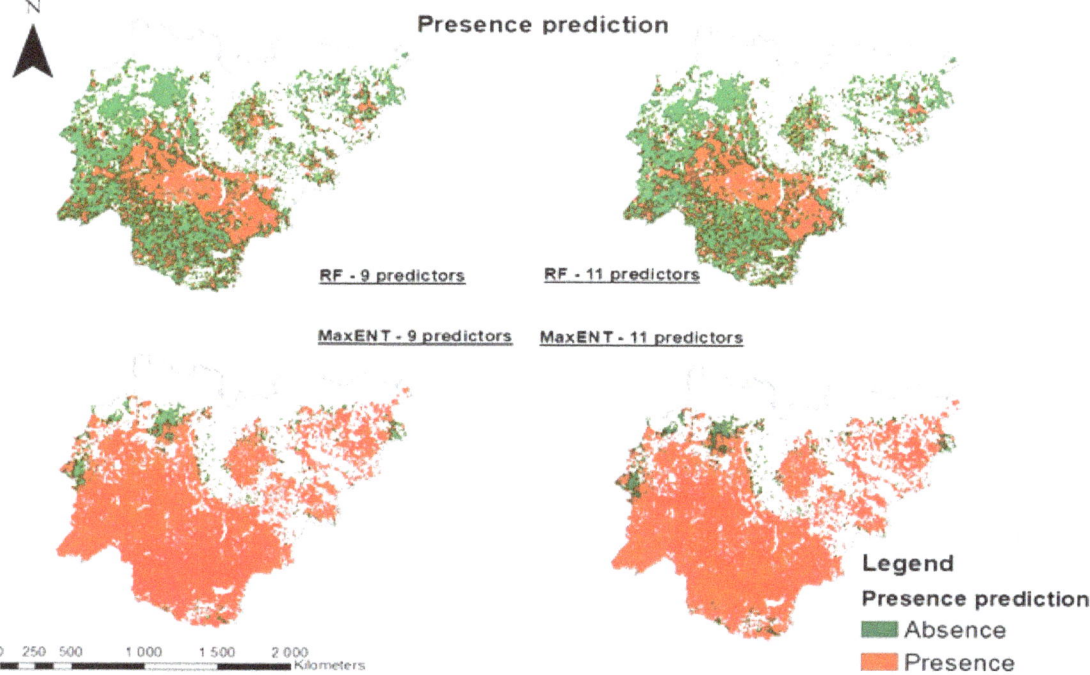

Figure 5. Results of long-term wildfire presence prediction modeling in Yakutia (macroscale).

Table 3. Comparison of the different types of long-term wildfire presence probability modeling methods in Yakutia (macroscale).

| Fire Probability | Percentage of Fire Points in Each Probability Class | | | |
| | Random Forest | | MaxENT | |
	9 Predictors	11 Predictors	9 Predictors	11 Predictors
Very Low	0.07	0.07	0.69	0.95
Low	1.74	1.65	1.57	2.30
Moderate	8.96	9.89	9.19	9.59
High	19.82	18.53	10.64	10.58
Very High	25.52	28.02	27.01	25.37
Extreme	43.89	41.84	50.89	51.21
R	0.97	0.98	0.91	0.90

Table 4. Comparison of the different types of long-term wildfire presence prediction modeling methods in Yakutia (macroscale).

| Presence Prediction | Percentage of Fire Points in Presence and Absence Class | | | |
| | Random Forest | | MaxENT | |
	9 Predictors	11 Predictors	9 Predictors	11 Predictors
Absence	0.20	0.14	12.50	14.13
Presence	99.80	99.86	87.50	85.87

When modeling using the maximum entropy prediction model, better results gave a dataset with nine variables. In both models, the highest number of fires between 2015–2018 is in the high, very high, and extreme probability classes. In the very low-risk class, there were less than 1% of fire points. Presence prediction models showed a superior random forest method over the maximum entropy method. In the random forest method, almost 100% of the fire points were in the presence class. The maximum entropy method was characterized by worse results. In the absence, the class located more than 10% of the fire points from the validation dataset.

Modeling carried out at the territory of the Nyurbinsky region showed slightly different results than in Yakutia (Figures 6 and 7). Correlation coefficients were not as high and did not exceed 0.9 (Table 5). According to the coefficient of correlation, the smallest error was characterized by a model of maximum entropy using six predictor variables. The coefficient of correlation for 11 predictors was only slightly lower and was equal to 0.89.

Table 5. Comparison of the different types of long-term wildfire presence probability modeling methods in Nyurbinsky (microscale).

| Fire Probability | Percentage of Fire Points in Each Probability Class | | | |
| | Random Forest | | MaxENT | |
	6 Predictors	11 Predictors	6 Predictors	11 Predictors
Very Low	498	4.97	0.00	0.00
Low	15.09	17.04	0.00	0.00
Moderate	20.40	16.06	1.92	12.81
High	15.60	20.25	32.28	31.23
Very High	24.94	21. 56	31.73	29.02
Extreme	19.00	20.13	34.07	26.94
R	0.75	0.82	0.90	0.89

A Comparison of Two Machine Learning Classification Methods for Remote Sensing Predictive Modeling...

33

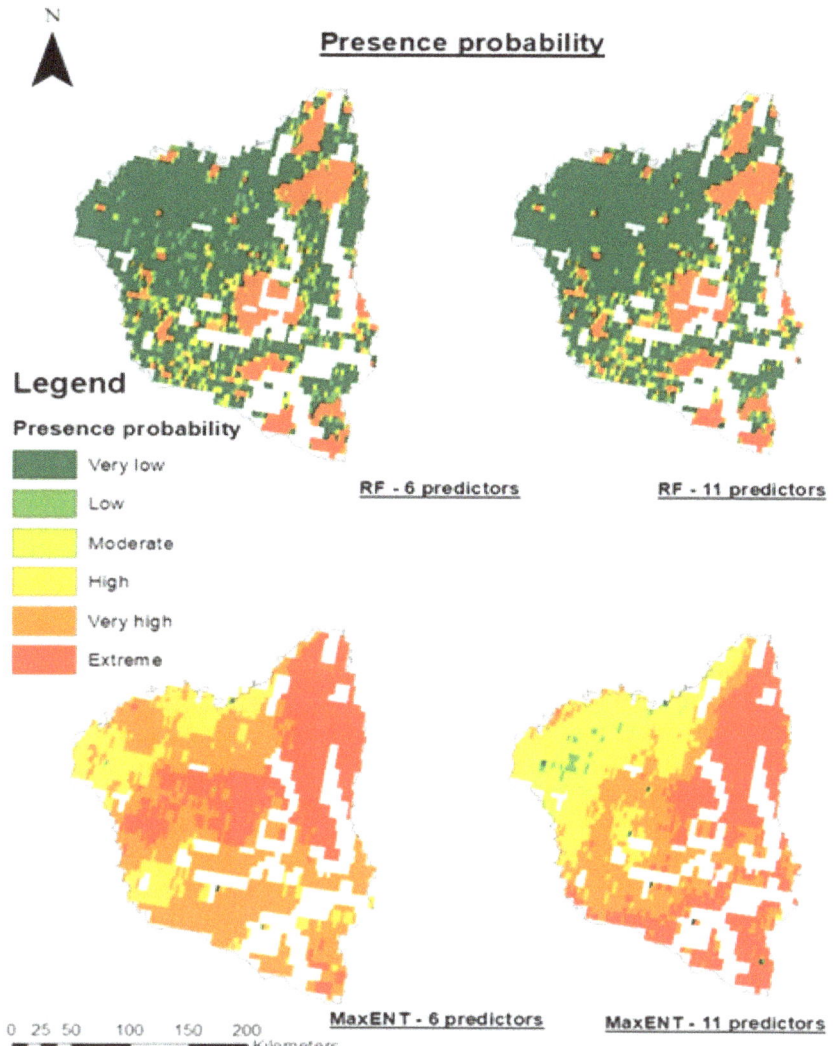

Figure 6. Results of the long-term wildfire presence probability modeling in Nyurbinksy (microscale).

It was observed that in the low and very low fire risk classes, there were no fire points. Significantly worse results were shown by the random forest prediction method. In both cases of the random forest method (six and 11 predictors), the coefficient of correlation did not exceed 0.82. Over 20% of all fire points were at low-risk classes. Similar results were observed by analyzing the presence prediction maps (Table 6). The random forest prediction method was characterized by more than 30% of fire points in the absence class. The results were distributed in a similar way in the maximum entropy prediction model, which is characterized by 11 predictors. The best results were observed in the maximum entropy classification using six predictive variables. In this model, more than 90% of the points were in the presence class.

Table 6. Comparison of the different types of long-term wildfire presence prediction modeling methods in Nyurbinsky (microscale).

Presence Prediction	Percentage of Fire Points in Presence and Absence Class			
	Random Forest		MaxENT	
	6 Predictors	11 Predictors	6 Predictors	11 Predictors
Absence	32.73	31.22	8.34	32.66
Presence	67.27	68.78	91.66	67.34

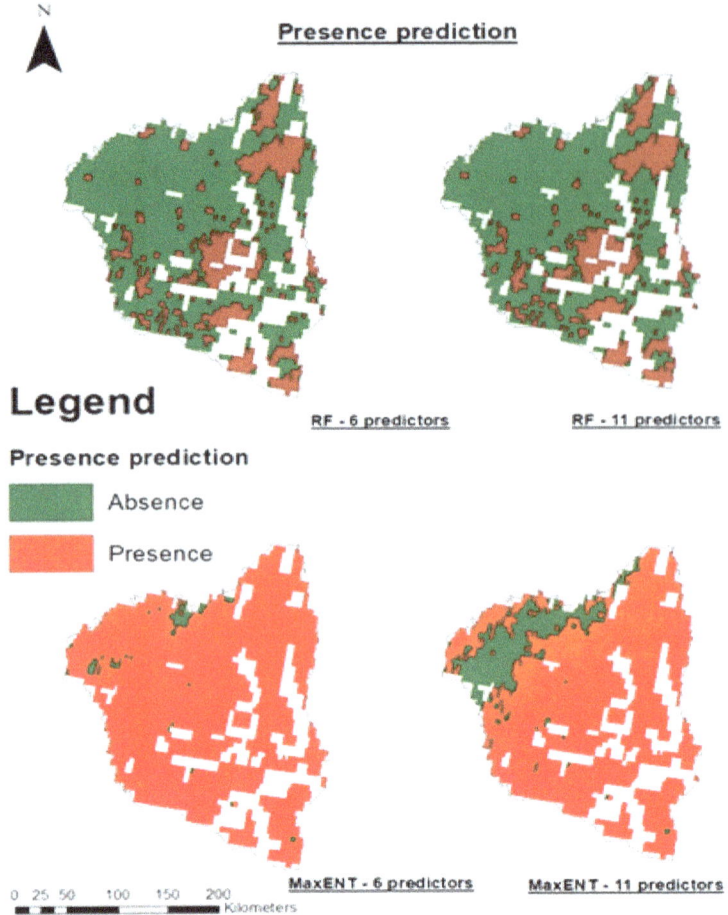

Figure 7. Results of the long-term wildfire presence prediction modeling in Nyurbinsky (macroscale).

It was observed that in both regions (Yakutia and Nyurbinsky), the random forest presence prediction method gave more accurate results. Even though the Nyurbinsky region incorrectly classified a larger number of fire points, it can be unambiguously stated that it is wrong. This is because (Figures 5 and 7) in the presence maximum entropy, presence prediction method, almost all of the territory of Yakutia and Nyurbinsky was classified as territory with the presence of fires. The results of the random forest presence prediction method showed a much narrower territory of the possibility of the occurrence of fires and yet it classified more points correctly in the territory of Yakutia.

4. Discussion

In fire studies, fire risk is one of the major topics and there are many different approaches to this subject. Blanchi et al., in their work about a methodological approach in fire risk studies, collected more than 50 works connected with fire probability cartography. Risk mapping has less than twenty years. Previously, there were rather preferred descriptive approaches. Since the 1990s, there has been an increasing interest in this field [36].

The study presented the possibilities of using different types of GIS and remote sensing data in modeling the wildfire risk. Results allow us to reflect various aspects of fire studies. The difficulties in fire risk assessment were to point out and clarify possibilities to define wildfire risk. There are many different approaches in hazard mapping based on different datasets, scales, and algorithms. The multi-approach is relevant in fire management studies [37,38]. Gai et al. [39] developed a spatially weighted index model for fire risk assessment. You et al. [40] integrated a Forest Resource Inventory Database based on four aspects of topographical, human activity, climate, and forest characteristic factors. Goldarag et al. [41] used neural networks and logistic regression for fire risk assessment.

The biggest challenge in the study was to collect the necessary data due to poor exploration of the area of studies. GIS technologies in Yakutia are under development [42]. For this purpose, we chose a semi-probabilistic mode of modeling, which gave possibilities of combining historical fire data and physical fire mechanisms [43].

The results of the analysis showed that the fire risk assessment in the Republic of Sakha (Yakutia) is not a problem that can be easily solved. The specificity of the studied territory is significantly different from other parts of the world that are facing the problem of wildfires. Boreal forests of Yakutia are perfectly adapted to extremely severe climatic conditions. Fires have been affecting the natural functioning of the forests of the region for centuries. Fires not only affect the climatic conditions, but also the formation of the terrain [36].

Results of the correlation analysis in macroscale showed that with increasing radiation, the risk of fires increases, and the same happened with increasing maximum summer temperature. Lim et al., in their study, highlighted that fire data showed a high correlation with climate factors [44]. NDVI was highly correlated with fires (R = 0.91). With increasing NDVI, the number of points of ignition increased. This situation was associated with the accumulation of combustible materials along with an increase in forest biomass. Elevation and slope correlated inversely proportional. According to Pourghasemi et al., slope and aspect are some of the most important factors controlling forest fire occurrence [45]. Kurbatsky et al. wrote that long drought conditions in the Yakutian valleys caused intense droughts and accumulation of fire fuels, which caused strong fires [46]. As the slope increased, we observed a decreasing number of wildfires. This situation may be due to the strongly inclined slopes that act as a barrier to the spread of fire. Additionally, on inclined slopes, many combustible materials cannot accumulate. Ghorbanzadeh et al. claimed that environmental variables are not the only reason for fire susceptibility and risk [47]. A very strong connection between the increase in the number of fire points and the human factors was observed. In places with proximity to transport routes, the number of fires increases, and the same thing happens if we consider the settlements. This situation indicates a strong anthropogenic impact on the risk of fire. There are not so many fires near water lines, which can be caused by the presence of moist areas close to the rivers.

The largest correlation at the microscale was observed when taking rainfall into account. As the rainfall increased, the number of fires increased. The other climatic factors did not show a strong correlation. NDVI, similar to the macroscale, strongly influences the number of fires. The properties of combustible materials relate to type, phytomass, condition, and moisture, among which the moisture content is the most important for fire protection [48]. In a completely different way, anthropogenic factors were distributed on the microscale than on the macroscale. The Nyurbinsky region has a relatively high population density compared to the entire territory of Yakutia, which may affect the differences between the results. Most fires continue to occur along roads, but the closer the settlements are, it occurs less frequently. This situation may be because fires that approach the villages pose a threat to people and fire-fighting starts quickly (fires do not reach large sizes), while those that are far from populated areas remain without any action as they do not pose a great threat. Along roads that are far from villages, fires can reach serious sizes. Additionally, villages in this region are usually not surrounded by forests, but by meadows called "alas", with lower susceptibility to ignition.

Human activity has a greater influence on the fire regimes and differs in the macro- and microscale. The results showed that distance from settlements (R = −0.97, and R = 0.86), distance from rivers (R = 0.91, and R = −0.54), and distance from roads (R = −0.80, and R = −0.82) had a great importance in fire risk assessment. Ajin et al. [49] showed a high correlation between fire occurrence and distance from roads and settlements. The study by Werf showed that in residential areas and near roads, more human activities are witnessed, and human activity is the most significant factor in the fire outbreak [50]. Studies of the territory of Yakutia in residential areas and near roads show that more human activities are witnessed, and the human activity is the most significant factor in fire outbreaks.

Very large differences were observed between the results of the research on the macro- and microscale. Often the results were quite the opposite. It is probably related to the specificity of the Yakutia region. The Niyurbinsky region is quite highly urbanized compared to other regions of Yakutia, which may affect the results. Additionally, the regional climatic conditions differed from the average for the entire region. The Nyuribnsky region is characterized by flat terrain, and most of Yakutia is mountainous, which can also affect the conditions for fires. To analyze the macroscale, it is necessary to take into account a wide variety of conditions: climatic, geographical, and human influence. However, on a microscale, these conditions are more homogeneous.

The analysis of the different methods of wildfire risk assessment allows for the identification of the high danger areas in Nyurbinsky and the entire Yakutia. The maximum accuracy was demonstrated by the random forest method with 11 predictors (R = 0.98) for the entire territory of Yakutia. The analysis showed that the random forest method gave more accurate results and a much narrower area of the possibility of fires than the MaxENT method, which allows us to propose that this model is preferable. A model created using the maximum entropy method has a small differentiation into zones, which does not allow for its use in practice as uninformative. Parisien used the MaxENT and random forest model to predict fire ignition across the USA [30] with satisfactory results. The MaxENT model was used to assess fire risk in India's Ghats Mountains [34]. In their studies, the authors showed a good correlation between the risk models obtained when using machine learning and forest fires. Due to their power, versatility, and ease of use, random forests are quickly becoming one of the most popular machine learning methods [48]. The studies confirmed that the use of GIS and remote sensing technologies can be used to assess the long-term risk and probability of fires in Yakutia. The results showed that reducing the number of factors during modeling did not show a significant impact on the results in both methods. In the few last years, an increase in the number and severity of fires has been observed as has more years with extreme fire seasons [9]. This situation may also be affected by climate change. On the other hand, a larger number of fires may have an impact on the climate, permafrost ecosystems, and the functioning of indigenous people. In connection with this situation, it is necessary to work on new fire risk assessment methods in the Arctic and subarctic zone of Yakutia. Methods should use global databases of fire, climate, and human activities. For building new models that include all these factors, it is necessary to use different types of datasets, like OSM data, climatic datasets, and remote sensing data from different sensors (MODIS, Landsat, Sentinel). According to Gigović et al., the random forest model could be used at the regional level for forest fire susceptibility mapping [51]. Banerjee et al. claim that the MaxENT method can be used as a decision support tool for stakeholders of forest resources [52]. Pham et al. claim that models that consider climate variables, vegetation, and human influences can explain fire risk better than those that only account for some of these factors [53].

The objective of work on modeling long-term wildfire risk in the Nyurbinsky region of Yakutia was accomplished. The accuracy of the results depends on the data used, and here available. The results were valid within the limits caused by the data used; the approach was chosen as was the research objective. Despite these limitations, it was possible to obtain interesting results that enabled us to answer the main questions posed by the problem.

5. Conclusions

The problem of the forest fires in Yakutia is not as well studied as in other countries. The results of the research have shown a strong human influence on the risk in this region despite the low population density. Anthropogenic factors showed a high correlation with the occurrence of wildfires more than climatic or topographical factors. Other factors affect the risk of fires at the macroscale, and others the microscale, which should be considered when modeling.

The random forest method showed better results in the macroscale, however, the maximum entropy model was better in the microscale. The exclusion of variables that do not show high correlation does not always improve the modeling results. The random forest presence prediction model is a more accurate method and significantly reduces the risk territory. The reverse is the method of maximum entropy, which is not as accurate and classifies very large areas as endangered.

Further study of this topic requires a clearer and conceptually developed approach to the application of remote sensing data. Therefore, this work makes sense to lay the foundations for the future of a completely automated fire risk assessment application in the Republic of Sakha. The results can be used for fire prophylactics and planning fire prevention. In the future, to determine the risk well, it is necessary to combine the obtained maps with the seasonal risk determined using indices (for example, the Nesterov index 1949) and the periodic dynamics of forest fires, which Isaev and Utkin studied in 1963. Such actions can help build an application with which it will be possible to determine the risk of wildfire and the spreading of fire during extreme events.

Author Contributions: S.G. and P.J. participated in the conceptualization of the manuscript. S.G. and P.J. participated in the methodology development and software development. P.J. conceived the designed the experiments. S.G. and P.J. wrote the manuscript. S.G. provided supervision on the design and implementation of the research. All authors contributed to the review and improvement of the manuscript. All authors have read and agreed to the published version of the manuscript.

References

1. Kasischke, E.S.; Stocks, B.J. *Fire, Climate Change, and Carbon Cycling in the Boreal Forest*; Springer Science & Business Media: Berlin/Heidelberg, Germany, 2012; pp. 203–204, ISBN 978-0-387-21629-4.
2. Troeva, E.I.; Isaev, A.P.; Cherosov, M.M.; Karpov, N.S. *The Far North: Plant Biodiversity and Ecology of Yakutia*; Springer Science & Business Media: Berlin/Heidelberg, Germany, 2010; pp. 203–204, ISBN 978-90-481-3774-9.
3. San-Miguel-Ayanz, J.; Carlson, J.D.; Alexander, M.; Tolhurst, K.; Morgan, G.; Sneeuwjagt, R.; Dudley, M. Current Methods to Assess Fire Danger Potential. In *Wildland Fire Danger Estimation and Mapping*; Series in Remote Sensing; World Scientific: Singapore, 2003; Volume 4, pp. 21–61, ISBN 978-981-238-569-7.
4. Жданко, В.А.; Гриценко, М.В. Метод анализа лесопожарных сезонов: Практические рекомендации. Л. ЛНИИЛХ **1980**. (In Russian)
5. Tedim, F.; Leone, V.; Amraoui, M.; Bouillon, C.; Coughlan, M.R.; Delogu, G.M.; Fernandes, P.M.; Ferreira, C.; McCaffrey, S.; McGee, T.K.; et al. Defining Extreme Wildfire Events: Difficulties, Challenges, and Impacts. *Fire* **2018**, *1*, 9. [CrossRef]
6. Oliveira, S.; Oehler, F.; San-Miguel-Ayanz, J.; Camia, A.; Pereira, J.M.C. Modeling spatial patterns of fire occurrence in Mediterranean Europe using Multiple Regression and Random Forest. *For. Ecol. Manag.* **2012**, *275*, 117–129. [CrossRef]
7. Cardille, J.A.; Ventura, S.J.; Monica, G.T. Environmental and social factors influencing wildfires in the Upper Midwest, United States. *Ecol. Appl.* **2001**, *11*, 111–127. [CrossRef]
8. González-Cabán, A. *Proceedings of the Second International Symposium on Fire Economics, Planning, and Policy: A Global View*; U.S. Department of Agriculture, Forest Service, Pacific Southwest Research Station: Albany, CA, USA, 2008; 720p.
9. Isaev, A.P.; Protopopov, A.V.; Protopopova, V.V.; Egorova, A.A.; Timofeyev, P.A.; Nikolaev, A.N.; Shurduk, I.F.; Lytkina, L.P.; Ermakov, N.B.; Nikitina, N.V.; et al. Vegetation of Yakutia: Elements of Ecology and Plant Sociology. In *The Far North: Plant Biodiversity and Ecology of Yakutia*; Troeva, E.I., Isaev, A.P., Cherosov, M.M., Karpov, N.S., Eds.; Plant and Vegetation; Springer: Dordrecht, The Netherlands, 2010; pp. 143–260, ISBN 978-90-481-3774-9.
10. Cherosov, M.M.; Isaev, A.P.; Mironova, S.I.; Lytkina, L.P.; Gavrilyeva, L.D.; Sofronov, R.R.; Arzhakova, A.P.; Barashkova, N.V.; Ivanov, I.A.; Shurduk, I.F.; et al. Vegetation and Human Activity. In *The Far North: Plant Biodiversity and Ecology of Yakutia*; Troeva, E.I., Isaev, A.P., Cherosov, M.M., Karpov, N.S., Eds.; Plant and Vegetation; Springer: Dordrecht, The Netherlands, 2010; pp. 261–295, ISBN 978-90-481-3774-9.
11. Peters, M.P.; Iverson, L.R.; Matthews, S.N.; Prasad, A.M. Wildfire hazard mapping: Exploring site conditions in eastern US wildland–urban interfaces. *Int. J. Wildland Fire* **2013**, *22*, 567–578. [CrossRef]
12. Phillips, S.J.; Dudík, M. Modeling of species distributions with Maxent: New extensions and a comprehensive evaluation. *Ecography* **2008**, *31*, 161–175. [CrossRef]

13.   Gull, S.F.; Skilling, J. Maximum entropy method in image processing. In *IEE Proceedings F (Communications, Radar and Signal Processing)*; IET Digital Library: London, UK, 1984; Volume 131, pp. 646–659.

14.   Атлас сельского хозяйства ЯАССР, А. под редакциейАГ Гущиной. *М. Главное УправлениеГеодезии и Картографии приСоветеМинистровСССР* **1989**. (In Russian)

15.   Воробьев, К.А.; Ким, А.Ю. ГеографическийАтлас *Республика Саха (Якутия)*; Роскартография: Якутск, Russia, 2000. (In Russian)

16.   Еловская, Л.Г. Классификация и диагностика мерзлотных почв Якутии. Якутский филиалСОАНСССР. 1987. Available online: https://search.rsl.ru/ru/record/01001419857 (accessed on 27 November 2020). (In Russian).

17.   Giglio, L. MODIS Collection 6 Active Fire Product User's Guide Revision A. Available online: /paper/MODIS-Collection-6-Active-Fire-Product-User%27s-Guide-Giglio/4aacae34ad3bcd557591067399ebc38580eb8286 (accessed on 14 September 2020).

18.   Prichard, S.J.; Kennedy, M.C. Fuel treatments and landform modify landscape patterns of burn severity in an extreme fire event. *Ecol. Appl.* **2014**, *24*, 571–590. [CrossRef]

19.   Martínez, J.; Vega-Garcia, C.; Chuvieco, E. Human-caused wildfire risk rating for prevention planning in Spain. *J. Environ. Manag.* **2009**, *90*, 1241–1252. [CrossRef]

20.   Syphard, A.D.; Radeloff, V.C.; Keuler, N.S.; Taylor, R.S.; Hawbaker, T.J.; Stewart, S.I.; Clayton, M.K. Predicting spatial patterns of fire on a southern California landscape. *Int. J. Wildland Fire* **2008**, *17*, 602–613. [CrossRef]

21.   Kwak, H.; Lee, W.-K.; Saborowski, J.; Lee, S.-Y.; Won, M.-S.; Koo, K.-S.; Lee, M.-B.; Kim, S.-N. Estimating the spatial pattern of human-caused forest fires using a generalized linear mixed model with spatial autocorrelation in South Korea. *Int. J. Geogr. Inf. Sci.* **2012**, *26*, 1589–1602. [CrossRef]

22.   Fick, S.E.; Hijmans, R.J. WorldClim 2: New 1-km spatial resolution climate surfaces for global land areas. *Int. J. Climatol.* **2017**, *37*, 4302–4315. [CrossRef]

23.   Onigemo, A.E.; Santos, S.A.; Pellegrin, L.A.; Abreu, U.G.P.; Silva, E.T.J.; Soriano, B.M.A.; Crispim, S.M.A. Application of vegetation index to assess fire risk in open grasslands with predominance of cespitous grasses in the Nhecolândia sub-region of the Pantanal. *Simpósio Bras. Sens. Remoto* **2007**, *13*, 4493–4500.

24.   Illera, P.; Fernández, A.; Delgado, J.A. Temporal evolution of the NDVI as an indicator of forest fire danger. *Int. J. Remote Sens.* **1996**, *17*, 1093–1105. [CrossRef]

25.   Goward, C.R.; Murphy, J.P.; Atkinson, T.; Barstow, D.A. Expression and purification of a truncated recombinant streptococcal protein G. *Biochem. J.* **1990**, *267*, 171–177. [CrossRef]

26.   Danielson, J.J.; Gesch, D.B. *Global Multi-Resolution Terrain Elevation Data 2010 (GMTED2010)*; US Department of the Interior, US Geological Survey: Newton, MA, USA, 2011.

27.   Pew, K.L.; Larsen, C.P.S. GIS analysis of spatial and temporal patterns of human-caused wildfires in the temperate rain forest of Vancouver Island, Canada. *For. Ecol. Manag.* **2001**, *140*, 1–18. [CrossRef]

28.   Zhang, Z.X.; Zhang, H.Y.; Zhou, D.W. Using GIS spatial analysis and logistic regression to predict the probabilities of human-caused grassland fires. *J. Arid Environ.* **2010**, *74*, 386–393.

29.   Jenks, G.F. The Data Model Concept in Statistical Mapping. *Int. Yearb. Cartogr.* **1967**, *7*, 186–190.

30.   Parisien, M.-A.; Snetsinger, S.; Greenberg, J.A.; Nelson, C.R.; Schoennagel, T.; Dobrowski, S.Z.; Moritz, M.A. Spatial variability in wildfire probability across the western United States. *Int. J. Wildland Fire* **2012**, *21*, 313–327. [CrossRef]

31.   Harte, J.; Newman, E.A. Maximum information entropy: A foundation for ecological theory. *Trends Ecol. Evol.* **2014**, *29*, 384–389. [CrossRef]

32.   Elith, J.; Graham, C.H.; Anderson, R.P.; Dudík, M.; Ferrier, S.; Guisan, A.; Hijmans, R.J.; Huettmann, F.; Leathwick, J.R.; Lehmann, A. Novel methods improve prediction of species' distributions from occurrence data. *Ecography* **2006**, *29*, 129–151. [CrossRef]

33.   Franklin, J. *Mapping Species Distributions: Spatial Inference and Prediction*; Cambridge University Press: Cambridge, UK, 2010; ISBN 978-1-139-48529-6.

34.   Renard, Q.; Pélissier, R.; Ramesh, B.R.; Kodandapani, N. Environmental susceptibility model for predicting forest fire occurrence in the Western Ghats of India. *Int. J. Wildland Fire* **2012**, *21*, 368–379. [CrossRef]

35.   Cutler, D.R.; Edwards, T.C., Jr.; Beard, K.H.; Cutler, A.; Hess, K.T.; Gibson, J.; Lawler, J.J. Random forests for classification in ecology. *Ecology* **2007**, *88*, 2783–2792. [CrossRef]

36.   Blanchi, R.; Jappiot, M.; Alexandrian, D. Forest fire risk assessment and cartography. A methodological approach. In Proceedings of the IV International Conference on Forest Fire Research, Luso, Portugal, 18–23 November 2002.

37. Tehrany, M.S.; Jones, S.; Shabani, F.; Martínez-Álvarez, F.; Bui, D.T. A novel ensemble modeling approach for the spatial prediction of tropical forest fire susceptibility using logitboost machine learning classifier and multi-source geospatial data. *Theor. Appl. Climatol.* **2019**, *137*, 637–653. [CrossRef]

38. Koetz, B.; Morsdorf, F.; van der Linden, S.; Curt, T.; Allgöwer, B. Multi-Source land cover classification for forest fire management based on imaging spectrometry and LiDAR data. *For. Ecol. Manag.* **2008**, *256*, 263–327. [CrossRef]

39. Gai, C.; Weng, W.; Yuan, H. GIS-Based forest fire risk assessment and mapping. In Proceedings of the 2011 Fourth International Joint Conference on Computational Sciences and Optimization, Kunming/Lijiang, China, 5–19 April 2011; pp. 1240–1244.

40. You, W.; Lin, L.; Wu, L.; Ji, Z.; Yu, J.; Zhu, J.; Fan, Y.; He, D. Geographical information system-based forest fire risk assessment integrating national forest inventory data and analysis of its spatiotemporal variability. *Ecol. Indic.* **2017**, *77*, 176–184. [CrossRef]

41. Goldarag, Y.J.; Mohammadzadeh, A.; Ardakani, A.S. Fire risk assessment using neural network and logistic regression. *J. Indian Soc. Remote Sens.* **2016**, *44*, 885–894. [CrossRef]

42. Andreev, D.V.; Makarova, M.E. GIS technologies application in useful fossils search in the territory of the Republic of Sakha (Yakutia). In Proceedings of the IOP Conference Series: Earth and Environmental Science, Changsha, China, 18–20 September 2020; IOP Publishing: Bristol, UK, 2020; Volume 548, p. 032006.

43. Boike, J.; Grau, T.; Heim, B.; Günther, F.; Langer, M.; Muster, S.; Gouttevin, I.; Lange, S. Satellite-derived changes in the permafrost landscape of central Yakutia, 2000–2011: Wetting, drying, and fires. *Glob. Planet. Chang.* **2016**, *139*, 116–127. [CrossRef]

44. Lim, C.-H.; Kim, Y.S.; Won, M.; Kim, S.J.; Lee, W.-K. Can satellite-based data substitute for surveyed data to predict the spatial probability of forest fire? A geostatistical approach to forest fire in the Republic of Korea. *Geomat. Nat. Hazards Risk* **2019**, *10*, 719–739. [CrossRef]

45. Pourghasemi, H.R.; Kariminejad, N.; Amiri, M.; Edalat, M.; Zarafshar, M.; Blaschke, T.; Cerda, A. Assessing and mapping multi-hazard risk susceptibility using a machine learning technique. *Sci. Rep.* **2020**, *10*, 1–11. [CrossRef]

46. Курбатский, Н.П. Техника и тактика тушения лесных пожаров. М. Гослесбумиздат **1962**. (In Russian)

47. Ghorbanzadeh, O.; Blaschke, T.; Gholamnia, K.; Aryal, J. Forest fire susceptibility and risk mapping using social/infrastructural vulnerability and environmental variables. *Fire* **2019**, *2*, 50. [CrossRef]

48. Dennison, P.E.; Roberts, D.A.; Reggelbrugge, J.C. Characterizing chaparral fuels using combined hyperspectral and synthetic aperture radar data. In Proceedings of the 9th AVIRIS Earth Science Workshop, Pasadena, CA, USA, 23–25 February 2000; Jet Propulsion Lab: Pasadena, CA, USA, 2000; Volume 6, pp. 23–25.

49. Ajin, R.S.; Loghin, A.-M.; Jacob, M.K.; Vinod, P.G.; Krishnamurthy, R.R. The risk assessment study of potential forest fire in Idukki Wildlife Sanctuary using RS and GIS techniques. *Int. J. Adv. Earth Sci. Eng.* **2016**, *5*, 308–318.

50. Werf, G.R.; Randerson, J.T.; Giglio, L.; Collatz, G.J.; Mu, M.; Kasibhatla, P.S.; Morton, D.C.; DeFries, R.S.; Jin, Y.; van Leeuwen, T.T. Global fire emissions and the contribution of deforestation, savanna, forest, agricultural, and peat fires (1997–2009). *Atmos. Chem. Phys.* **2010**, *10*, 11707–11735. [CrossRef]

51. Gigović, L.; Pourghasemi, H.R.; Drobnjak, S.; Bai, S. Testing a new ensemble model based on SVM and random forest in forest fire susceptibility assessment and its mapping in Serbia's Tara National Park. *Forests* **2019**, *10*, 408. [CrossRef]

52. Banerjee, P. Drivers and distribution of forest fires in Sikkim Himalaya: A maximum entropy-based approach to spatial modelling. **2020**, In Review.

53. Pham, B.T.; Jaafari, A.; Avand, M.; Al-Ansari, N.; Dinh Du, T.; Yen, H.P.H.; Phong, T.V.; Nguyen, D.H.; Le, H.V.; Mafi-Gholami, D. Performance evaluation of machine learning methods for forest fire modeling and prediction. *Symmetry* **2020**, *12*, 1022. [CrossRef]

3

Spatial Modelling of Gully Erosion using GIS and R Programing: A Comparison among Three Data Mining Algorithms

Alireza Arabameri [1], Biswajeet Pradhan [2,*], Hamid Reza Pourghasemi [3], Khalil Rezaei [4] and Norman Kerle [5]

[1] Department of Geomorphology, Tarbiat Modares University, Tehran 36581-17994, Iran; alireza.ameri91@yahoo.com

[2] Centre for Advanced Modelling and Geospatial Information Systems (CAMGIS), Faculty of Engineering and IT, University of Technology Sydney, Ultimo, NSW 2007, Australia

[3] Department of Natural Resources and Environmental Engineering, College of Agriculture, Shiraz University, Shiraz 71441-65186, Iran; hm_porghasemi@yahoo.com

[4] Faculty of Earth Sciences, Kharazmi University, Tehran 14911-15719, Iran; kh.rezaei@gmail.com

[5] Department of Earth Systems Analysis (ESA), Faculty of Geo-Information Science and Earth Observation (ITC), University of Twente, 7522 Enschede, The Netherlands; kerle@itc.nl

* Correspondence: biswajeet24@gmail.com or Biswajeet.Pradhan@uts.edu.au

Abstract: Gully erosion triggers land degradation and restricts the use of land. This study assesses the spatial relationship between gully erosion (GE) and geo-environmental variables (GEVs) using Weights-of-Evidence (WoE) Bayes theory, and then applies three data mining methods—Random Forest (RF), boosted regression tree (BRT), and multivariate adaptive regression spline (MARS)—for gully erosion susceptibility mapping (GESM) in the Shahroud watershed, Iran. Gully locations were identified by extensive field surveys, and a total of 172 GE locations were mapped. Twelve gully-related GEVs: Elevation, slope degree, slope aspect, plan curvature, convergence index, topographic wetness index (TWI), lithology, land use/land cover (LU/LC), distance from rivers, distance from roads, drainage density, and NDVI were selected to model GE. The results of variables importance by RF and BRT models indicated that distance from road, elevation, and lithology had the highest effect on GE occurrence. The area under the curve (AUC) and seed cell area index (SCAI) methods were used to validate the three GE maps. The results showed that AUC for the three models varies from 0.911 to 0.927, whereas the RF model had a prediction accuracy of 0.927 as per SCAI values, when compared to the other models. The findings will be of help for planning and developing the studied region.

Keywords: gully erosion; environmental variables; data mining techniques; SCAI; GIS

1. Introduction

Today, reducing natural resources, especially soil and water, is one of the major problems and major threats to human life and is one of the most important environmental problems worldwide that has intensified in recent years, with increasing population and the alternation of human activities [1]. According to the data from United Nations research, the world's population is growing at a rate of 1.8% per year and it is expected to rise from 8 billion in 2025 to 9.4 billion in 2050 [2]. This increase in world population would demand the need for food, water, forage, and others, which consequently would add huge pressure on land exploitation, non-standard exploitation, and eventually lead to an increase in erosion rates [1,3]. Soil erosion is one of the factors that endangers water and soil [1]. Soil erosion by water, such as GE, is considered as a major cause of land degradation around the world [4,5]. It leads to a range of problems, such as desertification, flooding and sediment deposition in reservoirs [6,7], the destructive effects on the ecosystem reducing soil fertility, and imposes huge economic costs [8]. GE is typically defined as a deep channel that has been eroded by concentrated water flow, removing surface soils and materials [9,10]. The amount of moisture and its changes as a result of the dry and wet seasons is a main parameter in creating cracks and grooves in fine-grained clay formations containing clay and silt, and ultimately developing rilled erosion and gullies [10]. The alternation of warm and dry seasons makes it possible to create cracks, in the formation of fine grains, in warm seasons with the drying of the land and the wilting of the vegetation, and these cracks at the time of the first sudden rainfall concentrate the runoff and therefore cause rill and GE to emerge [11]. GE occurs when the erosion of the water flow or the erodibility of the sediments or the formation of the area is higher than the geomorphological threshold of the area [11]. Mapping gully erosion systems is essential for implementing soil conservation measures [6]. GEVs that influence gully occurrence are rainfall, topography-derived factors such as elevation, slope degree, slope aspect, and plan curvature, lithology [12], soil properties [13], and LU/LC [14]. The distribution of precipitation affects the hydraulic flow and moisture content of the soil, and the erosion strength of the flow and soil resistance to erosion is different before and after erosion [11]. Generally, the amount and volume of flow are controlled by the topographic features of the area including slope, aspect, and drainage area of the area. Depth and morphology of the cross section of the gullies are controlled by soil erodibility features of the geological layers of the area. The characteristics of the region's soil affect the subsurface flow and the phenomenon of piping erosion, and the pipes cause a gully when their ceiling collapses [10].

Susceptibility maps of GE are essential for conservation of natural resources, and for evaluating the relationship among gully occurrence and relevant GEVs [12]. Several models have been applied to assess soil erosion and GE rate in a quantitative and qualitative way, such as the Universal Soil Loss Equation (USLE) [1,15], Erosion Potential Method, Modified Pacific Southwest Interagency Committee Model (MPSIAC) [16], Water Erosion Prediction Project (WEEP) [17], European Soil Erosion Model (EUROSEM) [18], Ephemeral Gully Erosion Model (EGEM) [19], and Chemicals, Runoff, and Erosion from Agricultural Management Systems (CREAMS) [20].

Within the soil conservation research field, the distribution of soil erosion is one of the primary sources of information. This is also relevant for GE; however, in the above mentioned methods, spatial distribution of gullies has not been addressed. Remote sensing-based methods to identify GE have been developed [21], including with RF machine learning, though they serve more to validate susceptibility models and to explain the actual erosion presence and distribution. In recent years, scientific research for susceptibility analysis of GE, and work on the statistical relationships between GEVs and the spatial distribution of gullies, have been addressed using various statistical and machine learning methods including bivariate statistics (BS) [1], weights-of-evidence (WoE) [13], index of entropy (IofE) [8], logistic regression (LR) [22–26], information value (IV) [24,25], random forest (RF) [27], bivariate statistical models [28,29], maximum entropy (ME) [30,31], frequency ratio (FR) [28], analytical hierarchy processes (AHP) [29], artificial neural network (ANN) [12,31], support vector machine (SVM) [31], and boosted regression trees (BRT) [12]. For this purpose, various GEVs such as topography (e.g., elevation, slope, aspect, plan curvature, profile curvature, slope length), lithology, land use, soil properties (e.g., soil texture, soil type, erosivity, soil water content), land use, climate (rainfall intensity, rainfall period, and spatial distribution of rainfall), infrastructures (road, bridge) and hydrology (e.g., TWI, SPI, drainage density) were used.

A comprehensive literature review shows that there are still dimensions that require further research, and that a large number of potentially useful methods have not yet been fully implemented to provide GE susceptibility maps. The main objectives of this study are: (i) To determine the relationship between gully occurrence and conditioning factors using Weights-of-Evidence Bayes theory, (ii) assessing the capability of RF, MARS, and BRT data mining/machine learning models to predict GE susceptibility; and (iii) validation of models using the AUC curve and SCAI methods. Study of the research background showed that using MARS, BRT, and RF data mining models in GE zonation is very new. It will help managers in future planning to prevent human intervention in sensitive areas.

2. Materials and Methods

2.1. Study Area

The Shahroud watershed, with an area of about 848 km² and elevation range from 1084 to 2131 m a.s.l., is located in the northeastern part of Semnan Province, Iran (Figure 1). The study area receives an average rainfall of less than 250 mm has an arid and semi-arid climate [32]. Various types of lithological formations cover this watershed, and the landforms are mainly low level pediment fans and valley terrace deposits. The dominant land use is rangelands, but irrigation farming and bare lands are also present.

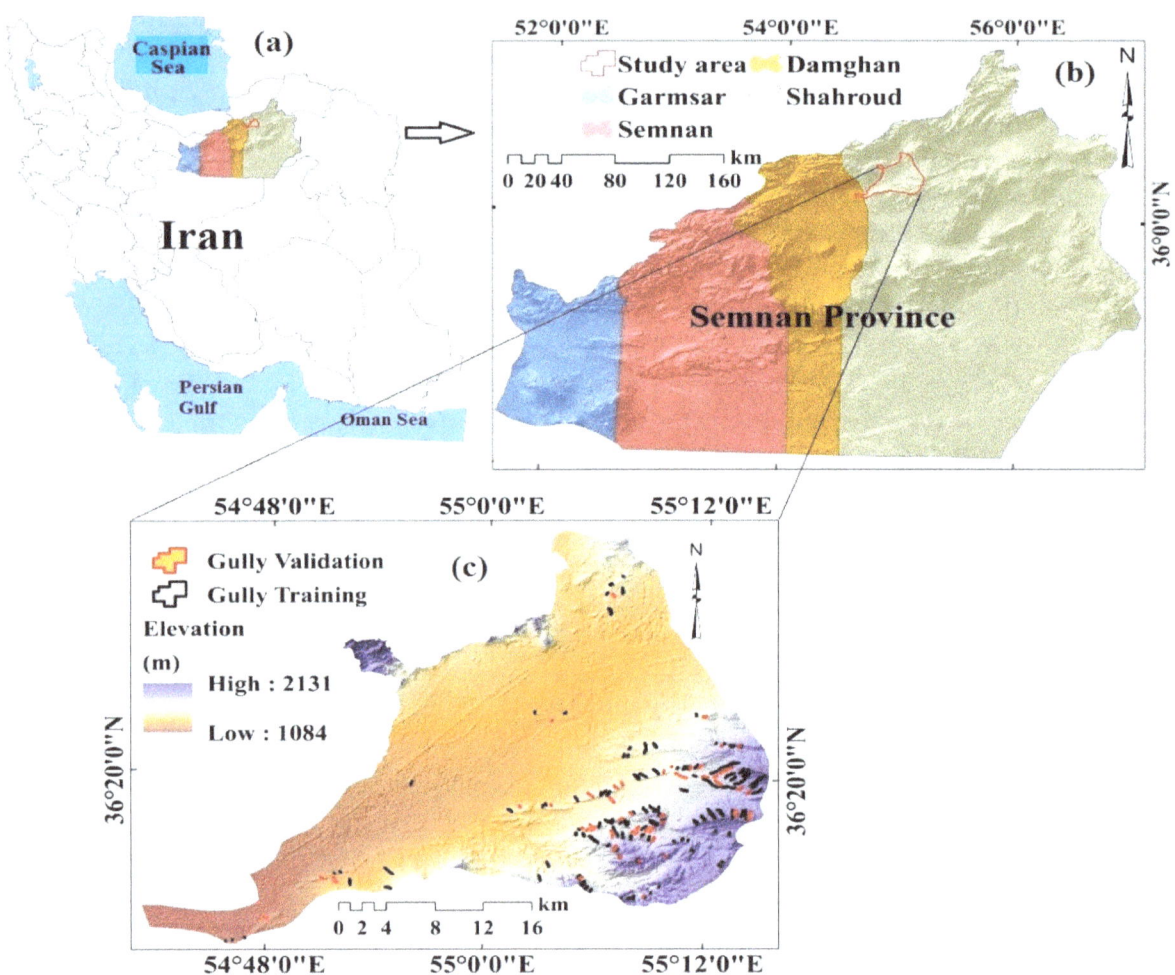

Figure 1. (**a**) Location of the Semnan provinces in Iran, (**b**) location of study area, and (**c**) gully erosion locations with the digital elevation model map of the Shahroud watershed.

2.2. Data and Method

Figure 2 shows the methodological approach applied to map GE susceptibility in the Shahroud watershed using BRT, MARS, and RF models. For preparing an accurate and reliable gully inventory map, extensive field surveys with a DGPS device were performed in the study area to determine the location of the Gullies [27,28]. Then, among 172 detected gully locations, randomly (70/30 ratio), 121 gully locations (70%) and 51 gully locations (30%) in the polygon format were used for training the testing models [28]. The locations of training and testing gullies are shown in Figure 1. Interventionary studies involving animals or humans, and other studies require ethical approval must list the authority that provided approval and the corresponding ethical approval code.

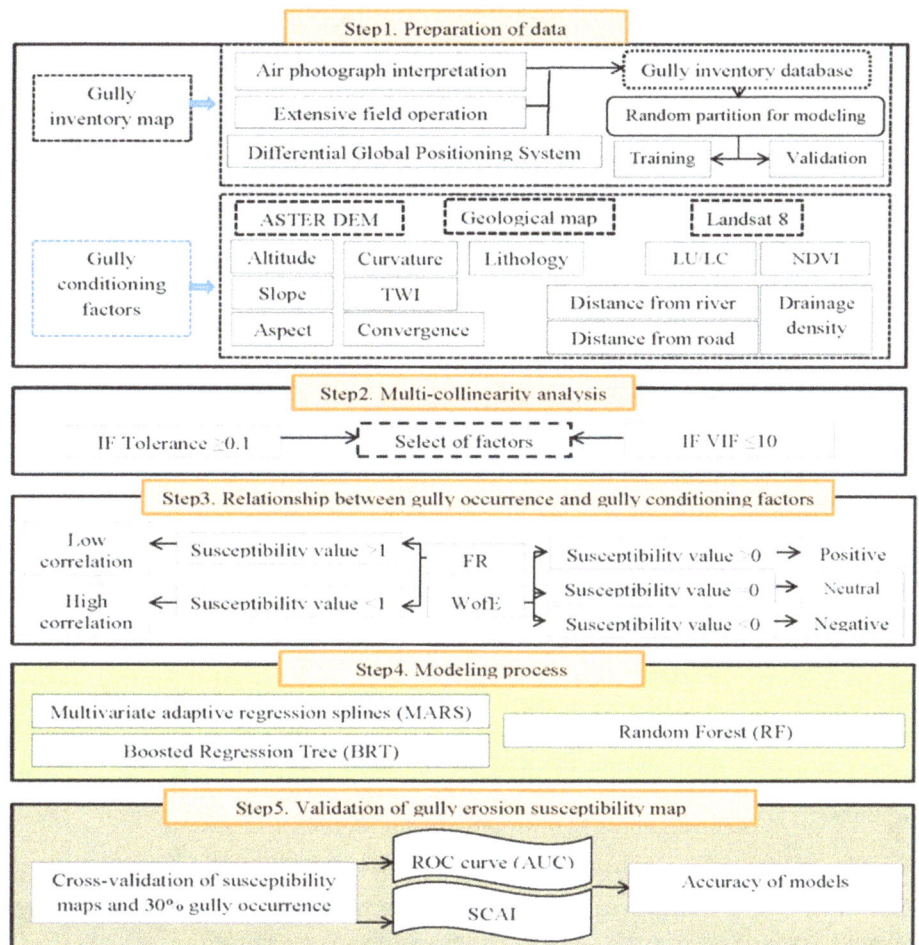

Figure 2. Flowchart of research methodology.

The tools used in present study are ArcGIS10.5, ENVI 4.8, SAGA-GIS 2.1.1, and a DGPS. The basic maps used were geological maps [33], at a scale of 1:100,000, topographic maps, at a scale of 1:50,000, satellite images acquired by Landsat8, and ASTER GDEM with spatial resolution of 30 m [34]. In this study, based on literature review [24,26,31] and local conditions of the study area, twelve factors were selected. Elevation map was divided into six classes: <1200 m, 1200–<1350 m, 1350–<1450 m, 1450–1600 m, and >1600 m (Figure 3a). Slope degree affects surface runoff [35], soil erosion, and pattern of drainage density. Slope degree map was classified into six classes [24,26]: <5°, 5–<10°, 10–<15°, 15–<20°, 20–<25°, 25–30°, and >30° (Figure 3b).

The aspect map was classified into nine classes (Figure 3c). Positive and negative values of plan curvatures define convexity and concavity of slope curvature, whereas zero is flat surface. The plan curvature map was divided into 3 categories: Concave, Flat, and Convex. The TWI indicator is important for identifying prone areas to GE [36]. TWI is calculated by Equation (1):

$$TWI = \ln\left(\frac{S}{\tan \alpha}\right) \qquad (1)$$

TWI map of study area is divided into four classes [24,26,37] including <5, 5–<7.5, 7.5–11, and >11 (Figure 3e). The convergence index (CI) gives a measure of how flow in a cell diverges (convergence index in negative and positive values) [38]. The CI map was prepared in SAGA-GIS 2.1.1 and divided into 3 classes: <0, 0–10, and >10 (Figure 3f). In this research, for the computation of the effect of drainage network and infrastructures on GE, the distance from rivers and roads was considered [14] and divided into four classes: <170 m, 170–<370 m, 370–650 m, and >650 m for rivers (Figure 3g) and <500 m, 500–<1500 m, 1500–3000 m, and >3000 m for roads (Figure 3h). The line density tool in ArcGIS 10.5 was used for calculating drainage density and then its map was divided into four categories: <1.4, 1.4–<2.4, 2.4–3.7, and >3.7 km/km^2 (Figure 3i). A geological map at a 1:100,000 scale was used to prepare the lithological unit layer. The lithological units were classified into ten categories based on their sensitivity to gully occurrence using expert knowledge method (Figure 3j and Table 1). The advantage of this method is it is easy to use, however this method has certain disadvantages, such as the possibility of a mistake by the expert.

Table 1. Lithology of the study area.

Code	Lithology	Geological Age
Murmg	Gypsiferous marl	Miocene
Qft2	Low level piedment fan and vally terrace deposits	Quaternary
Ku	Upper cretaceous, undifferentiated rocks	Cretaceous
Jd	Well—bedded to thin—bedded, greenish—grey argillaceous limestone with intercalations of calcareous shale (DALICHAI FM)	Jurassic
PeEz	Reef-type limestone and gypsiferous marl (ZIARAT FM)	Paleocene-Eocene
PlQc	Fluvial conglomerate, Piedmont conglomerate and sandstone.	Pliocene-Quaternary
Jl	Light grey, thin—bedded to massive limestone (LAR FM)	Jurassic-Cretaceous
E2c	Conglomerate and sandstone	Eocene
PlQc	Fluvial conglomerate, Piedmont conglomerate and sandstone.	Pliocene-Quaternary
E1c	Pale-red, polygenic conglomerate and sandstone	Paleocene-Eocene

The LU/LC map was obtained using Landsat 8 images [39–41]. The main LU/LC types identified in the study area were range, irrigation farming, and bare lands (Figure 3k). The NDVI map was also produced using Landsat 8 images and classified into 3 categories: <0.11, 0.11–0.25, and >0.25 (Figure 3l).

For multi-collinearity checking, the tolerance (TOL) and variance inflation factor (VIF) were used. If during modeling there is collinearity among the variables, the accuracy of the model's prediction decreases. Values of TOL and VIF were ≤0.1 and ≥10, respectively, indicating that multi-collinearity among parameters [28].

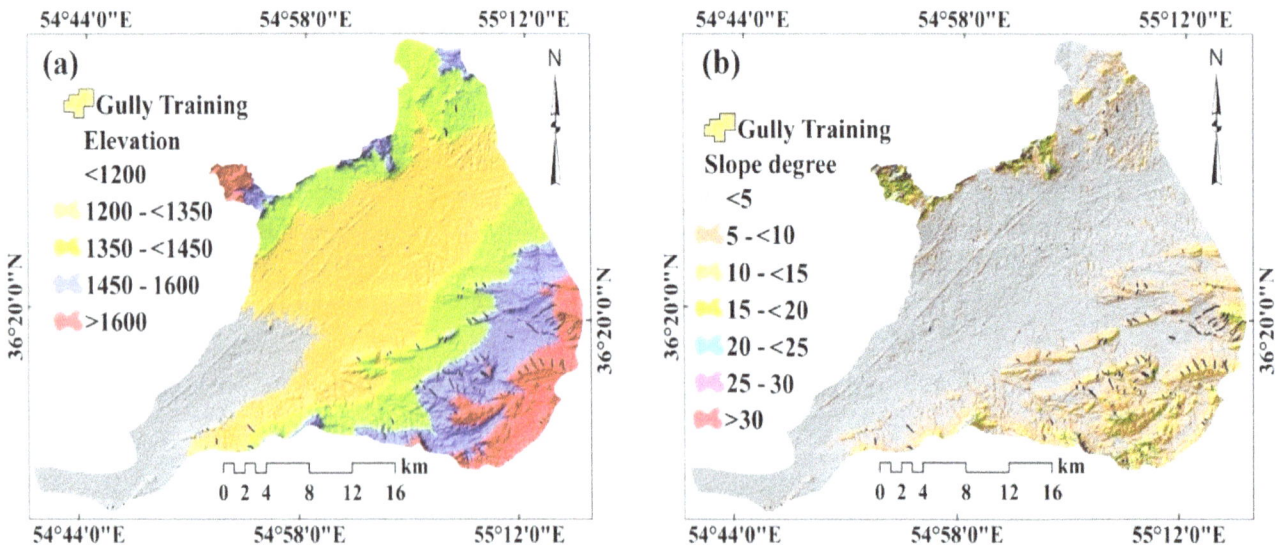

Figure 3. *Cont.*

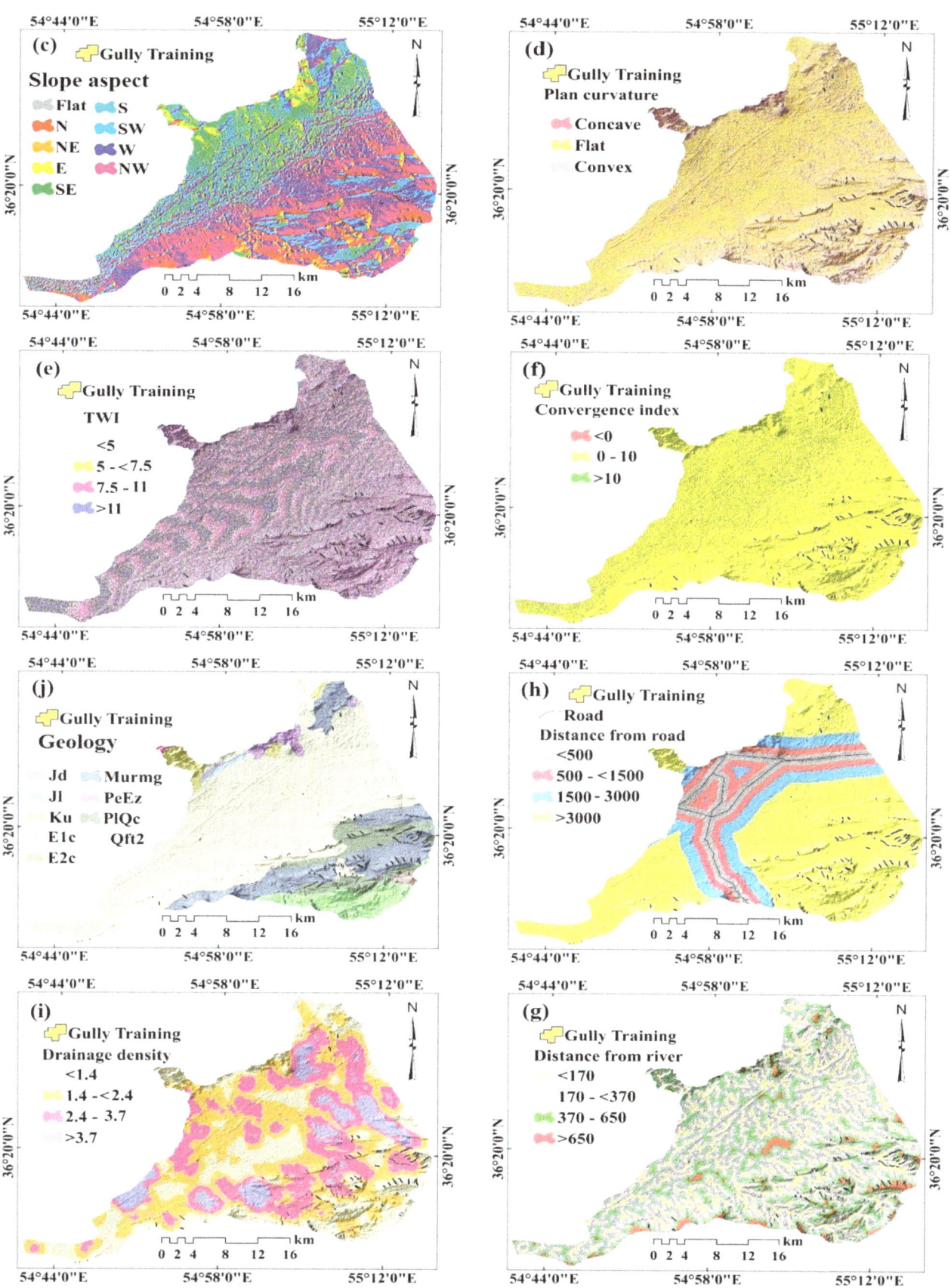

Figure 3. *Cont.*

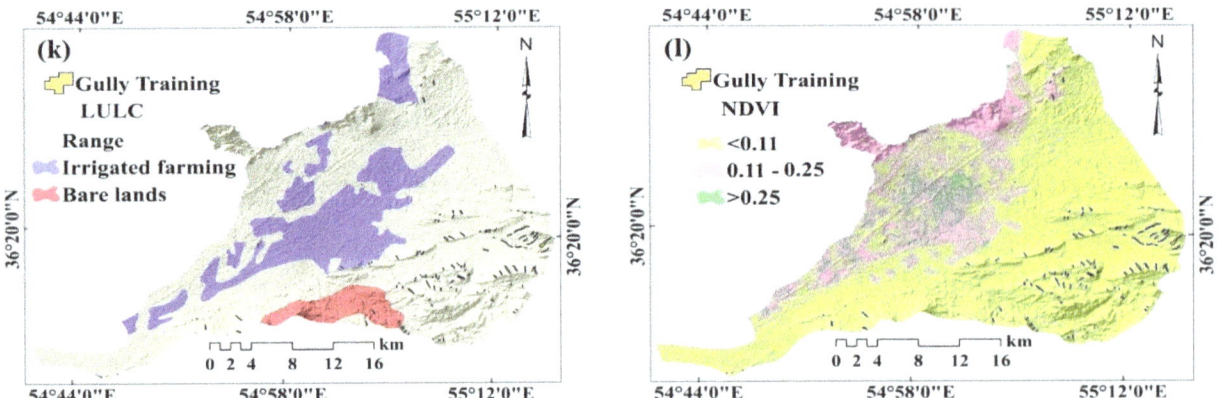

Figure 3. Gully erosion conditioning factors: (**a**) Elevation, (**b**) slope, (**c**) aspect, (**d**) plan curvature, (**e**) TWI, (**f**) convergence index, (**j**) geology, (**h**) distance from road, (**i**) drainage density, (**g**) distance from road, (**k**) land use/land cover (LU/LC), and (**l**) NDVI.

2.3. Gully Erosion Modelling

2.3.1. WoE Model

WoE is according to the Bayesian probability framework, to predict the significance of effective factor classes through a statistical approach [42–51]. In this method, the spatial relationship between GE areas and GEVs are identified. The WoE model is based on the calculation of positive (W^+) and negative (W^-) weights. This model computes the weight of each GEVs according to the existence or absence of the gully inventory [52] as follows:

$$W_i^+ = \ln\left(\frac{P\{B|L\}}{P\{B|\overline{L}\}}\right) \tag{2}$$

$$W_i^- = \ln\left(\frac{P\{\overline{B}|L\}}{P\{\overline{B}|\overline{L}\}}\right) \tag{3}$$

$$C = W^+ + W^- \tag{4}$$

$$S(C) = \sqrt{S^2(W^+) + S^2(W^-)} \tag{5}$$

$$S^2(W^+) = \frac{1}{N\{B \cap L\}} + \frac{1}{\{B \cap L\}} \tag{6}$$

$$S^2(W^-) = \frac{1}{\{\overline{B} \cap L\}} + \frac{1}{\{\overline{B} \cap \overline{L}\}} \tag{7}$$

$$W = \left(\frac{C}{S(C)}\right) \tag{8}$$

where ln is the natural log function and P is the probability, B and $\overline{B}$ indicate the presence and absence of the gully geo-environmental factor, respectively, L is the presence of gully, and $\overline{L}$ is the absence of a gully. W^+ and W^- are positive and negative weights, with W^+ indicating that a geo-environmental factor is present in the gully inventory. $S^2(W^+)$ is the variance of the W^+ and $S^2(W^-)$ is the variance of W^-. C indicates the overall association between GEVs and gully occurrence. S(C) is the standard deviation of the contrast and W is final weight of each class factor.

2.3.2. RF Model

RF is a controlled learning method that uses multiple trees in the classification [21]. The RF algorithm, by replacing and continuously changing the factors that affect the target, leads to the creation of a large number of decision trees, then all trees are combined to make decisions [21]. The RF consists of 3 user-defined parameters, which include: (1) The number of variables used in the construction of each tree, which expresses the power of each independent tree; (2) number of trees in RF; and (3) minimum number of nodes [43]. RF prediction power increases with the increasing strength of independent trees and reducing the correlation between them [44]. This algorithm uses 66% of the data to grow a tree called Bootstrap, and then a predictor variable is introduced randomly during the growing process to split a node in the tree construction. The remaining 33% of the data is also used to evaluate the fitted tree [45]. This process is repeated several times and the average of all predicted values is used as the final prediction of the algorithm. In this model, two factors, including the mean decrease accuracy and mean decrease Gini, are used to prioritize of each of the effective factors. The use of the mean decrease accuracy in comparison to mean decrease Gini index is more effective in determining the priority of effective factors, especially in the context of the relationship between environmental factors [46]. The RF analyses were carried out in R 3.3.1, using the "Randomforest" package [21].

2.3.3. BRT Model

BRT is one of several techniques that can help improve the performance of a single model by combining multiple models [47]. BRT uses two algorithms for modeling: Boosting and regression [48]. Boosting is a way to increase the accuracy of the model, and based on this, the construction, combination, and averaging of a large number of models are better and more accurate than an individual model on its own [49]. BRT overcomes the greatest weakness of the single decision tree, which is relatively weak in data processing. In BRT, only the first tree of all the training data is constructed, the next trees are grown on the remaining data from the tree before it; trees are not built on all data and only use a number of data [50]. The main idea in this method is to combine a set of weak predictor models (high predictive error) to arrive at strong prediction (low predictive error) [51]. Thus, in this study, BRT was used for GE spatial modeling using GMB (Generalized Boosted Models) and dismo (Species Distribution Modeling) packages in R 3.3.1.3.

2.3.4. MARS Model

The MARS model is a form of regression algorithm that was introduced by Friedman in 1991 to predict continuous numerical outputs [52]. This technique generates flexible regression models for predicting the target variable by means of dividing the problem space into intervals of input variables and processing a basic function in each interval.

The base function represents information in relation to one or more independent variables. A base function is defined in a given interval, in which the primary and end points are called knote. The knote is the key concept in this method and represents the point at which the behavior of the function changes at that point. The base function expresses the relationship between the input variables and the target variable and is in the form of $Max\,(0, X - c)$ or $Max\,(0, c - X)$, in which c is threshold value and X is the impute variable. The general form of the MARS model is as follows:

$$f\,(x) = \beta_0 + \sum_{j=1}^{P} \sum_{b=1}^{B} \left[\beta_{jb}(+)Max\left(0, x_i - H_{bj}\right) + \beta_{jb}(-)\,Max\left(0, H_{bj} - x_J\right)\right] \tag{9}$$

where x = input, $f(x)$ = output, P = predictor variables, and B = basis function. $Max\,(0, x - H)$ and $Max\,(0, H - x)$ are basis function and do not have to be present if their coefficients are 0. β_0 is constant, β_{jb} is the coefficient of the jth base function (BF), and the H values are called knots. The MARS model includes three main steps: (1) A forward stepwise algorithm to select certain spline basis functions, (2) a backward stepwise algorithm to delete base functions (BFs) until the best set is found, and (3) a smoothing method which gives the final MARS approximation a certain degree of continuity [52]. First, the MARS model estimates the value of the target function with a constant value, and then generates the best processing in the forward direction by searching among the variables.

The search process continues as long as all possible (BFs) are added to the model. At this stage, a very complex model with a large number of knotes is obtained. In the next step, through the process of pruning backward, BFs that are less important are identified and deleted by using the generalized cross-validation (*GCV*) criterion [27]. *GCV* is a criterion for data fitting and eliminates a large number of BFs and reduces the probability of overfitting. This indicator is obtained by using Equation (10):

$$GCV = \frac{1}{2} \sum_{i=1}^{N} [y_i - f(x_i)]^2 / \left[1 - \frac{C(B)}{N} \right]^2 \qquad (10)$$

where N is the number of data and $C(B)$ is a complexity penalty that increases with the number of BF in the model and which is defined as:

$$C(B) = (B = 1) + dB \qquad (11)$$

where d is a penalty for each BF included into the model. This process continues until a complete review of all the basic functions, and at the end of the optimal model is obtained by applying base functions [52]. MARS model is an adaptive approach, since the selection of BFs and node locations is based on the data and type of purpose. After determining the optimal MARS model, the analysis of variance (ANOVA) method can be used to estimate the participation rate of each of the input variables and BFs. A detailed description of the MARS model can be found in [45]. MARS was run with R 3.3.1 and the "Earth" package [53].

2.4. Validation of GESMs Using Three Data Mining Models

A single criterion is not enough to select the best model among a large number of models, and judging about choosing a superior model by one criteria. It is not a suitable approach and it raises the chance of mistake in choosing the suitable model [27,37,54]. In this study, to compare the performance between data mining models and select the appropriate model, AUC and SCAI were used [28,36,55]. For calculating AUC, different thresholds were considered from 0 to 1, and for each threshold, the number of cells detected by the model as gully erosion was compared with observed gully erosion cells and positive and negative ratio indicators was calculated. After calculating these two indicators, we arranged them in ascending order, then they were plotted to calculate AUC. The AUC values range from 0.5 to 1. If a model cannot estimate the occurrence of an event better than a probable or random viewpoint, its AUC is 0.5 and therefore it will have the least accuracy, while if the AUC is equal to one, the model will have the highest accuracy [56,57]. The quantitative–qualitative relationship between AUC value and prediction accuracy can be classified as follows: 0.5–0.6, poor; 0.6–0.7, average; 0.7–0.8, good; 0.8–0.9, very good; and 0.9–1, excellent. SCAI is the ratio of the percentage area of each of the zoning classes to the percentage of gullies occurring on each class. Based on the SCAI indicator, the values of SCAI in very high sensitivity class are lower than very low sensitivity class.

3. Results

3.1. Multi-Collinearity Analysis

Multicollinearity is a condition of very high inter-correlations or inter-associations among the independent variables. Therefore, it is a type of disturbance in the data, and if present in the data, the statistical conclusions of the data may not be reliable [27]. A TOL value less than 0.1 or a VIF value larger than 10 indicates a high multicollinearity [56]. The outcomes of the coherent analysis among the 12 GEVs are shown in Table 2. The outcomes showed that the TOL and VIF of all GEVs were $\geq$0.1 and $\leq$5, respectively. As a result, no multi-collinearity is seen among the GEVs.

Table 2. Multi-collinearity of effective factors using tolerance (TOL) and variance inflation factor (VIF).

Conditioning Factors	Collinearity Statistics	
	Tolerance	VIF
Constant Coefficient	-	-
Slope degree	0.998	1.002
Distance from road	0.672	1.489
Distance from river	0.323	3.094
Plan curvature	0.674	1.483
Lithology	0.945	1.058
LU	0.864	1.158
Drainage density	0.826	1.211
Elevation	0.920	1.087
Convergence index	0.666	1.503
Aspect	0.299	3.343
TWI	0.942	1.062
NDVI	0.941	1.063

3.2. Spatial Relationship Using WoE Model

The outcomes of WoE model are shown in Table 3. In elevation, the results of WoE indicate that there is a direct correlation between classes of elevation and GE, and with an increase in elevation, GE also increases. Therefore, the class of >1600 m with WoE 47.95 had the greatest impact on gully occurrence. The result of slope degree indicate that classes 5–<10 with WoE 34.96 had a strong relation with GEIM. For slope aspect, NE–facing slopes with a value of 19.46 show high probability of gully occurrence. In the case of plan curvature, among the three classes of concave, flat, and convex, the concave class had the highest value (78.04), and thus a positive correlation with GE. This result is in line with [11,50]. In TWI, the class of >11 has the strongest relationship with GE with the highest value (78.04). In the case of the convergence index, the class of 0–10 with values of 13.18 has a positive relation with gully occurrence. With respect to distance from river, class of >650 with value of 25.86 and regarding distance from road the class of >3000 m with values of 16.25 had the greatest effect on gully occurrence. For the drainage density factor, the class of <1.4 km/km^2 showed the highest value (14.23) and thus high correlation with gully occurrence. According to the lithology factor, Gypsiferous marl with greatest value (51.23) is more prone to GE than other lithology units. Concerning LU/LC, most gullies are located in the range land use type and this class with the highest value (21.02) has the strongest relationship with gully occurrence. In NDVI, results indicated that all gullies are located in the class of <0.11, showing that very low vegetation density renders slopes susceptible to GE.

Table 3. Relationship between conditioning factors and gully erosion using weights-of-evidence (WoE) model.

Factor	Class	Number of Pixels in Domain	Pixels of Gullies	Weights-of-Evidence (WoE)				
				C	S2 (w$^+$)	S2 (w$^-$)	S	W
	<1200	144,200	21	−3.16	0.05	0.00	0.22	−14.41
	1200–<1350	348,463	89	−2.87	0.01	0.00	0.11	−26.60
1	1350–<1450	230,735	502	−0.37	0.00	0.00	0.05	−7.52
	1450–1600	133,305	1057	0.33	0.00	0.00	0.00	0.00
	>1600	85,376	1074	1.88	0.00	0.00	0.04	47.95
	<5	705,163	896	−1.83	0.00	0.00	0.04	−44.90
	5–<10	171,923	1259	1.34	0.00	0.00	0.04	34.96
	10–<15	38,854	397	1.38	0.00	0.00	0.05	25.36
2	15–<20	13,936	121	1.13	0.01	0.00	0.09	12.13
	20–<25	6223	50	1.03	0.02	0.00	0.14	7.22
	25–30	3396	15	0.42	0.07	0.00	0.26	1.62
	>30	2584	5	−0.41	0.20	0.00	0.45	−0.92

Table 3. *Cont.*

Factor	Class	Number of Pixels in Domain	Pixels of Gullies	Weights-of-Evidence (WoE)				
				C	S2 (w$^+$)	S2 (w$^-$)	S	W
3	Flat	16,770	2	−3.22	0.50	0.00	0.71	−4.55
	N	72,345	208	−0.01	0.00	0.00	0.07	−0.19
	NE	79,383	209	−0.11	0.00	0.00	0.07	−1.52
	E	72,794	43	−1.66	0.02	0.00	0.15	−10.81
	SE	91,567	54	−1.68	0.02	0.00	0.14	−12.24
	S	114,731	246	−0.34	0.00	0.00	0.07	−5.13
	SW	119,263	396	0.15	0.00	0.00	0.05	2.81
	W	142,533	459	0.12	0.00	0.00	0.05	2.35
	NW	232,693	1126	0.76	0.00	0.00	0.04	19.46
4	Concave	54,613	1493	2.99	0.00	0.00	0.04	78.04
	Flat	574,180	749	−1.43	0.00	0.00	0.04	−33.33
	Convex	313,286	501	−0.80	0.00	0.00	0.05	−16.27
5	<7	24,272	63	0.00	0.00	0.00	0.02	−0.15
	5–<7.5	42,453	91	−0.32	0.01	0.00	0.11	−3.00
	7.5–11	89,328	225	−0.16	0.00	0.00	0.07	−2.29
	>11	786,026	2364	0.21	0.00	0.00	0.06	3.87
6	<0	75,370	26	−2.21	0.04	0.00	0.20	−11.21
	0–10	776,920	2534	0.95	0.00	0.00	0.07	13.18
	>10	89,789	183	−0.39	0.01	0.00	0.08	−5.08
7	<170	382,383	522	−1.07	0.00	0.00	0.05	−21.99
	170–<370	329,586	835	−0.21	0.00	0.00	0.04	−4.99
	370–650	179,671	914	0.75	0.00	0.00	0.04	18.62
	>650	50,444	472	1.31	0.00	0.00	0.05	25.86
8	<500	90,285	0	−0.10	0.00	0.00	0.02	−5.29
	500–<1500	102,453	0	−0.12	0.00	0.00	0.02	−6.05
	1500–3000	113,685	12	−3.44	0.08	0.00	0.29	−11.91
	>3000	635,661	2731	4.70	0.00	0.08	0.29	16.25
9	<1.4	277,251	1150	0.55	0.00	0.00	0.04	14.23
	1.4–<2.4	353,215	1000	−0.04	0.00	0.00	0.04	−1.12
	2.4–3.7	231,503	573	−0.21	0.00	0.00	0.05	−4.49
	>3.7	80,115	20	−2.54	0.05	0.00	0.22	−11.32
10	Murmg	144,412	1544	1.97	0.00	0.00	0.04	51.23
	Qft2	617,176	417	−2.36	0.00	0.00	0.05	−44.47
	Ku	23,972	0	−0.03	0.00	0.00	0.02	−1.35
	Jd	18,232	0	−0.02	0.00	0.00	0.02	−1.03
	PeEz	1449	0	0.00	0.00	0.00	0.02	−0.08
	PlQc	71,058	600	1.24	0.00	0.00	0.05	26.85
	Jl	3,274	0	0.00	0.00	0.00	0.02	−0.18
	E2c	58,380	174	0.03	0.01	0.00	0.08	0.33
	E1c	4,820	8	−0.56	0.13	0.00	0.35	−1.60
11	Range	708,879	2669	2.48	0.00	0.01	0.12	21.02
	Farming	193,682	33	−3.06	0.03	0.00	0.18	−17.47
	Bare land	39,523	41	−1.06	0.02	0.00	0.16	−6.75
12	<0.11	863,198	2743	0.09	0.00	0.00	0.02	4.59
	0.11–0.25	56,745	0	−0.06	0.00	0.00	0.02	−3.26
	>0.25	22,140	0	−0.02	0.00	0.00	0.02	−1.25

1. Elevation, 2. Slope degree, 3. Slope aspect, 4. Plan curvature, 5. topographic wetness index (TWI), 6. Convergence index, 7. Distance from river, 8. Distance from road, 9. Drainage density, 10. Lithology, 11. land use (LU), 12. NDVI.

3.3. *Applying RF Model*

The outcomes of the confusion matrix for RF model are shown in Table 4. The result shows that the model predicted 2487 non-gully pixels as non-gullies and 256 non-gullies as gully. On the other hand, the RF model predicted 2677 gullies as gullies and 66 gullies as non-gullies. Moreover, the out-of-bag error (OOB) for RF was 5.82%. This means that the model has a precision of 94.18%, which expresses the excellent accuracy of the model in predicting gully erosion.

Table 4. Confusion matrix from the random forest (RF) model (0 = no gully, 1 = gully).

	0	1	Class Error
0	2487	256	0.0933
1	66	2677	0.0240

Prioritization results of RF are shown in Table 5 and Figure 4. The results show that the distance from roads (381.67, 22%), elevation (335.06, 19%), and lithology (234.21, 14%) had the highest values, followed by slope degree, drainage density, distance from river, NDVI, convergence index, slope aspect, TWI, plan curvature, and LU/LC.

Table 5. Relative influence of effective conditioning factors in the RF model.

Conditioning Factors	Weight
Distance from road	381.67
Elevation	335.06
Lithology	234.21
Slope degree	153.85
Drinage density	126.72
Distance from river	106.84
NDVI	105.26
Convergence index	73.97
Slope aspect	72.41
TWI	71.3
Plan curvature	42.43
LU	25.38

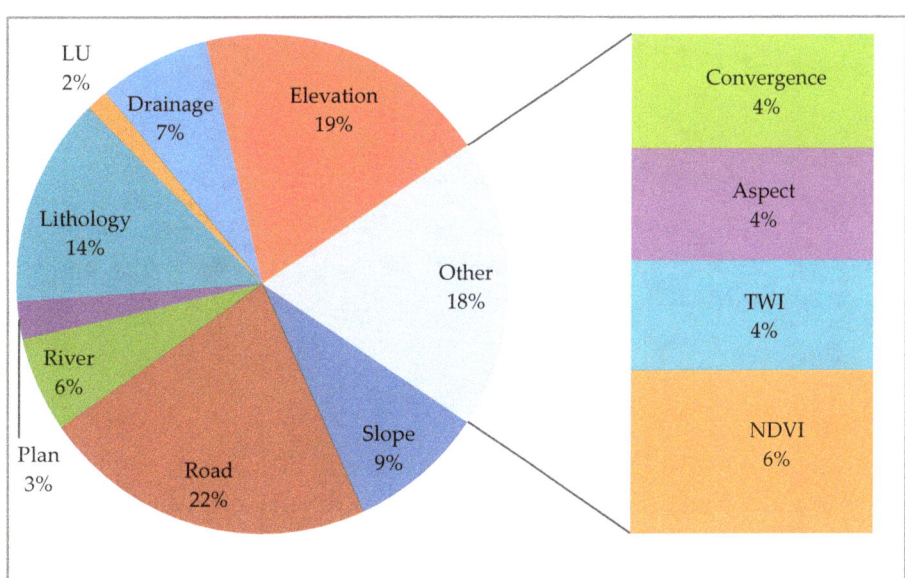

Figure 4. Relative influence of effective conditioning factors in the random forest (RF) model.

Finally, the GESM by the RF model was prepared in ArcGIS 10.5 and divided into five classes from very low to very high (Figure 5a), using a natural break classification [8]. According to the results, of the entire study area (847.87 km^2), 525.97 km^2 (62.03%) are located in the very low susceptibility class, 148.28 km^2 (17.49%) in the low susceptibility, 79.42 km^2 (9.37%) in the moderate class, 56.34 km^2 (6.64%) in the high class, and 37.88 km^2 (4.47%) are located in the very high susceptibility class. Of the total area of GE (0.729 km^2) in the study area, 0.86% (0.01 km^2) are located in the very low susceptibility class,

5.67% (0.04 km^2) in the low susceptibility, 14.80% (0.11 km^2) in the moderate susceptibility, 21.95% (0.16 km^2) in the high susceptibility, and 56.72% (0.41 km^2) in the very high susceptibility classes.

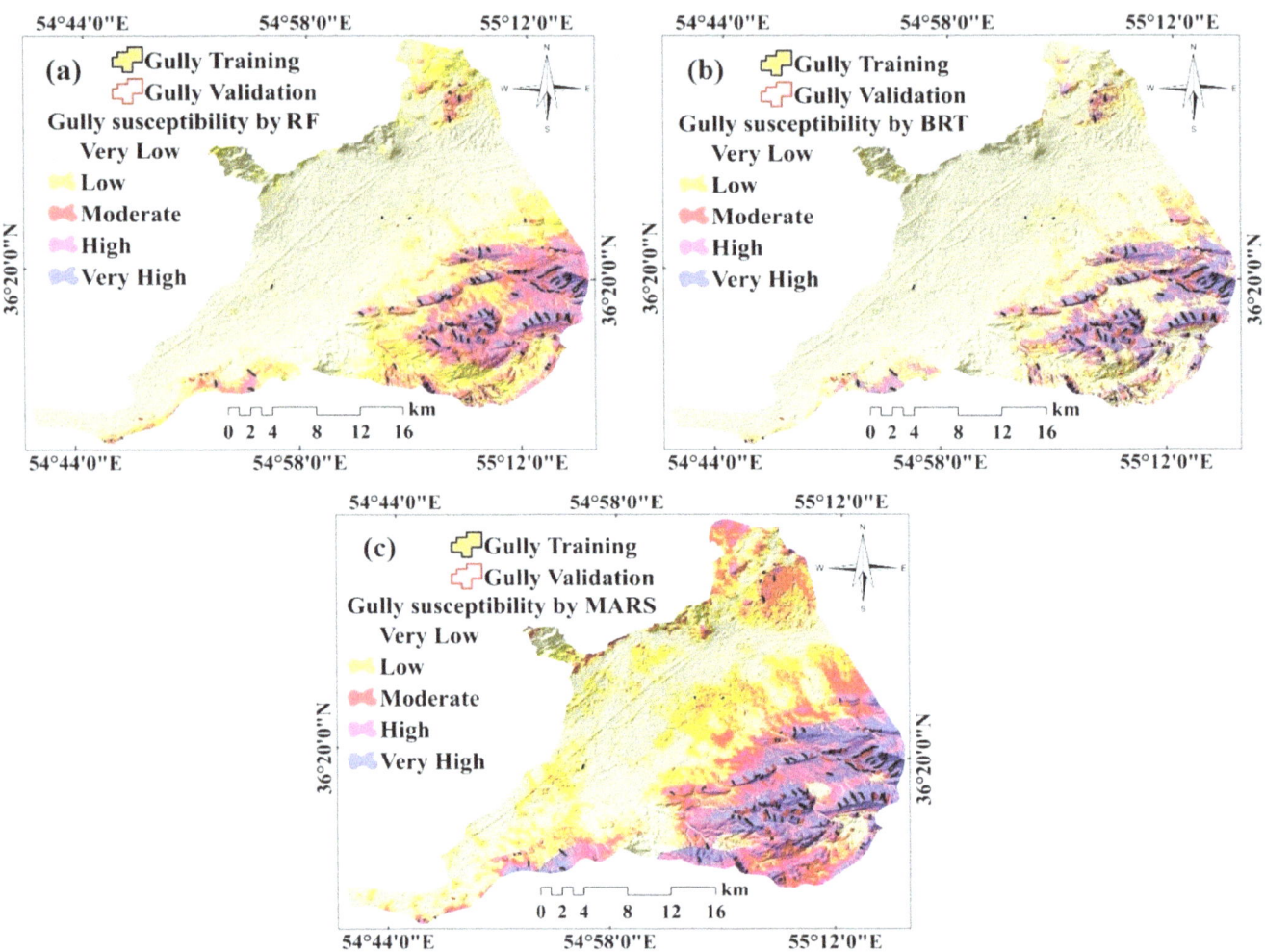

Figure 5. Gully erosion susceptibility maps using: (**a**) RF model, (**b**) BRT model, and (**c**) multivariate adaptive regression spline (MARS) model.

3.4. Applying BRT Model

The BRT model was used to reveal the spatial correlation between the existing GE and the GEVs in the study area. The results of the model are shown in Figure 6. They indicate that the factors distance from roads (31.1%), elevation (27.2%), and lithology (11%) had the highest importance on GE, mirroring the outcomes of the RF model, followed by slope degree (7%), drainage density (6.7%), distance from river (5.1%), slope aspect (3.8%), convergence index (2.4%), NDVI (2.2%), plan curvature (1.6%), TWI (1.6%), and LU/LC (0.3%). The gully susceptibility map by the BRT model was also prepared in ArcGIS 10.5 and divided into five classes of very low to very high (Figure 6c). The results of the GE susceptibility class by the BRT model covered 847.87 km^2 of the study area an area distribution in the very low, low, moderate, high, and very high susceptibility classes are 605.37 km^2, 88.38 km^2, 52.01 km^2, 34.13 km^2, and 67.98 km^2, and percentage distribution in the susceptibility classes of are 71.40, 10.42, 6.13, 4.03, and 8.02, respectively. Of the actual GE area of 0.729 km^2, 0.04 (5.55%), 0.03 (4.56%), 0.06 (8.26%), 0.08 (11.34%), and 0.51 km^2 (70.28%) are located in the very low to very high susceptibility classes, respectively.

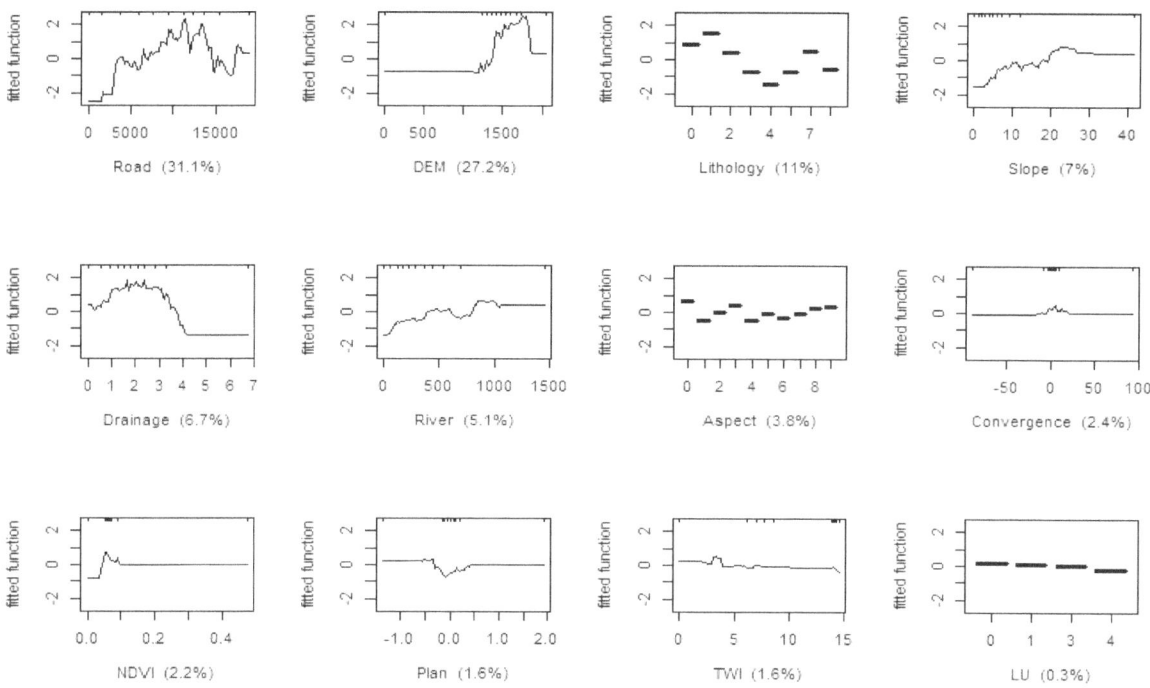

Figure 6. Relative influence of effective conditioning factors in boosted regression tree (BRT) model.

3.5. Applying MARS Model

The optimal MARS model included 28 terms, and the GCV was 0.157. MARS model provides the optimal model only by selecting the necessary parameters. In this research, nine GEVs including lithology, distance from road, distance from river, drainage density, elevation, aspect, convergence index, slope, and NDVI were used to construct the optimal model from the 12 GEVs. The GESM by the MARS model was implemented in ArcGIS 10.5 using Equation (12). According to Equation (12), distance from roads, elevation, and lithology were the most important variables. Values of GESM by MARS model varies from −9.8 to 7.3. At first, GESM classified using quantile, equal interval, natural break, and geometrical interval classification techniques, then, by comparatively analyses of the distribution of training and validation gullies in high and very high classes, the natural break classification technique was most accurate. As a result, GESM by MARS were classified into very low (−9.86–−6.24), low (−6.24–−2.31), moderate (−2.3–0.04), high (0.04–0.38), and very high (0.38–7.32) gully erosion susceptibility zones by natural break classification technique (Figure 5c). The results indicate that 0.02 km² (2.10%) of GE in the study area are located in the very low susceptibility class, with 339.01 km² (39.98% of total study area) and 0.58 km² (79.16%) located in the very high susceptibility class with 105.50 km² (0.58%) (Table 6). In general, the results indicate that for all three models with increasing susceptibility (from very low to very high), the area of the respective classes decreased, while in contrast the areas of GE increased. These results is in line with Youssef et al. (2015).

Table 6. Area under the curve (AUC) values of RF, MARS, and BRT data mining models.

Models	AUC	Standard Error	Asymptotic Significant	Asymptotic 95% Confidence Interval	
				Lower Bound	Upper Bound
RF	0.927	0.007	0.000	0.914	0.941
MARS	0.911	0.008	0.000	0.896	0.926
BRT	0.919	0.007	0.000	0.905	0.933

3.6. Validation of Models

The results of the validation of the models using the AUC curve and SCAI indicator are shown in Figure 7, and in Tables 6 and 7. The results show that the values of the AUC for the three models vary from 0.911 to 0.927, indicating very good prediction accuracy for all models, with RF resulting in the highest value. In addition, the SCAI values for the three models, RF (61.08–0.00), MARS (10.45–0.03), BRT (12.59–0.01), show that the RF model has higher SCAI values compared to the other models in the very low, low, and very high susceptibility classes (Figure 7). In spite of the high efficiency and accuracy of the RF model for GE sensitivity mapping, so far this model has not been used by the research community.

$$
\begin{aligned}
GESP_{MARS} = \ & 0.74 + (0.659 \times Lithology1) + (0.656 \times Lithology7) - 0.0001 \\
& \times max(0, 13445 - Distance\ from\ road) + 0.0001 \\
& \times max(0, Distance\ from\ road - 13445) - 0.0002 \\
& \times max(0, 2907.97 - Distance\ from\ River) - 0.087 \\
& \times max(0, 2.377 - Drainage\ density) - 0.106 \\
& \times max(0, Drainage\ density - 2.377) + 0.001 \times max(0, 1793 \\
& - Elevation) - 0.002 \times max(0, Elevation - 1793) - 0.605 \\
& \times Lithology7 \times Aspect4 - 0.0001 \times max(0, 7355.32 \\
& - Distance\ from\ road) \times Lithology1 - 0.0001 \\
& \times max(0, 11249.2 - Distance\ from\ road) \times Lithology7 \\
& - 0.00002 \times max(0, Distance\ from\ road - 11249.2) \\
& \times Lithology7 + 0.0001 \times max(0, 13445 \\
& - Distance\ from\ road) \times Lithology10 - 0.005 \\
& \times max(0, 84.853 - Distance\ from\ River) \times Lithology1 \\
& - 0.0003 \times max(0, Distance\ from\ River - 84.853) \\
& \times Lithology1 + 0.001 \times Lithology2 \times max(0, Elevation \\
& - 1249) - 0.001 \times Lithology2 \times max(0, 1249 - Elevation) \\
& - 0.019 \times Lithology7 \times max(0, 0.772 - Convergence) - 23.54 \\
& \times Lithology7 \times max(0, NDVI - 0.055) - 22.23 \times Lithology7 \\
& \times max(0, 0.055 - NDVI) - 0.00001 \times max(0, 7.65 - Slope) \\
& \times max(0, Distance\ from\ road - 13445) + 0.00001 \\
& \times max(0, Slope - 7.65) \times max(0, Distance\ from\ road - 13445) \\
& - 0.0001 \times max(0, 8.186 - Slope) \times max(0, 1793 - Elevation) \\
& + 0.00004 \times max(0, Slope - 8.19) \times max(0, 1793 - Elevation) \\
& - 0.0000001 \times max(0, Distance\ from\ road - 3877.78) \\
& \times max(0, 907.97 - Distance\ from\ River) + 0.00000004 \\
& \times max(0, 8861.03 - Distance\ from\ road) \times max(0, 907.97 \\
& - Distance\ from\ River) + 0.0000001 \times max(0, Road \\
& - 8861.03) \times max(0, 907.97 - Distance\ from\ River) - 0.001 \\
& \times max(0, Distance\ from\ road - 13445) \\
& \times max(0, Drainage\ density - 1.821) - 0.000001 \\
& \times max(0, Distance\ from\ road - 11435.4) \times max(0, 1793 \\
& - Elevation) - 0.000001 \times max(0, Distance\ from\ road \\
& - 13238.3) \times max(0, 1793 - Elevation)
\end{aligned}
\tag{12}
$$

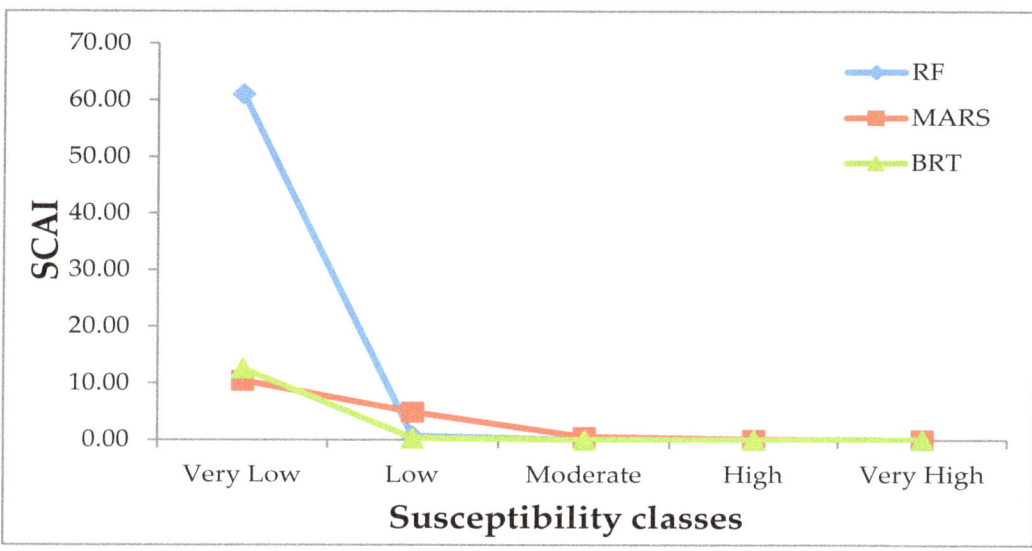

Figure 7. Seed cell area index (SCAI) values for different susceptibility classes in RF, MARS, and BRT data mining models.

Table 7. Seed cell area index (SCAI) values in RF, multivariate adaptive regression spline (MARS), and boosted regression tree (BRT) data mining models.

Model	Susceptibility Classes	Total Area of Classes		Gully in Classes		No Gully Area (km)	Seed Cell (%)	SCAI
		Area (km)	%	Area (km)	%			
RF	Very Low	525.97	62.03	0.01	0.86	525.96	0.01	61.08
	Low	148.28	17.49	0.04	5.67	148.24	0.24	0.74
	Moderate	79.42	9.37	0.11	14.80	79.31	1.15	0.08
	High	56.34	6.64	0.16	21.95	56.18	2.41	0.03
	Very High	37.88	4.47	0.41	56.72	37.46	9.27	0.00
MARS	Very Low	339.01	39.98	0.02	2.10	339.00	0.04	10.45
	Low	194.83	22.98	0.01	1.48	194.82	0.05	4.89
	Moderate	131.17	15.47	0.04	5.67	131.13	0.27	0.58
	High	77.35	9.12	0.08	11.59	77.26	0.93	0.10
	Very High	105.50	12.44	0.58	79.16	104.92	4.64	0.03
BRT	Very Low	605.37	71.40	0.04	5.55	605.33	0.06	12.59
	Low	88.38	10.42	0.03	4.56	88.34	0.32	0.33
	Moderate	52.01	6.13	0.06	8.26	51.95	0.98	0.06
	High	34.13	4.03	0.08	11.34	34.05	2.06	0.02
	Very High	67.98	8.02	0.51	70.28	67.46	6.40	0.01

4. Discussion

Determining effective parameters in GE and providing a GESM are the first steps in risk management. In regards to this, prediction of areas susceptible to erosion is associated with uncertainty, various models can be used to predict it accurately. Over the past decades, numerous statistical and empirical models have been developed to predict environmental hazards, such as GE, by various researchers around the world [12,14,28,30,31,45]. Due to some of the limitations of the aforementioned models such as time consuming, complexity, costly, and need a lot of data, in recent years data mining methods have been presented. Data mining is a process of discovery of relationships, patterns, and trends that consider the vast amount of information stored in databases with template recognition technology [51,58,59]. The most important applications of data mining are categorization, estimation, forecasting, group dependency, clustering, and descriptions. The results of data mining models show that in RF, BRT, and MARS mode, distance from roads had the highest impact in the occurrence of gully erosion in the study area. This result is in line with [10,49]. If the engineering measures are not considered in site selection and construction of roads as anthropogenic structures in nature, they can act as a causative factor in environmental hazards such as landslide and gully erosion. The construction of roads in bare lands with erosion-sensitive formations has led to the expansion of gully erosion in the study area, so that the construction of a road without proper culverts causes disrupted of natural drainage and runoff concentrations, thus eroding the bare lands and resulting in the formation of a gully. The results of the validation of data mining models showed that the RF model more accurately predicted areas that are sensitive to gully erosion. These results are consistent with the results of [36,43,46,59], which introduced the RF model as a strong and high-performance model. One of the most widely used data mining methods is the RF model. The advantages of the RF method over other models is that this model can apply several input factors without eliminating any factors, and return a very small set of categories that support high prediction accuracy [6]. The classification accuracy of this model is affected by many factors such as the number, scale, type, and precision of input data. Thus, in the processing, the use of all suitable factors causes the accuracy of the model to increase. Compared with other models, RF has higher sufficiency to apply a very high number of datasets [6]. The RF model has the potential as a tool of spatial model for assessing environmental issues and environmental hazards. The RF model combines several tree algorithms to generate a repeated prediction of each phenomenon. This method can learn complicated patterns and consider the nonlinear relationship between explanatory variables and dependent variables. It can also incorporate and combine different types of data in the analysis, due to the lack of distribution of assumptions about the data used. This model can use and apply thousands of input variables without deleting one of them. This method is less sensitive to artificial neural networks, in case of noise data, and can better estimate the parameters [60]. The greatest advantages of RF model are high predictive accuracy, the ability to learn nonlinear relationships, the ability to determine the important variables in prediction, its nonparametric nature, and in dealing with distorted data, it works better than other algorithms for categorization. The main disadvantages of this algorithm include high memory occupation, hard and time-consuming in implementation for large datasets, high cost of pruning, high number of end nodes in case of overlap, and the accumulation of layers of errors in the case of the tree growing. [15,61] stated that the CART, BRT, and RF models showed better accuracy compared to bivariate and multivariate methods. Pourghasemi et al. concluded that the RF and maximum entropy (ME), models have high performance and precision in modeling [31]. Mojaddadi et al. showed that BRT, CART, and RF methods are suitable for modelling [55]. Chen et al. indicates that the MARS and RF models are good estimators for mapping [36]. Lai et al. indicated that the RF model has significant potential for weight determination on landslide modelling [62]. Kuhnert et al stated that RF with AUC = 97.0 is suitable for landslide susceptibility [27]. Lee et al stated that the prediction accuracy of RF model is high (90.8) and that this model had a high capability for landslide prediction [43]. They applied RF and boosted-tree models for spatial prediction of flood susceptibility in Seoul metropolitan city, Korea [43]. They stated that the RF model has better performance compared to boosted-tree. As a scientific achievement, the methodology framework used in this research has shown that the proper selection of effective variables in gully erosion, along with the use of modern data mining models and Geography Information System (GIS) technique, are able to successfully identify areas susceptible to gully erosion. The susceptibility map prepared using this methodology is a suitable tool for sustainable planning to protect the land against gully erosion processes. Therefore, this methodology can be used to assess gully erosion in other similar areas, especially in arid and semi-arid regions.

5. Conclusions

GE is one of the main processes causing soil degradation and there is a need to improve methods to predict susceptible areas and responsible environmental factors, to allow early intervention to prevent, limit, or reverse gully formation. The utility of three data mining models, RF, BRT, and MARS, to predict GE in the Shahroud watershed, Iran, was assessed. For this purpose, twelve causative factors and 121 gully locations (70%) are used for applying the models. In addition, 51 gully locations (30%) are used for validation of models. The correlation between GE and conditioning factor classes was researched with a WoE Bayes theory. Distance to roads, elevation, and lithology were the key factors. Validation of the models showed that all three models have high accuracy for GE mapping. Data mining/machine learning methods have a unique ability and accuracy for GESM. The results also showed that the southwestern part of the study region has a high susceptibility to GE.

Therefore, it is recommended that the following suggestions should be made to prevent and reduce soil erosion and its subsequent risks in the Sharoud watershed: (1) Control of gullies by restoration of vegetation adaptable with the natural conditions of the area; (2) gully controlling by building dams that could prevent soil erosion by slowing down the flow of water and aggravation of sedimentation; (3) awareness of farmers by environmental officials of the region, in terms of the type and principles of proper cultivation and prevention of overgrazing and destruction of vegetation; (4) correction of land use based on natural ability and restrictions related to geomorphologic and physiographic soil characteristics of the area.

Author Contributions: Conceptualization, A.A., B.P., and H.R.P.; Data curation, A.A.; Formal analysis, A.A., and H.R.P.; Investigation, A.A., B.P., and H.R.P.; Methodology, B.P., A.A., H.R.P., and K.R.; Resources, B.P. and A.A.; Software, H.R.P., and A.A.; Supervision, B.P., N.K. and H.R.P.; Validation, H.R.P., and A.A.; Writing–original draft, A.A.; Writing–review and editing, B.P., A.A., H.R.P., N.K. and K.R.

References

1. Magliulo, P. Assessing the susceptibility to water-induced soil erosion using a geomorphological, bivariate statistics-based approach. *Environ. Earth Sci.* **2012**, *67*, 1801–1820. [CrossRef]

2. UNEP. The Emissions Gap Report. United Nations Environment Programme (UNEP). Nairobi, 2017. Available online: www.unenvironment.org/resources/emissions-gap-report (accessed on 13 January 2018).

3. Haregeweyn, N.; Tsunekawa, A.; Poesen, J.; Tsubo, M.; Meshesha, D.T.; Fenta, A.A.; Nyssen, J.; Adgo, E. Comprehensive assessment of soil erosion risk for better land use planning in river basins: Case study of the Upper Blue Nile River. *Sci. Total Environ.* **2017**, *574*, 95–108. [CrossRef] [PubMed]

4. Nampak, H.; Pradhan, B.; Mojaddadi Rizeei, H.; Park, H.-J. Assessment of Land Cover and Land Use Change Impact on Soil Loss in a Tropical Catchment by Using Multi-Temporal SPOT-5 Satellite Images and RUSLE model. *Land Degrad. Dev.* **2018**. [CrossRef]

5. Rizeei, H.M.; Saharkhiz, M.A.; Pradhan, B.; Ahmad, N. Soil erosion prediction based on land cover dynamics at the Semenyih watershed in Malaysia using LTM and USLE models. *Geocarto Int.* **2016**, *31*, 1158–1177. [CrossRef]

6. Zhang, X.; Fan, J.; Liu, Q.; Xiong, D. The contribution of gully erosion to total sediment production in a small watershed in Southwest China. *Phys. Geogr.* **2018**, *39*, 246–263. [CrossRef]

7. Mojaddadi, H.; Habibnejad, M.; Solaimani, K.; Ahmadi, M.; Hadian-Amri, M. An Investigation of Efficiency of Outlet Runoff Assessment. *J. Appl. Sci.* **2009**, *9*, 105–112.

8. Zabihi, M.; Mirchooli, F.; Motevalli, A.; Darvishan, A.K.; Pourghasemi, H.R.; Zakeri, M.A.; Sadighi, F. Spatial modelling of gully erosion in Mazandaran Province, northern Iran. *Catena* **2018**, *161*, 1–13. [CrossRef]

9. Kirkby, M.; Bracken, L. Gully processes and gully dynamics. *Earth Surf. Process. Landf. J. Br. Geomorphol. Res. Group* **2009**, *34*, 1841–1851. [CrossRef]

10. Torri, D.; Poesen, J.; Borselli, L.; Bryan, R.; Rossi, M. Spatial variation of bed roughness in eroding rills and gullies. *Catena* **2012**, *90*, 76–86. [CrossRef]

11. Mccloskey, G.; Wasson, R.; Boggs, G.; Douglas, M. Timing and causes of gully erosion in the riparian zone of the semi-arid tropical Victoria River, Australia: Management implications. *Geomorphology* **2016**, *266*, 96–104. [CrossRef]

12. Rahmati, O.; Tahmasebipour, N.; Haghizadeh, A.; Pourghasemi, H.R.; Feizizadeh, B. Evaluating the influence of geo-environmental factors on gully erosion in a semi-arid region of Iran: An integrated framework. *Sci. Total Environ.* **2017**, *579*, 913–927. [CrossRef] [PubMed]

13.   Dube, F.; Nhapi, I.; Murwira, A.; Gumindoga, W.; Goldin, J.; Mashauri, D. Potential of weight of evidence modelling for gully erosion hazard assessment in Mbire District–Zimbabwe. *Phys. Chem. Earth Part A/B/C* **2014**, *67*, 145–152. [CrossRef]

14.   Zakerinejad, R.; Maerker, M. An integrated assessment of soil erosion dynamics with special emphasis on gully erosion in the Mazayjan basin, southwestern Iran. *Nat. Hazards* **2015**, *79*, 25–50. [CrossRef]

15.   Pham, T.G.; Degener, J.; Kappas, M. Integrated universal soil loss equation (USLE) and Geographical Information System (GIS) for soil erosion estimation in A Sap basin: Central Vietnam. *Int. Soil Water Conserv. Res.* **2018**, *6*, 99–110. [CrossRef]

16.   Pournader, M.; Ahmadi, H.; Feiznia, S.; Karimi, H.; Peirovan, H.R. Spatial prediction of soil erosion susceptibility: An evaluation of the maximum entropy model. *Earth Sci. Inform.* **2018**, *11*, 389–401. [CrossRef]

17.   Althuwaynee, O.F.; Pradhan, B.; Park, H.-J.; Lee, J.H. A novel ensemble bivariate statistical evidential belief function with knowledge-based analytical hierarchy process and multivariate statistical logistic regression for landslide susceptibility mapping. *Catena* **2014**, *114*, 21–36. [CrossRef]

18.   Morgan, R.; Quinton, J.; Smith, R.; Govers, G.; Poesen, J.; Auerswald, K.; Chisci, G.; Torri, D.; Styczen, M. The European Soil Erosion Model (EUROSEM): A dynamic approach for predicting sediment transport from fields and small catchments. *Earth Surf. Process. Landf. J. Br. Geomorphol. Res. Group* **1998**, *23*, 527–544. [CrossRef]

19.   Barber, M.; Mahler, R. Ephemeral gully erosion from agricultural regions in the Pacific Northwest, USA. *Ann. Wars. Univ. Life Sci.-SGGW. Land Reclam.* **2010**, *42*, 23–29. [CrossRef]

20.   Leonard, R.; Knisel, W.; Still, D. GLEAMS: Groundwater loading effects of agricultural management systems. *Trans. ASAE* **1987**, *30*, 1403–1418. [CrossRef]

21.   Liaw, A.; Breiman, W.M. Cutler's Random Forests for Classification and Regression. Available online: https://www.rdocumentation.org/packages/randomForest (accessed on 1 April 2018).

22.   Akgün, A.; Türk, N. Mapping erosion susceptibility by a multivariate statistical method: A case study from the Ayvalık region, NW Turkey. *Comput. Geosci.* **2011**, *37*, 1515–1524. [CrossRef]

23.   Conoscenti, C.; Angileri, S.; Cappadonia, C.; Rotigliano, E.; Agnesi, V.; Märker, M. Gully erosion susceptibility assessment by means of GIS-based logistic regression: A case of Sicily (Italy). *Geomorphology* **2014**, *204*, 399–411. [CrossRef]

24.   Conforti, M.; Aucelli, P.P.; Robustelli, G.; Scarciglia, F. Geomorphology and GIS analysis for mapping gully erosion susceptibility in the Turbolo stream catchment (Northern Calabria, Italy). *Nat. Hazards* **2011**, *56*, 881–898. [CrossRef]

25.   Lucà, F.; Conforti, M.; Robustelli, G. Comparison of GIS-based gullying susceptibility mapping using bivariate and multivariate statistics: Northern Calabria, South Italy. *Geomorphology* **2011**, *134*, 297–308. [CrossRef]

26.   Meyer, A.; Martınez-Casasnovas, J. Prediction of existing gully erosion in vineyard parcels of the NE Spain: A logistic modelling approach. *Soil Tillage Res.* **1999**, *50*, 319–331. [CrossRef]

27.   Kuhnert, P.M.; Henderson, A.K.; Bartley, R.; Herr, A. Incorporating uncertainty in gully erosion calculations using the random forests modelling approach. *Environmetrics* **2010**, *21*, 493–509. [CrossRef]

28.   Rahmati, O.; Haghizadeh, A.; Pourghasemi, H.R.; Noormohamadi, F. Gully erosion susceptibility mapping: The role of GIS-based bivariate statistical models and their comparison. *Nat. Hazards* **2016**, *82*, 1231–1258. [CrossRef]

29.   Svoray, T.; Michailov, E.; Cohen, A.; Rokah, L.; Sturm, A. Predicting gully initiation: Comparing data mining techniques, analytical hierarchy processes and the topographic threshold. *Earth Surf. Proc. Land.* **2012**, *37*, 607–619. [CrossRef]

30.   Zakerinejad, R.; Märker, M. Prediction of Gully erosion susceptibilities using detailed terrain analysis and maximum entropy modeling: A case study in the Mazayejan Plain, Southwest Iran. *Geogr. Fis. Din. Quat.* **2014**, *37*, 67–76.

31.   Pourghasemi, H.R.; Yousefi, S.; Kornejady, A.; Cerdà, A. Performance assessment of individual and ensemble data-mining techniques for gully erosion modeling. *Sci. Total Environ.* **2017**, *609*, 764–775. [CrossRef] [PubMed]

32.   I.R. of Iran Meteorological Organization. 2012. Available online: http://www.mazandaranmet.ir/ (accessed on 12 October 2017).

33. Geological Survey Department of Iran (GSDI). 2012. Available online: http://www.mazandaranmet.ir/ (accessed on 12 October 2017).

34. Althuwaynee, O.F.; Pradhan, B.; Lee, S. Application of an evidential belief function model in landslide susceptibility mapping. *Comput. Geosci.* **2012**, *44*, 120–135. [CrossRef]

35. Rizeei, H.M.; Pradhan, B.; Saharkhiz, M.A. Surface runoff prediction regarding LULC and climate dynamics using coupled LTM, optimized ARIMA, and GIS-based SCS-CN models in tropical region. *Arab. J. Geosci.* **2018**, *11*, 53. [CrossRef]

36. Chen, W.; Xie, X.; Wang, J.; Pradhan, B.; Hong, H.; Bui, D.T.; Duan, Z.; Ma, J. A comparative study of logistic model tree, random forest, and classification and regression tree models for spatial prediction of landslide susceptibility. *Catena* **2017**, *151*, 147–160. [CrossRef]

37. Tehrany, M.S.; Pradhan, B.; Mansor, S.; Ahmad, N. Flood susceptibility assessment using GIS-based support vector machine model with different kernel types. *Catena* **2015**, *125*, 91–101. [CrossRef]

38. Claps, P.; Fiorentino, M.; Oliveto, G. Informational entropy of fractal river networks. *J. Hydrol.* **1996**, *187*, 145–156. [CrossRef]

39. Aal-Shamkhi, A.D.S.; Mojaddadi, H.; Pradhan, B.; Abdullahi, S. Extraction and modeling of urban sprawl development in Karbala City using VHR satellite imagery. In *Spatial Modeling and Assessment of Urban Form*; Springer: Berlin, Germany, 2017; pp. 281–296.

40. Abdullahi, S.; Pradhan, B.; Mojaddadi, H. City compactness: Assessing the influence of the growth of residential land use. *J. Urban Technol.* **2018**, *25*, 21–46. [CrossRef]

41. Rizeei, H.M.; Shafri, H.Z.; Mohamoud, M.A.; Pradhan, B.; Kalantar, B. Oil palm counting and age estimation from WorldView-3 imagery and LiDAR data using an integrated OBIA height model and regression analysis. *J. Sensors* **2018**, *2018*, 2536327. [CrossRef]

42. Xie, Z.; Chen, G.; Meng, X.; Zhang, Y.; Qiao, L.; Tan, L. A comparative study of landslide susceptibility mapping using weight of evidence, logistic regression and support vector machine and evaluated by SBAS-InSAR monitoring: Zhouqu to Wudu segment in Bailong River Basin, China. *Environ. Earth Sci.* **2017**, *76*, 313. [CrossRef]

43. Lee, S.; Kim, J.-C.; Jung, H.-S.; Lee, M.J.; Lee, S. Spatial prediction of flood susceptibility using random-forest and boosted-tree models in Seoul metropolitan city, Korea. *Geomat. Nat. Hazards Risk* **2017**, *8*, 1185–1203. [CrossRef]

44. Cutler, D.R.; Edwards Jr, T.C.; Beard, K.H.; Cutler, A.; Hess, K.T.; Gibson, J.; Lawler, J.J. Random forests for classification in ecology. *Ecology* **2007**, *88*, 2783–2792. [CrossRef] [PubMed]

45. Simpson, G.L.; Birks, H.J.B. Statistical learning in palaeolimnology. In *Tracking Environmental Change Using Lake Sediments*; Springer: Berlin, Germany, 2012; pp. 249–327.

46. Nicodemus, K.K. Letter to the editor: On the stability and ranking of predictors from random forest variable importance measures. *Brief. Bioinform.* **2011**, *12*, 369–373. [CrossRef] [PubMed]

47. Bui, D.T.; Bui, Q.-T.; Nguyen, Q.-P.; Pradhan, B.; Nampak, H.; Trinh, P.T. A hybrid artificial intelligence approach using GIS-based neural-fuzzy inference system and particle swarm optimization for forest fire susceptibility modeling at a tropical area. *Agr. For. Meteorol.* **2017**, *233*, 32–44.

48. Elith, J.; Leathwick, J.R.; Hastie, T. A working guide to boosted regression trees. *J. Anim. Ecol.* **2008**, *77*, 802–813. [CrossRef] [PubMed]

49. Regmi, A.D.; Devkota, K.C.; Yoshida, K.; Pradhan, B.; Pourghasemi, H.R.; Kumamoto, T.; Akgun, A. Application of frequency ratio, statistical index, and weights-of-evidence models and their comparison in landslide susceptibility mapping in Central Nepal Himalaya. *Arab. J. Geosci.* **2014**, *7*, 725–742. [CrossRef]

50. Aertsen, W.; Kint, V.; Van Orshoven, J.; Özkan, K.; Muys, B. Comparison and ranking of different modelling techniques for prediction of site index in Mediterranean mountain forests. *Ecol. Model.* **2010**, *221*, 1119–1130. [CrossRef]

51. Krishnaiah, V.; Narsimha, G.; Chandra, N.S. Heart disease prediction system using data mining techniques and intelligent fuzzy approach: A review. *Heart Dis.* **2016**, *136*, 43–51. [CrossRef]

52. Oh, H.-J.; Pradhan, B. Application of a neuro-fuzzy model to landslide-susceptibility mapping for shallow landslides in a tropical hilly area. *Comput. Geosci.* **2011**, *37*, 1264–1276. [CrossRef]

53. Torgo, L. *Data Mining with R: Learning with Case Studies*; Chapman and Hall/CRC: Boca Raton, FL, USA, 2016.

54. Umar, Z.; Pradhan, B.; Ahmad, A.; Jebur, M.N.; Tehrany, M.S. Earthquake induced landslide susceptibility mapping using an integrated ensemble frequency ratio and logistic regression models in West Sumatera Province, Indonesia. *Catena* **2014**, *118*, 124–135. [CrossRef]

55. Mojaddadi, H.; Pradhan, B.; Nampak, H.; Ahmad, N.; Ghazali, A.H.B. Ensemble machine-learning-based geospatial approach for flood risk assessment using multi-sensor remote-sensing data and GIS. *Geomat. Nat. Hazards Risk* **2017**, *8*, 1080–1102. [CrossRef]

56. Pourghasemi, H.R.; Beheshtirad, M.; Pradhan, B. A comparative assessment of prediction capabilities of modified analytical hierarchy process (M-AHP) and Mamdani fuzzy logic models using Netcad-GIS for forest fire susceptibility mapping. *Geomat. Nat. Hazards Risk* **2016**, *7*, 861–885. [CrossRef]

57. Hong, H.; Naghibi, S.A.; Dashtpagerdi, M.M.; Pourghasemi, H.R.; Chen, W. A comparative assessment between linear and quadratic discriminant analyses (LDA-QDA) with frequency ratio and weights-of-evidence models for forest fire susceptibility mapping in China. *Arab. J. Geosci.* **2017**, *10*, 167. [CrossRef]

58. Mezaal, M.R.; Pradhan, B.; Shafri, H.; Mojaddadi, H.; Yusoff, Z. Optimized Hierarchical Rule-Based Classification for Differentiating Shallow and Deep-Seated Landslide Using High-Resolution LiDAR Data. In *Global Civil Engineering Conference*; Springer: Berlin, Germany, 2017.

59. Rizeei, H.M.; Pradhan, B.; Saharkhiz, M.A. An integrated fluvial and flash pluvial model using 2D high-resolution sub-grid and particle swarm optimization-based random forest approaches in GIS. *Complex Intell. Syst.* **2018**, 1–20. [CrossRef]

60. Kantardzic, M. *Data mining: Concepts, Models, Methods, and Algorithms*; John Wiley & Sons: Hoboken, NJ, USA, 2011.

61. Pham, B.T.; Pradhan, B.; Bui, D.T.; Prakash, I.; Dholakia, M. A comparative study of different machine learning methods for landslide susceptibility assessment: A case study of Uttarakhand area (India). *Environ. Model. Softw.* **2016**, *84*, 240–250. [CrossRef]

62. Lai, C.; Chen, X.; Wang, Z.; Xu, C.-Y.; Yang, B. Rainfall-induced landslide susceptibility assessment using random forest weight at basin scale. *Hydrol. Res.* **2017**, nh2017044. [CrossRef]

4

Landslide Susceptibility Modeling using Integrated Ensemble Weights of Evidence with Logistic Regression and Random Forest Models

Wei Chen [1,2,3,*,†], Zenghui Sun [1,2,†] and Jichang Han [1,2,*]

[1] Key Laboratory of Degraded and Unused Land Consolidation Engineering, Ministry of Land and Resources of China, Xi'an 710075, China

[2] Shaanxi Provincial Land Engineering Construction Group Co. Ltd., Xi'an 710075, China; sunzenghui061@126.com

[3] College of Geology & Environment, Xi'an University of Science and Technology, Xi'an 710054, Shaanxi, China

* Correspondence: chenwei0930@xust.edu.cn (W.C.); hanjc_sxdj@126.com (J.H.)

† These authors contribute equally to this paper.

Abstract: The main aim of this study was to compare the performances of the hybrid approaches of traditional bivariate weights of evidence (WoE) with multivariate logistic regression (WoE-LR) and machine learning-based random forest (WoE-RF) for landslide susceptibility mapping. The performance of the three landslide models was validated with receiver operating characteristic (ROC) curves and area under the curve (AUC). The results showed that the areas under the curve obtained using the WoE, WoE-LR, and WoE-RF methods were 0.720, 0.773, and 0.802 for the training dataset, and were 0.695, 0.763, and 0.782 for the validation dataset, respectively. The results demonstrate the superiority of hybrid models and that the resultant maps would be useful for land use planning in landslide-prone areas.

Keywords: landslide; weights of evidence; logistic regression; random forest; hybrid model

1. Introduction

Landslides are common geological hazards caused by multiple factors including landform [1,2], geological evolution [3], groundwater [4], land use type [5], precipitation [6,7], irrigation [8], earthquake [9], engineering construction [10], and climate change [11–13]. To avoid casualties caused by landslides and guarantee the stable development of mountainous areas, it is critical to determine a control and prevention scheme for landslides in a region. Generally, regional landslide susceptibility maps are beneficial to mitigate the effects of landslide hazards.

At present, various methods have been proposed and introduced into landslide susceptibility mapping. The existing modeling approaches can be put into two categories: qualitative approaches and quantitative approaches [14,15]. In recent years, conventional qualitative approaches have been gradually abandoned by many researchers due to the risk that expert opinion can make the results stray from objective reality [16]. Compared with qualitative approaches, quantitative approaches are mainly based on the hidden information of objective data instead of subjective experience. Additionally, quantitative approaches mainly include traditional mathematical statistic methods, deterministic models, and some state-of-the-art machine learning algorithms.

For traditional statistical methods, the probability-frequency ratio (FR) [17,18], weight of evidence (WoE) [19,20], statistical index (SI) [21,22], index of entropy (IoE) [23,24], certainty factors (CF) [25–27], evidential belief function (EBF) [28–30], and logistic regression (LR) [31,32] models have been extensively adopted in landslide susceptibility mapping. However, one limitation for all traditional statistical methods is that some hypotheses exist [33]. In deterministic models, the detail characteristics of slopes are necessary to construct the calculation model [34]. Although deterministic models conform to basic physical laws of landslide, these models are not very suitable for regional landslide susceptibility assessments due to the complex process of modeling and computing [34].

In the past decade, with the rise of machine learning and data mining, a number of relevant algorithms have been developed for landslide susceptibility zonation [35–39]. For instance, the logistic regression model (LRM), artificial neural network (ANN), support vector machine (SVM), and decision tree (DT) were the top four machine learning algorithms in landslide susceptibility mapping during the period of 2005–2016 [16]. It is clear that machine learning algorithms improve the prediction accuracy of regional landslide occurrence, but the generalization performance of single classifiers still needs to be promoted [40]. In this way, a series of ensemble approaches have recently become more and more popular in geo-hazard susceptibility mapping [37,41–43].

In terms of ensemble approaches, several single classifiers have been combined using ensemble frameworks including random subspace [44], random forest [45,46], Bagging [47], AdaBoost [48], MultiBoost [49], and so on [37,50–52]. Currently, some novel ensemble techniques have been proposed and applied in landslide susceptibility assessment, flood susceptibility mapping, and groundwater potential analysis [41,53,54]. Additionally, the excellent performance of ensemble algorithms on predictive ability and generalization capacity has also been proven. For example, Kadavi et al. [55] compared four ensemble-based machine learning models (AdaBoost, LogitBoost, Multiclass Classifier, and Bagging) with the traditional frequency ratio model (FRM) in the task of landslide susceptibility mapping. Furthermore, the results demonstrated that all of the AUC values of the four ensemble-based machine learning models were higher than that of FRM. In addition, many scholars preferred to construct ensemble learning models by integrating machine learning algorithms with bivariate statistical models because some of the hypotheses of the conventional models can be weakened through hybrid models [56]. Meanwhile, part of the merits of bivariate statistical models and machine learning models can remain by integrating together. Weights of evidence models, as a classic bivariate statistical approach, can calculate the weights of various categories of a conditioning factor based on sturdy mathematical theories [57]. Furthermore, the weights of evidence models can be integrated with other machine learning approaches to reveal the hidden correlations between different conditioning factors and landslide occurrence. Therefore, in the present study, based on GIS tools, the integrated ensemble weights of evidence with logistic regression and random forest models were employed to map landslide susceptibility, and the results were compared and analyzed quantitatively by receiver operating characteristic curves (ROC) and area under the curve (AUC).

2. Study Area

The study area was located in Shaanxi Province, China (Figure 1) where the average annual temperature is 14.2 °C, the average annual rainfall is 909.8 mm, and the evaporation is 1537.1 mm. Topographically, the study area is part of the Qinba Mountain. The general trend is high in the south and low in the north. Elevation ranges from 442 m to 2410 m above sea level, with an average elevation of 1171 m. Slope angles in the study area are in the range of 0 to 70°. Most of the slope angles are in classes of 10–20° (29.27%), followed by 20–30° (26.29%), 0–10° (23.64%), and 30–40° (14.99%). Only 5.81% of slope angles are higher than 40°.

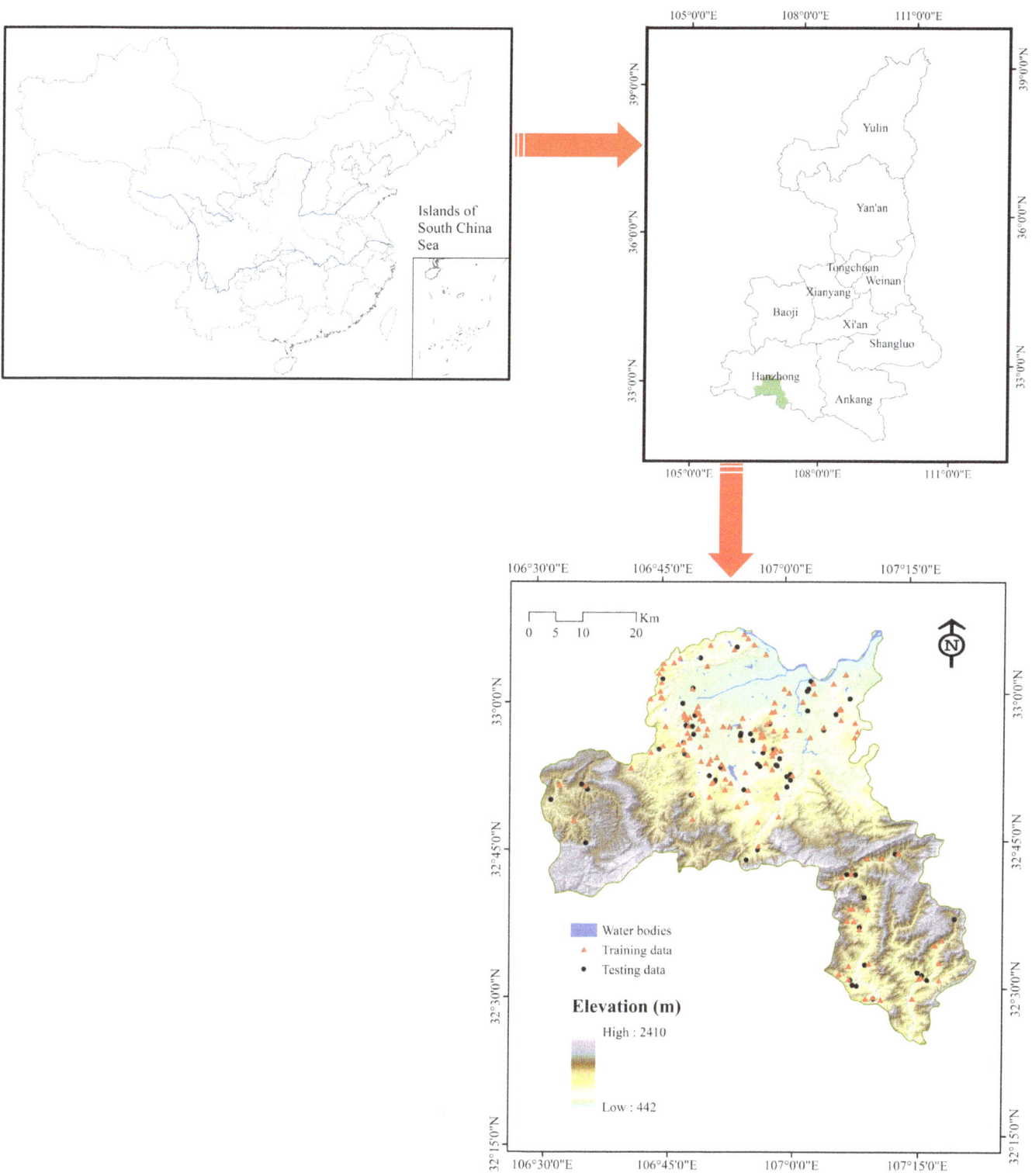

Figure 1. Location of the study area.

Geologically, the study area is located at the northern margin of the Yangtze plate. There are five major faults crossing the area including (1) the Gangchang fault (SW–NE direction), (2) the Xiaolengba–Qinjiaba fault (NW–SE direction), (3) the Xiaoba–Haitang fault (SW–NE direction), (4) the Moujiaba–Shuimohe fault (W–E direction), and (5) the Jiangjiawan–Zhujiaba–Tuqiangping fault (SW–NE direction) (Figure 2).

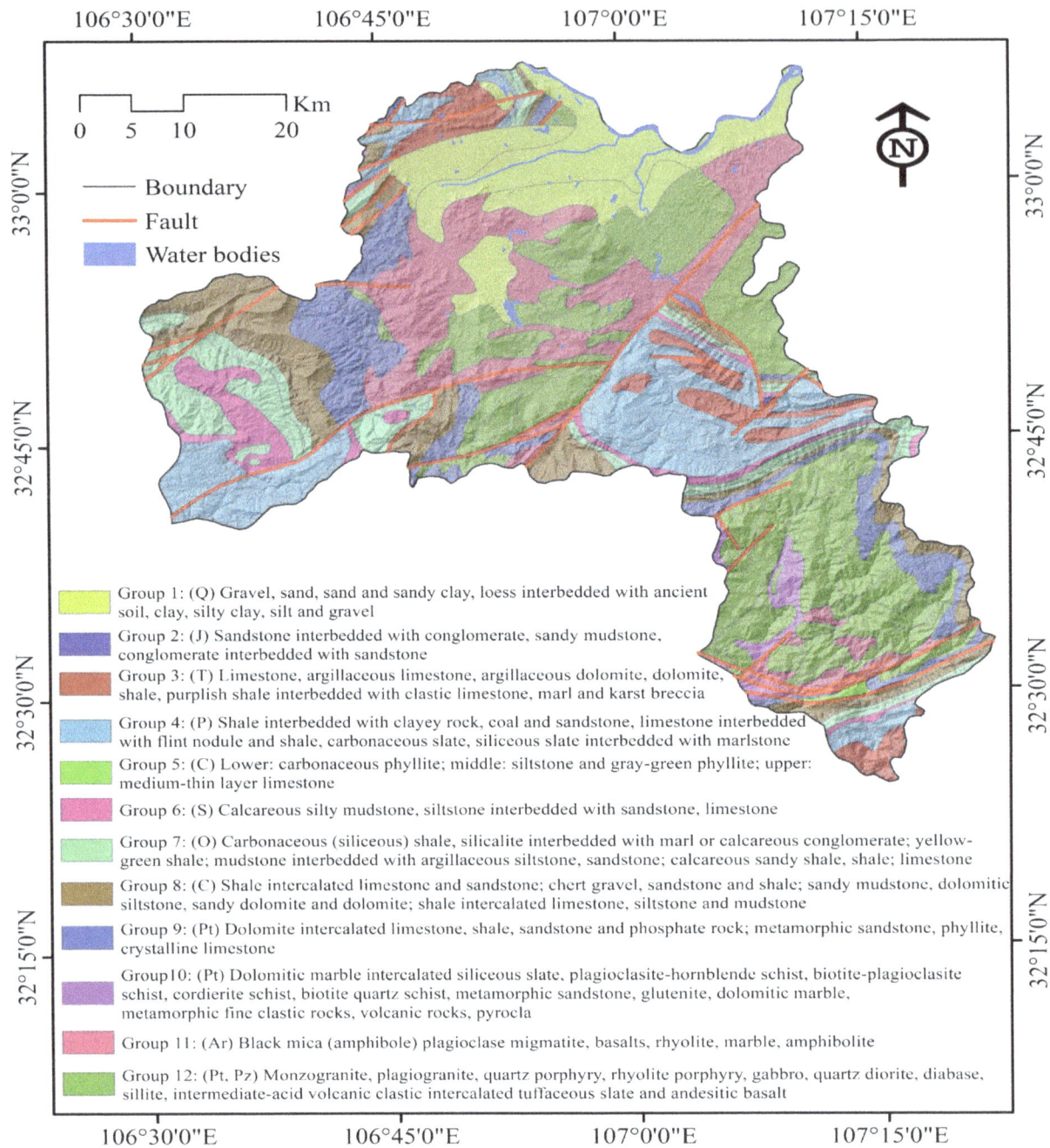

Figure 2. Geological map of the study area.

3. Materials and Methods

3.1. Data Preparation

A landslide inventory includes the locations of the past and recent landslides [21]. A landslide inventory can give insight into landslide location, dates, type, frequency of occurrence, state of activity, magnitude or size, failure mechanisms, causal factors, and damage caused [58,59]. In the present study, the landslide inventory map was prepared on the basis of satellite images (Google Earth and ZY03 images) and historical landslide records of the area, which were verified by GPS. A total of 202 landslides were identified to prepare the landslide susceptibility map, of which most of the landslides were slides (190), the others included 12 rock falls [60]. According to an analysis in the GIS environment, the smallest landslide was nearly 160 m³, the largest landslide was more than 1,000,000 m³, while the average was 33,000 m³. Finally, 141 landslides were randomly selected as training data and rest of them were used for the verification of the landslide susceptibility map (Figure 1).

There are no universal guidelines for selecting landslide conditioning factors [33,61]. A total of 16 landslide conditioning factors were used for landslide susceptibility mapping including slope angle, slope aspect, elevation, plan curvature, profile curvature, topographic wetness index (TWI), stream power index (SPI), sediment transport index (STI), distance to rivers, distance to roads, distance to faults, soil, land use, normalized difference vegetation index (NDVI), lithology, and rainfall, which are considered as controlling factors in the occurrence of landslides in the study area.

Slope angle is an important factor that affects the stress state of slope mass, and these positions where stress exceeds failure strength may contribute to landslide hazards [62,63]. In this case, as shown in Figure 3a, the thematic data layer of the slope angle was reclassified into seven categories with an interval of $10°$, namely, $(0–10°)$, $(10–20°)$, $(20–30°)$, $(30–40°)$, $(40–50°)$, $(50–60°)$, and $(60–72.83°)$.

Slope aspect is another common conditioning factor for the task of landslide susceptibility mapping [64,65]. It has been proven that most landslides usually occur at a certain slope aspect for a given study area, but the mechanism has not been revealed clearly [66]. Therefore, slope aspect was also employed as a conditioning factor. Here, slope aspect categories include flat, north, northeast, east, southeast, south, southwest, west, and northwest (Figure 3b).

Generally, it is considered that elevation has a firm relationship with landslide occurrence [67]. There is no denying that elevation can influence the topography, vegetation, temperature, humidity, human activities, and many other conditions that have a connection with slope stability [30,68]. In Figure 3c, the elevation of the study area was divided into ten classes with an interval of 200 m, i.e., (442–600 m), (600–800 m), (800–1000 m), (1000–1200 m), (1200–1400 m), (1400–1600 m), (1600–1800 m), (1800–2000 m), (2000–2200 m), and (2200–2410 m).

Plan curvature and profile curvature are two quantitative indices that embody topographic characteristics and trend from different perspectives [69]. Various curvature values indicate different runoff and erosion conditions of water. For instance, the upwardly convex surfaces have positive curvature values while negative curvature values mean upwardly concave surfaces [30]. In this study, the plan curvature and profile curvature values were both reclassified into three groups (Figure 3d,e).

TWI was proposed to indicate the local groundwater potential by Moore [70] in 1991. Currently, TWI is regarded as an extensively-used causative factor in landslide susceptibility assessment [71]. It is expressed as $TWI = \ln(\frac{\alpha}{\tan \beta})$, where β is the slope angle (radian), and α is the flow accumulation through a point [72]. The TWI values of the study area can be calculated by GIS software and reclassified as (<4), $(4–5)$, $(5–6)$, $(6–7)$, and (>7) with an interval of 1 (Figure 3f).

SPI can directly measure the erosion capacity of the stream. A higher SPI value indicates that the stream has more powerful erosion on the slope surface [55]. The SPI values are mainly determined as $SPI = \alpha \tan \beta$ [54,70]. In this study, the SPI values were identified as five categories with an interval of 20, namely, (<20), $(20–40)$, $(40–60)$, $(60–80)$, and (>80) (Figure 3g).

As another topographic index, STI has also been considered to construct the landslide susceptibility model [73]. Similar to SPI, STI can quantitatively reflect the regional topographic features and erosion conditions [74]. For the present study, STI values contained five categories with an interval of 10: (<10), $(10–20)$, $(20–30)$, $(30–40)$, (>40) (Figure 3h).

Rivers can not only affect the moisture distribution in slopes, but can also erode the toes of slopes, which cause slope deformation and failure [75]. Thus, it is necessary to consider the river effects when producing landslide susceptibility maps. In this study, based on the distance to rivers, five buffer zones with an interval of 200 m were generated for each river: $(<200 m)$, $(200–400 m)$, $(400–600 m)$, $(600–800 m)$, and $(>800 m)$ (Figure 3i).

Generally speaking, road construction in mountainous areas, which always produce an engineering load and destroy the integrity of slope structure, have significant negative impacts on the slope stability [76]. Hence, the distance to roads is usually selected as a conditioning factor to embody the influence of road engineering activities on landslide occurrence [77]. Here, values of the distance to roads were divided into five groups with an interval of 300 m, i.e., (<300 m), (300–600 m),(600–900 m), (900–1200 m), and (>1200 m) (Figure 3j).

Fault structures affect the spatial distribution and characteristics of landslides in a certain region [50]. According to relevant studies [30,78], the integrity of rock and soil mass generally decrease as the distance to the faults shorten. In this way, landslide hazards are more likely to occur in the neighboring area of faults. Ultimately, buffers of various faults in the study area were obtained and reclassified into five categories with an interval of 1000 m: (<1000 m), (1000–2000 m), (2000–3000 m),(3000–4000 m), and (>4000 m) (Figure 3k).

In terms of soil, this is an essential factor that has a strong correlation with landslide occurrence [79]. To a great extent, the strength, root cohesion, permeability, and vegetation coverage of the soil mass depend on the soil type [80,81], which can impact the failure characteristics of slopes [82,83]. In this study area, a total of nine soil types were identified including cumulic anthrosol, dystric cambisol, eutric cambisol, calcaric fluvisol, haplic luvisol, chromic luvisol, eutric planosol, calcaric regosol, and eutric regosol (Figure 3l).

Land use is one of the most frequently used conditioning factors, and the correlation between landslides and land use has been confirmed [84]. For instance, in some farmland regions, landslides are frequent and common under long-term irrigation [85]. For the study area, the types of land use mainly consist of farmland, forestland, grassland, water, residential areas, and bareland (Figure 3m).

NDVI is a very popular index to measure the degree of vegetation in a region. NDVI values can be figured out by the formula $NDVI = (I - R)/(IR + R)$, where IR is the infrared band and R is the red band of the electromagnetic spectrum [86]. The range of NDVI values is from $-1 to 1$, and a positive value means that the local ground is covered by vegetation. Five categories of NDVI values were generated based on the natural break method [87], namely (−0.21–0.21), (0.21–0.36), (0.36–0.44),(0.44–0.52), and (0.52–0.65) (Figure 3n).

Like soil, lithology is one of the most important factors that directly determines slope stability. According to many existing studies, the physical and mechanical properties of rock mass usually change dramatically with lithological units [88]. Therefore, most landslides occur in the sliding-prone lithological units that have lower strength and a higher moisture content. For this study area, the strata were mainly reclassified into twelve lithological units based on the lithofacies and geological ages, and the specific distribution of various lithologies was illustrated in Figure 3o.

Rainfall is a crucial triggering factor that causes massive landslides by means of raising the groundwater level and increasing pore water pressure [89]. It can be observed that the probability of landslide occurrence indeed grows under the actions of long-term or heavy rainfall. Based on the meteorological data of the study area, the corresponding rainfall map with an interval of 100 mm/yr was produced, i.e., (<900 mm/yr), (900–1000 mm/yr), (1000–1100 mm/yr), (1100–1200 mm/yr),(1200–1300 mm/yr), (1300–1400 mm/yr), (1400–1500 mm/yr), (1500–1600 mm/yr), (1600–1700 mm/yr), and (>1700 mm/yr) (Figure 3p).

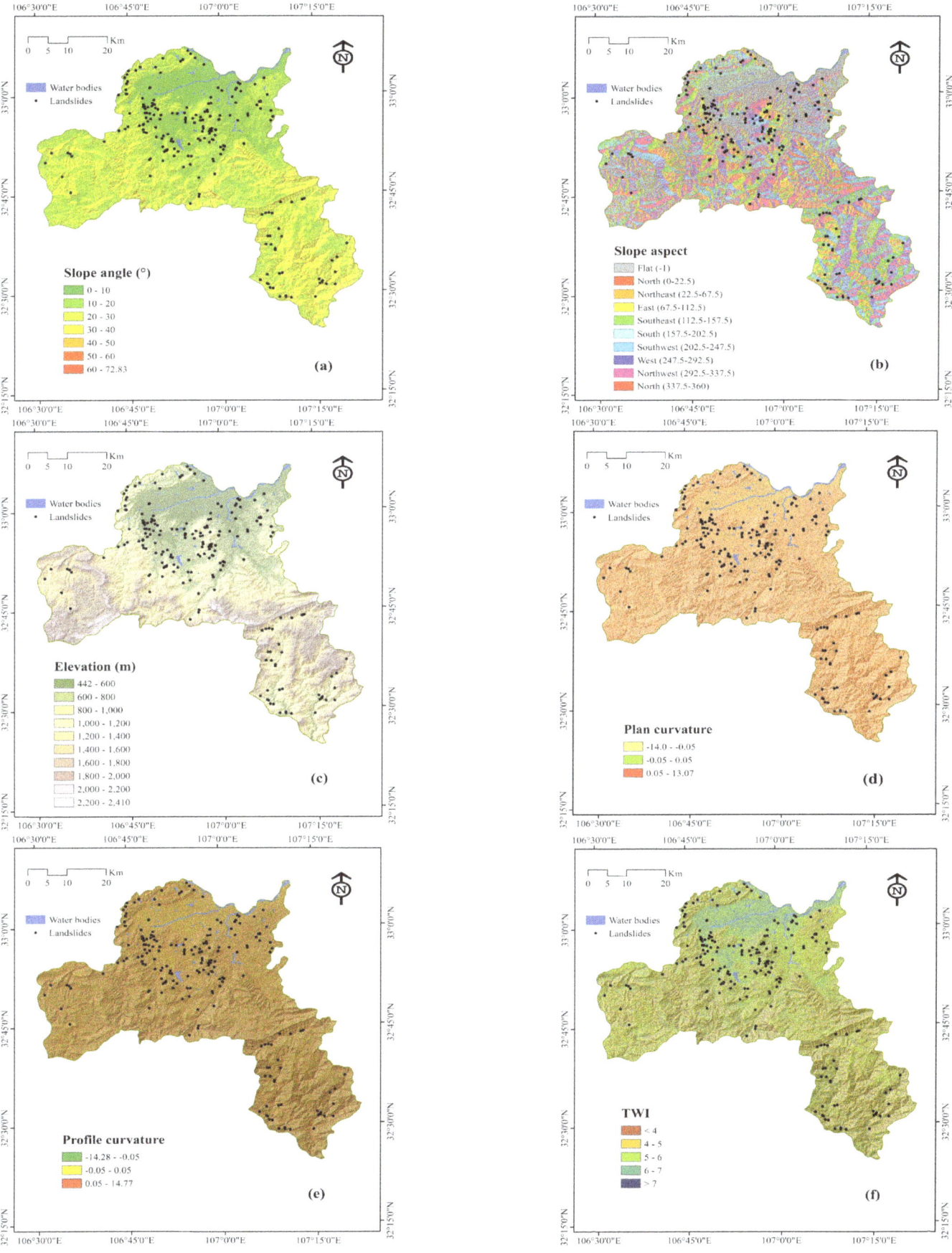

Figure 3. *Cont.*

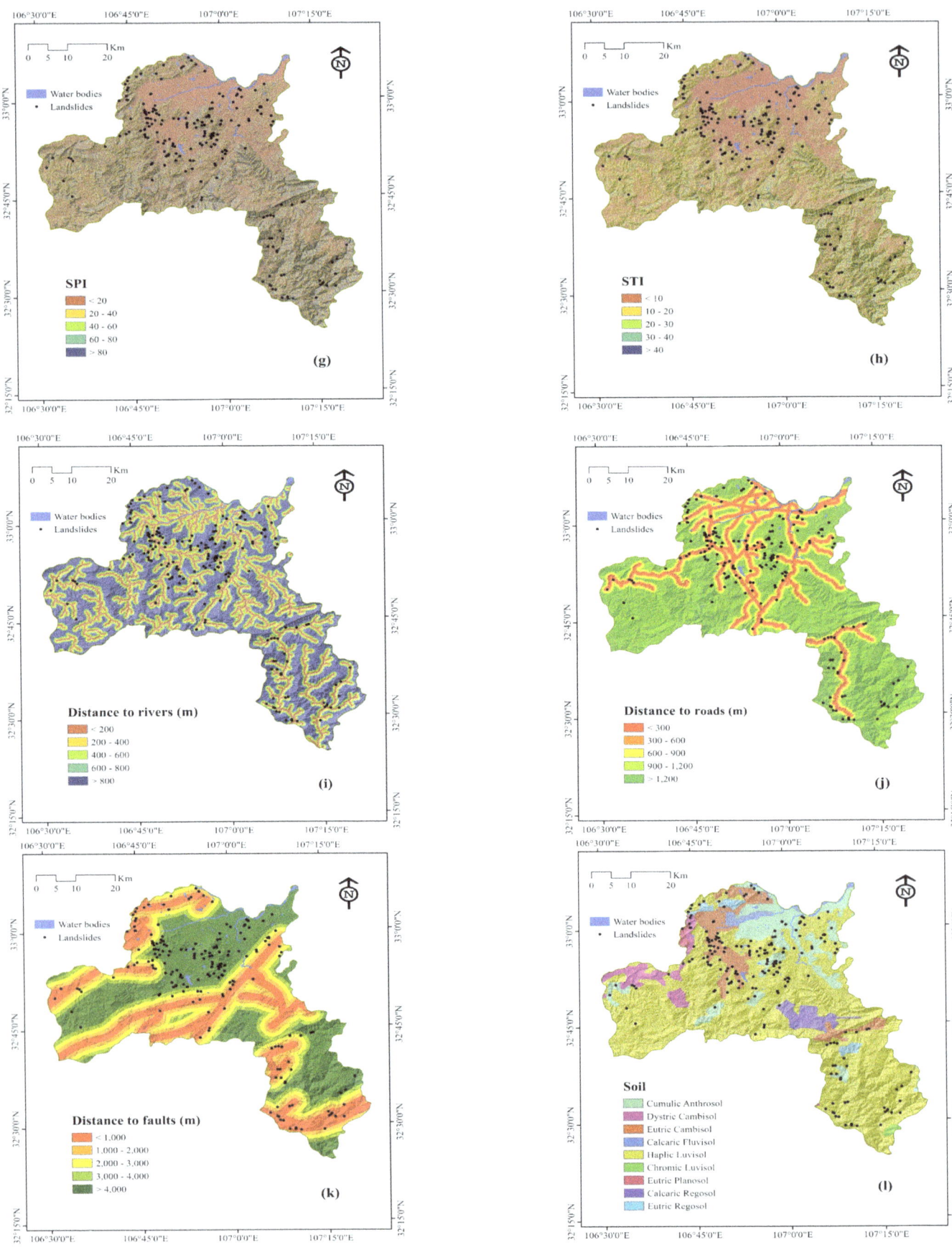

Figure 3. *Cont.*

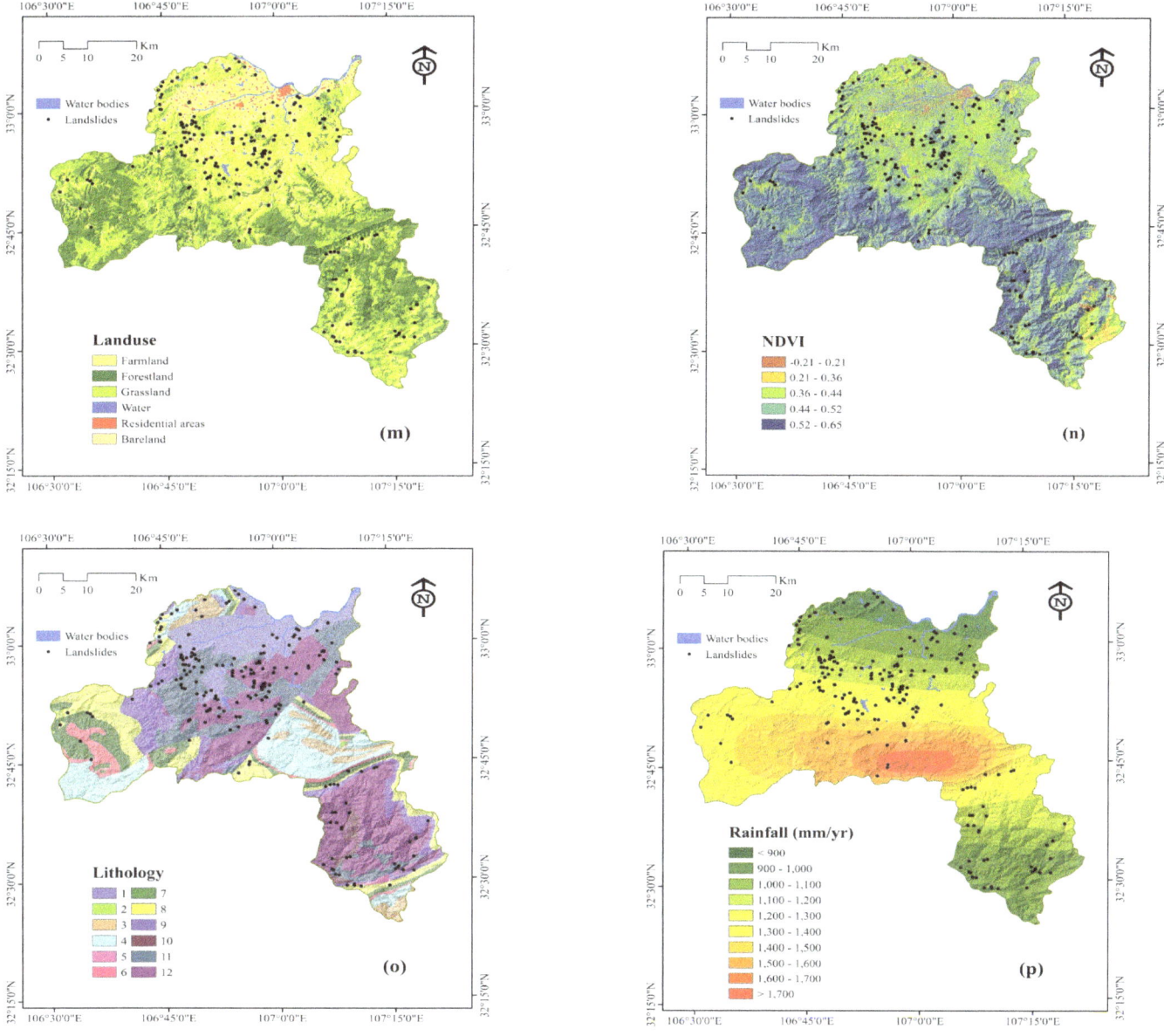

Figure 3. Thematic maps: (**a**) Slope angle; (**b**) Slope aspect; (**c**) Elevation; (**d**) Plan curvature; (**e**) Profile curvature; (**f**) TWI; (**g**) SPI; (**h**) STI; (**i**) Distance to rivers; (**j**) Distance to roads; (**k**) Distance to faults; (**l**) Soil; (**m**) Land use; (**n**) NDVI; (**o**) Lithology; and (**p**) Rainfall.

3.2. Weight of Evidence

Weight of evidence (WoE) is one of the most popular models that uses the Bayesian theory of conditional probability to quantify spatial associations between evidence layers and known mineral occurrences [90]. In the WoE method, conditional independence is the most important issue that should be considered. The WoE is based on the calculation of positive weight W^+ and negative weight W^- as follows:

$$W^+ = \ln \frac{p\{B|A\}}{p\{B|\overline{A}\}} \tag{1}$$

$$W^- = \ln \frac{p\{\overline{B}|A\}}{p\{\overline{B}|A\}} \tag{2}$$

where B is the presence predictive factor; $\overline{B}$ is the absence of the predictive factor; A is the presence of landslide; and $\overline{A}$ is the absence of landslide. In landslide susceptibility prediction, the weight contrast $W_f = W^+ - W^-$ was used to measure and reflect the spatial association between the landslide conditioning factors and landslide occurrence [91].

3.3. Logistic Regression

Logistic regression (LR) is one type of regression analysis where categorical outcomes can be predicted based on a certain predictor [92]. By using the logistic functions, probabilities of the possible outcomes can be modeled [93].

The logistic regression model is useful for two-class classification. Assuming there are n samples of the pairs, (x_i, y_i), $i = 1, 2, \ldots, n$, $y_i \in \{-1, +1\}$ is a binary class label for each sample $i = 1, 2, \ldots, n$ and weights (w, b). In the logistic regression for binary classification, the occurrence probability of the class is modeled with the below function:

$$P(y = \pm 1 | x, w) = \frac{1}{1 + \exp(-y(w^T x + b))} \tag{3}$$

where b is the intercept; T is the matrix transposition; and the k-dimensional coefficient vector, $w = (w_1, w_2, \ldots w_k)^T$ are parameters to be estimated.

3.4. Random Forest

Random forests (RF) are an ensemble of separately trained binary decision trees [94]. In the random forest algorithm, a random vector i_k is naturally produced, independent from the previous random vectors and distributed to all trees, and each tree is grown using the training dataset and random vector i_k, and outcomes are in the collection of tree-structured classifiers $h(x, i_k), k = 1, 2, \ldots n$ at input vector x. In this study, i_k is the landslide conditioning factors. The random forest consisted of two trees, namely, landslide and non-landslide, each constructed while considering sixteen random features.

Generally, in a random forest algorithm, the generalization error is described as below [95]:

$$GE = P_{x,y}(mg(x, y) < 0) \tag{4}$$

where x and y are the landslide conditioning factors indicating the probability over the x, y space, and mg is the margin function, which is defined as below:

$$mg(x, y) = av_k I(h_k(x) = y) - \max_{j \neq y} av_k I(h_k(x) = j) \tag{5}$$

Which measures the extent to which the average number of votes at random vectors for the right output exceeds the average vote for any other output. The $I(*)$ is the indicator function [96].

4. Results

4.1. Correlation Analysis

The correlation between the conditioning factors and probability of landslides occurrence was measured by the weight contrast W_f, and the calculation results of the WoE model are listed in Table 1. The LR method was employed to produce the landslide susceptibility map, and one of the most critical applicable conditions of LR is that the landslide conditioning factors are mutually independent [97]. Therefore, it is necessary to diagnose the multicollinearity of various conditioning factors when evaluating landslide susceptibility [98]. Currently, the tolerance (TOL) (TOL $= 1 - R^2$, and R is the coefficient of determination of the regression equation) and variance inflation factor (VIF) (VIF $= 1/TOL$) have been applied in multicollinearity diagnosis [99–101].

Table 1. Correlation between landslides and conditioning factors using the WoE model.

Factors	Class	No. of Landslide	No. of Pixels in Domain	W^+	W^-	W_f
Slope angle (°)	0–10	36	738,360	0.077	−0.025	0.102
	10–20	63	914,163	0.423	−0.246	0.669
	20–30	29	821,142	−0.246	0.075	−0.320
	30–40	11	468,077	−0.653	0.081	−0.734
	40–50	2	155,377	−1.255	0.037	−1.292
	50–60	0	24,357	0.000	0.008	0.000
	60–72.83	0	1710	0.000	0.001	0.000
Slope aspect	Flat	0	874	0.000	0.000	0.000
	North	16	443,863	−0.225	0.033	−0.258
	Northeast	16	405,251	−0.134	0.019	−0.153
	East	17	376,207	0.001	0.000	0.001
	Southeast	23	390,547	0.266	−0.044	0.310
	South	32	374,222	0.639	−0.130	0.769
	Southwest	13	344,928	−0.181	0.020	−0.201
	West	9	354,647	−0.576	0.055	−0.631
	Northwest	15	432,647	−0.264	0.037	−0.301
Elevation (m)	442–600	28	413,571	0.405	−0.079	0.485
	600–800	48	512,157	0.730	−0.237	0.968
	800–1000	31	377,619	0.598	−0.119	0.717
	1000–1200	17	326,381	0.143	−0.018	0.161
	1200–1400	15	398,407	−0.182	0.024	−0.206
	1400–1600	2	385,439	−2.163	0.117	−2.281
	1600–1800	0	376,083	0.000	0.128	0.000
	1800–2000	0	247,350	0.000	0.083	0.000
	2000–2200	0	78,216	0.000	0.025	0.000
	2200–2410	0	7963	0.000	0.003	0.000
Plan curvature	−14.0– −0.05	58	144,0116	−0.114	0.088	−0.203
	−0.05–0.05	13	215,290	0.291	−0.025	0.316
	0.05–13.07	70	1,467,780	0.055	−0.051	0.106
Profile curvature	−14.28—−0.05	66	1,428,952	0.023	−0.020	0.042
	−0.05–0.05	16	177,891	0.689	−0.062	0.751
	0.05–14.77	59	1,516,343	−0.149	0.123	−0.271
TWI	<4	11	558,428	−0.829	0.116	−0.945
	4–5	50	1,000,955	0.101	−0.052	0.153
	5–6	48	746,522	0.354	−0.143	0.497
	6–7	20	393,490	0.119	−0.018	0.137
	>7	12	423,791	−0.467	0.057	−0.523
SPI	<20	88	1,740,663	0.113	−0.164	0.277
	20–40	20	497,521	−0.116	0.021	−0.137
	40–60	12	231,236	0.139	−0.012	0.151
	60–80	5	133,800	−0.189	0.008	−0.197
	>80	16	519,966	−0.383	0.062	−0.445
STI	<10	90	1,722,652	0.146	−0.215	0.361
	10–20	32	702,426	0.009	−0.003	0.012
	20–30	6	295,062	−0.798	0.056	−0.853
	30–40	5	141,300	−0.244	0.010	−0.254
	>40	8	261,746	−0.390	0.029	−0.419
Distance to rivers (m)	<200	27	521,129	0.138	−0.030	0.168
	200–400	22	463,390	0.050	−0.009	0.059
	400–600	18	427,717	−0.070	0.011	−0.081
	600–800	19	374,831	0.116	−0.017	0.133
	>800	55	1,336,119	−0.092	0.064	−0.156
Distance to roads (m)	<300	33	343,852	0.754	−0.150	0.904
	300–600	16	279,559	0.237	−0.027	0.264
	600–900	8	245,226	−0.325	0.023	−0.348
	900–1200	15	219,752	0.413	−0.040	0.453
	>1200	69	2,034,797	−0.286	0.382	−0.668

Table 1. *Cont.*

Factors	Class	No. of Landslide	No. of Pixels in Domain	W^+	W^-	W_f
Distance to faults (m)	<1000	32	671,796	0.054	−0.015	0.069
	1000–2000	18	503,008	−0.232	0.039	−0.271
	2000–3000	21	412,189	0.121	−0.020	0.141
	3000–4000	8	348,794	−0.677	0.060	−0.737
	>4000	62	1,187,399	0.145	−0.101	0.246
Soil	Cumulic Anthrosol	20	360,361	0.206	−0.030	0.237
	Dystric Cambisol	4	113,893	−0.251	0.008	−0.259
	Eutric Cambisol	31	249,592	1.012	−0.165	1.177
	Calcaric Fluvisol	0	37,035	0.000	0.012	0.000
	Haplic Luvisol	80	2,211,459	−0.222	0.393	−0.615
	Chromic Luvisol	0	10,045	0.000	0.003	0.000
	Eutric Planosol	3	14,836	1.500	−0.017	1.516
	Calcaric Regosol	1	82,141	−1.311	0.020	−1.330
	Eutric Regosol	2	43,824	0.011	0.000	0.011
Land use	Farmland	86	90,0284	0.750	−0.601	1.351
	Forestland	4	96,7369	−2.390	0.342	−2.732
	Grassland	51	1,202,442	−0.062	0.037	−0.100
	Water	0	18,838	0.000	0.006	0.000
	Residential areas	0	33,563	0.000	0.011	0.000
	Bareland	0	690	0.000	0.000	0.000
NDVI	−0.21–0.21	4	67,502	0.272	−0.007	0.279
	0.21–0.36	10	207,991	0.063	−0.005	0.068
	0.36–0.44	63	651,020	0.762	−0.358	1.121
	0.44–0.52	56	1,089,392	0.130	−0.077	0.207
	0.52–0.65	8	1,107,281	−1.832	0.379	−2.212
Lithology	1	27	363,139	0.499	−0.089	0.588
	2	0	1694	0.000	0.001	0.000
	3	2	136,901	−1.128	0.031	−1.159
	4	6	398,403	−1.098	0.093	−1.191
	5	0	7470	0.000	0.002	0.000
	6	0	107,848	0.000	0.035	0.000
	7	5	225,834	−0.713	0.039	−0.751
	8	10	319,450	−0.366	0.034	−0.401
	9	9	276,290	−0.326	0.027	−0.353
	10	1	39,158	−0.570	0.005	−0.575
	11	32	435,539	0.487	−0.107	0.594
	12	49	811,460	0.291	−0.126	0.417
Rainfall (mm/yr)	<900	8	189,533	−0.067	0.004	−0.071
	900–1000	29	582,217	0.098	−0.024	0.122
	1000–1100	23	282,006	0.591	−0.083	0.675
	1100–1200	35	329,319	0.856	−0.174	1.030
	1200–1300	16	271,086	0.268	−0.030	0.298
	1300–1400	18	629,601	−0.457	0.089	−0.545
	1400–1500	7	351,254	−0.818	0.068	−0.886
	1500–1600	3	270,784	−1.405	0.069	−1.474
	1600–1700	1	135,625	−1.812	0.037	−1.849
	>1700	1	81,761	−1.306	0.019	−1.325

Generally, a TOL value less than 0.1 or a VIF value larger than 10 is regarded as a symbol of multicollinearity [61]. In this study, the results of the WoE model were used as inputs to calculate the TOL and VIF values of all of the conditioning factors. In accordance with the calculated results, there was no multicollinearity among the landslide conditioning factors (Table 2).

Table 2. Multicollinearity analysis.

Landslide Conditioning Factors	Collinearity Statistics	
	Tolerance (TOL)	Variance inflation factors (VIF)
Slope angle	0.761	1.315
Slope aspect	0.883	1.133
Elevation	0.650	1.539
Plan curvature	0.714	1.400
Profile curvature	0.855	1.170
TWI	0.828	1.208
SPI	0.434	2.303
STI	0.402	2.489
Distance to rivers	0.946	1.057
Distance to roads	0.779	1.284
Distance to faults	0.908	1.101
NDVI	0.774	1.292
Soil	0.642	1.557
Land use	0.627	1.595
Lithology	0.765	1.308
Rainfall	0.664	1.507

4.2. Application of the WoE Model

In terms of slope angle, the slope angle between $10°$–$20°$ (0.669) is more prone to landslide occurrence. Additionally, the W_f values of the region where slope angles larger than $50°$ are 0. For the slope aspect factor, W_f was the highest for south-facing (0.769). Furthermore, southeast-facing (0.310) and east-facing (0.001) also had a positive correlation with landslide occurrence. In the case of elevation, most landslides were distributed in the classes of 442–600 m (0.485), 600–800 m (0.968), and 800–1000 m (0.717). When the elevation was larger than 1200 m, elevation had an inhibitory effect on landslides.

In the case of plan curvature, flat areas had a more important impact on landslides, whereas the W_f values of convex areas and concave areas were 0.106 and −0.203, respectively. In the case of profile curvature, the W_f values of the concave class, flat class, and convex class ere 0.042, 0.751, and −0.271, respectively. For TWI, the highest W_f value was observed for the interval of 5–6 (0.497) while the class <4 (−0.945) had the lowest value. For SPI, the class <20 (0.277) had the highest W_f value, and the areas of SPI 20–40 and >60 were negative for landslides. For STI, the class <10 had the highest W_f value of 0.361, while the class 20–30 had the lowest value of −0.853. In the case of distance to rivers, the classes of <200 m (0.168) and 600–800 m (0.133) occupied higher W_f values when compared to the other classes. In the case of distance to roads, the class of <300 m (0.904) had a more intimate correlation with landslide occurrence. In the case of distance to faults, it can be seen that the class >4000 m had the highest W_f value of 0.246. For soil, eutric cambisol (1.177) and eutric planosol (1.516) were more likely to induce landslides due to the dramatic falling of soil strength under saturated conditions [85]. For land use, farmland (1.351) had the highest probability of landslide occurrence, which may be essentially caused by irrigation. According to the W_f values of NDVI, the class of 0.36–0.44 (1.121) mainly contributed to landslide occurrence, while the lowest value was for the class of 0.52–0.65 (−2.212), which indicates that high vegetation coverage can restrain landslides. In the case of lithology, the W_f values of group 1 (Q) (0.588), group 11 (Ar) (0.594), and group 12 (Pt, Pz) (0.417) were larger than 0, indicating that these lithological groups had the highest susceptibility to landslide. In the case of rainfall, the range between 1100–1200 mm/yr (1.030) showed high susceptibility for landslide occurrence.

The calculated W_f values for all landslide conditioning factors were summed using the following equation to construct the landslide susceptibility map (LSM):

$$
\begin{aligned}
LSM_{WoE} = \quad & \text{Slope angle}_{Wf} + \text{Slope aspect}_{Wf} + \text{Elevation}_{Wf} + \text{Plan curvature}_{Wf} \\
& + \text{Profile curvature}_{Wf} + \text{TWI}_{Wf} + \text{SPI}_{Wf} + \text{STI}_{Wf} + \text{Distance to rivers}_{Wf} \\
& + \text{Distance to roads}_{Wf} + \text{Distance to faults}_{Wf} + \text{NDVI}_{Wf} + \text{Soil}_{Wf} + \text{Landuse}_{Wf} \\
& + \text{Lithology}_{Wf} + \text{Rainfall}_{Wf}
\end{aligned} \tag{6}
$$

The integrated result of the WoE model is shown in Figure 4. The LSM was reclassified into five classes based on the natural break method: very low, low, moderate, high, and very high.

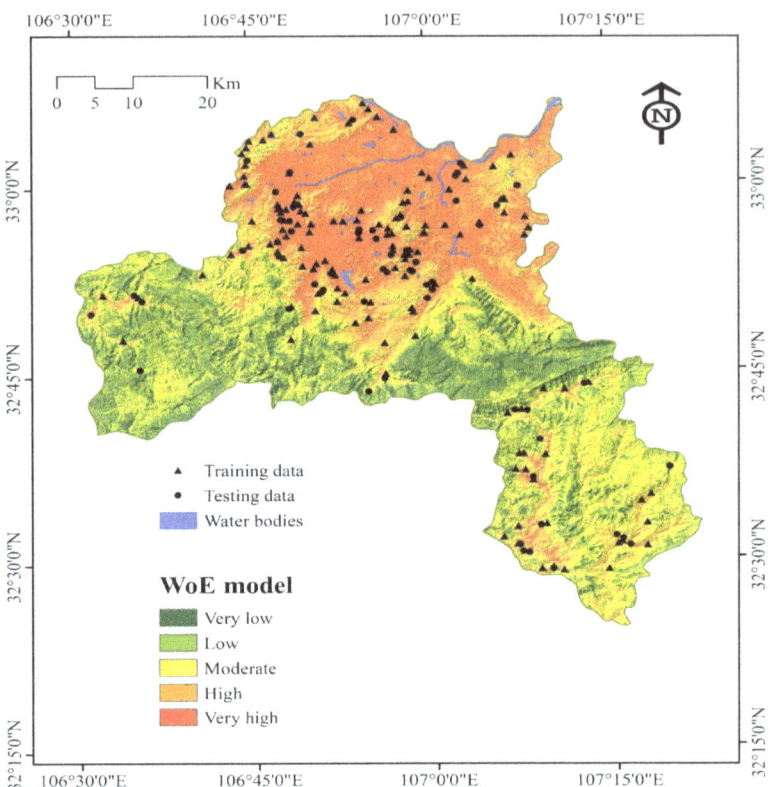

Figure 4. Landslide susceptibility map using the WoE model.

4.3. Application of the WoE-LR Model

In this case, SPSS 18.0 software was applied to build a landslide susceptibility model with the WoE-LR model. The input table of the LR model can be generated by the determined class values of variables based on the WoE model [54]. In the analysis process, a forward stepwise LR was adopted, and the analysis results are given in Tables 3 and 4. The Cox and Snell R Square (0.245) and Nagelkerke R Square (0.326) are two pseudo determined coefficients that are used to reflect the degree of independent variables explaining dependent variables [102,103]. According to Table 4, the LR equation and landslide occurrence probability P can be expressed as Equations (7) and (8), respectively.

$$
\begin{aligned}
y = \quad & 1.122 \times \text{Slope angle} + 2.157 \times \text{Slope aspect} + 0.986 \times \text{Elevation} \\
& + 2.505 \times \text{Plan curvature} + 0.868 \times \text{Profile curvature} + 1.764 \times \text{TWI} \\
& + 1.427 \times \text{SPI} + 1.142 \times \text{STI} + 0.512 \times \text{Distance to rivers} \\
& + 1.445 \times \text{Distance to roads} + 0.972 \times \text{Distance to faults} \\
& + 0.859 \times \text{NDVI} + 1.392 \times \text{Soil} + 1.634 \times \text{Landuse} + 1.032 \times \text{Lithology} \\
& + 1.594 \times \text{Rainfall} + 0.806
\end{aligned} \tag{7}
$$

$$
P = \frac{e^y}{1 + e^y} \tag{8}
$$

Table 3. Maximum likelihood estimation and Cox and Snell's and Nagelkerke's R-square.

—2 Log Likelihood	Cox & Snell R Square	Nagelkerke R Square
311.780	0.245	0.326

Table 4. Coefficients of WoE-LR model.

Landslide Conditioning Factors	Coefficients
Slope angle	1.122
Slope aspect	2.157
Elevation	0.986
Plan curvature	2.505
Profile curvature	0.868
TWI	1.764
SPI	1.427
STI	1.142
Distance to rivers	0.512
Distance to roads	1.445
Distance to faults	0.972
NDVI	0.859
Soil	1.392
Land use	1.634
Lithology	1.032
Rainfall	1.594
Constant	0.806

Ultimately, the landslide susceptibility index (LSI) for the LR model were obtained based on Equation (8), moreover, the LSI values were reclassified into five categories by the natural break method: very low, low, moderate, high, and very high (Figure 5).

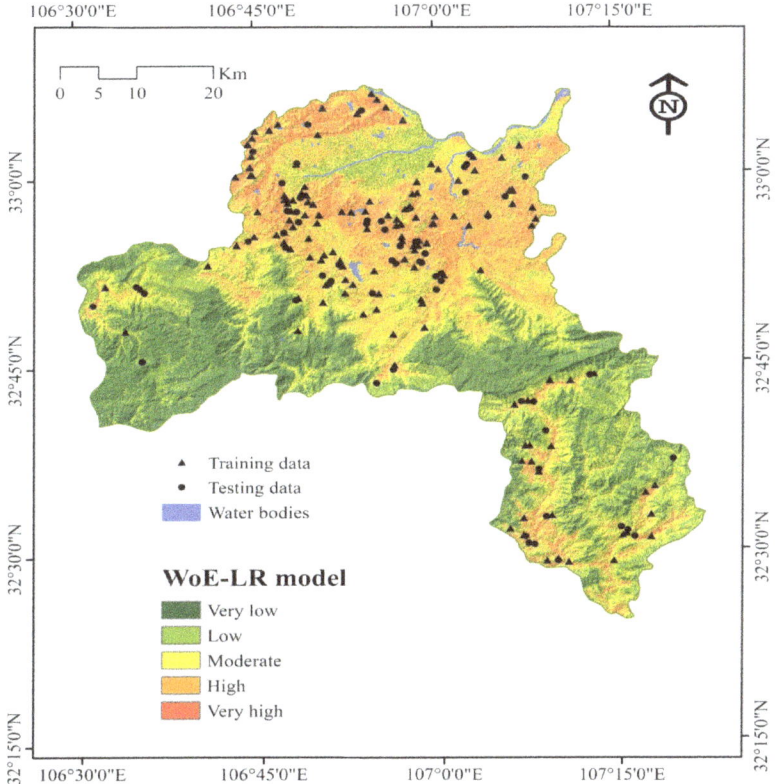

Figure 5. Landslide susceptibility map using the WoE-LR model.

4.4. Application of the WoE-RF Model

Similarly, the calculated results of the WoE model can be used as input for the RF model. In this study, training of the RF model was implemented by WEKA software. During the analyzing process, the importance of various conditioning factors can be measured quantitatively and ordered by MDA (mean decrease accuracy) and MDG (mean decrease Gini). Generally, MDA is determined during the Out-Of-Bag error calculation phase, while MDG is a measure of how each variable contributes to the homogeneity of the nodes and leaves [104]. The values of the above-mentioned two metrics of the conditioning factors are illustrated in Figure 6, and a larger value of MDA or MDG means a higher importance of the corresponding variable. Accordingly, in terms of MDA, land use is the most critical factor in the RF model, while soil is second in importance only to land use. For the MDG, the importance of elevation was first, followed by rainfall and land use. Finally, based on ArcGIS software, the landslide susceptibility map using the WoE-RF was generated and is shown in Figure 7.

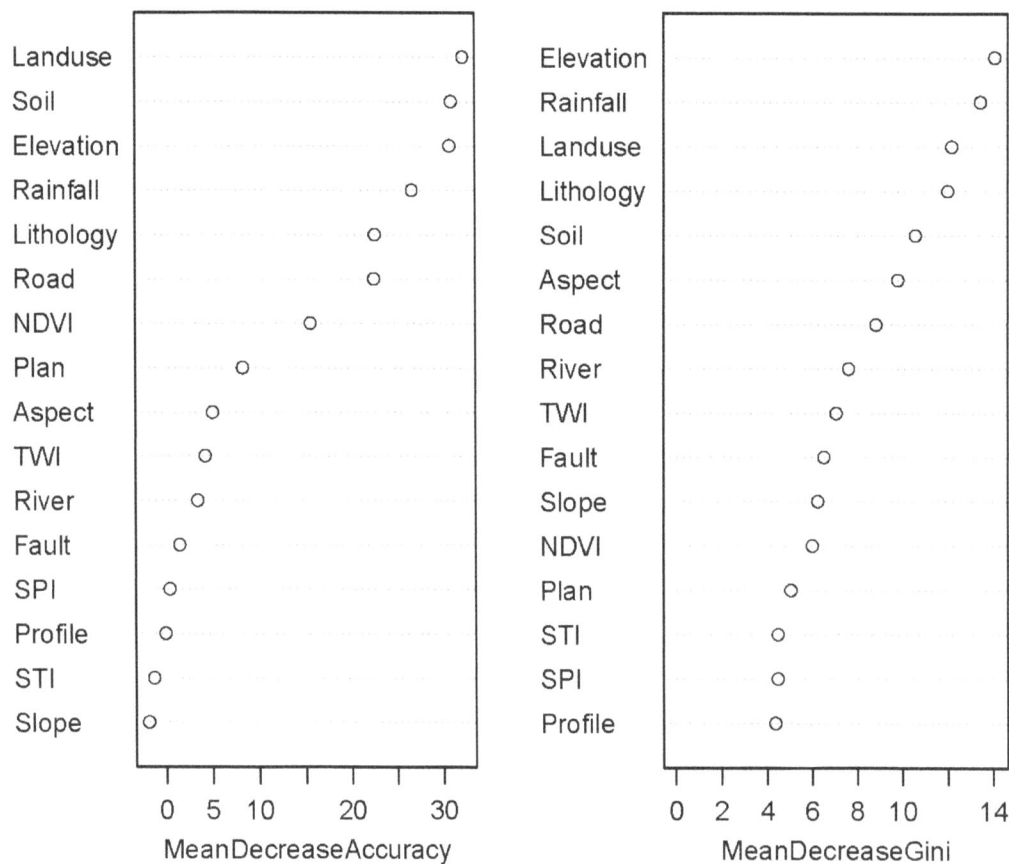

Figure 6. Mean decrease accuracy and mean decrease Gini.

4.5. Validation of Landslide Models

Currently, the ROC and AUC have been widely applied to validate the performance of determined landslide susceptibility models [64,69,105]. The ROC curve can be generated by plotting the false positive rate (100-specificity) in the x-axis versus the sensitivity in the y-axis [71]. The area under the ROC curve (AUC) is an indicator of the global summary measure of the performance of a model [106–108]. In the present study, to assess the validation of the WoE, WoE-LR, and WoE-RF models, the ROC curves of three models with training and validation datasets are described in Figures 8 and 9.

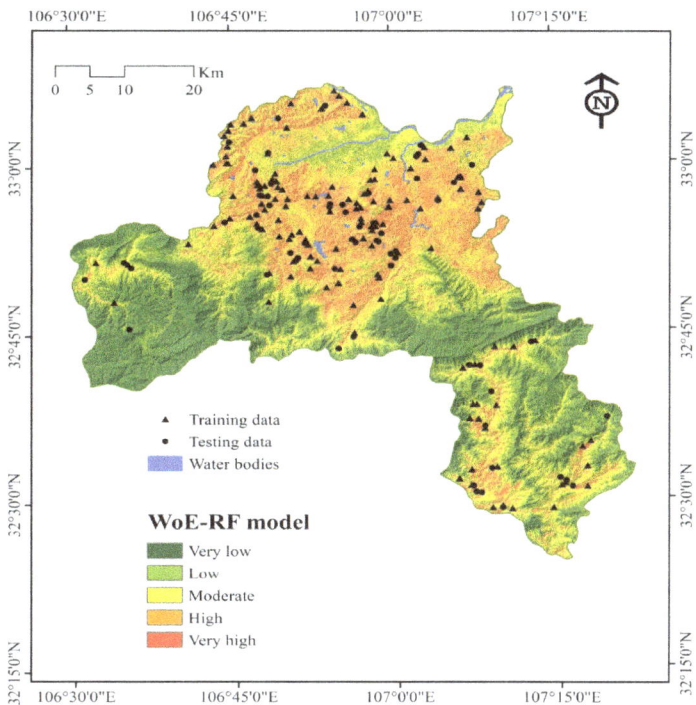

Figure 7. Landslide susceptibility map using the WoE-RF model.

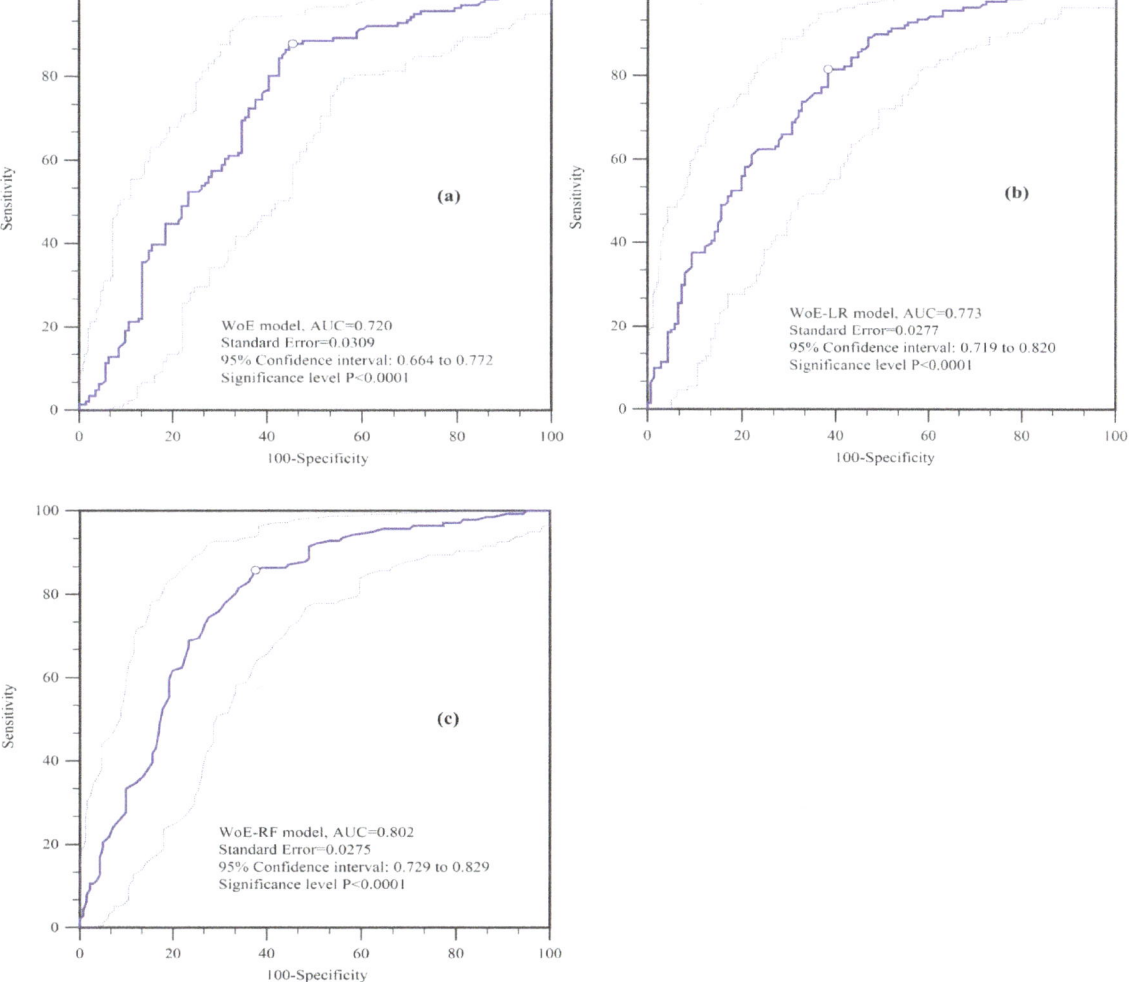

Figure 8. ROC curves using the training dataset.

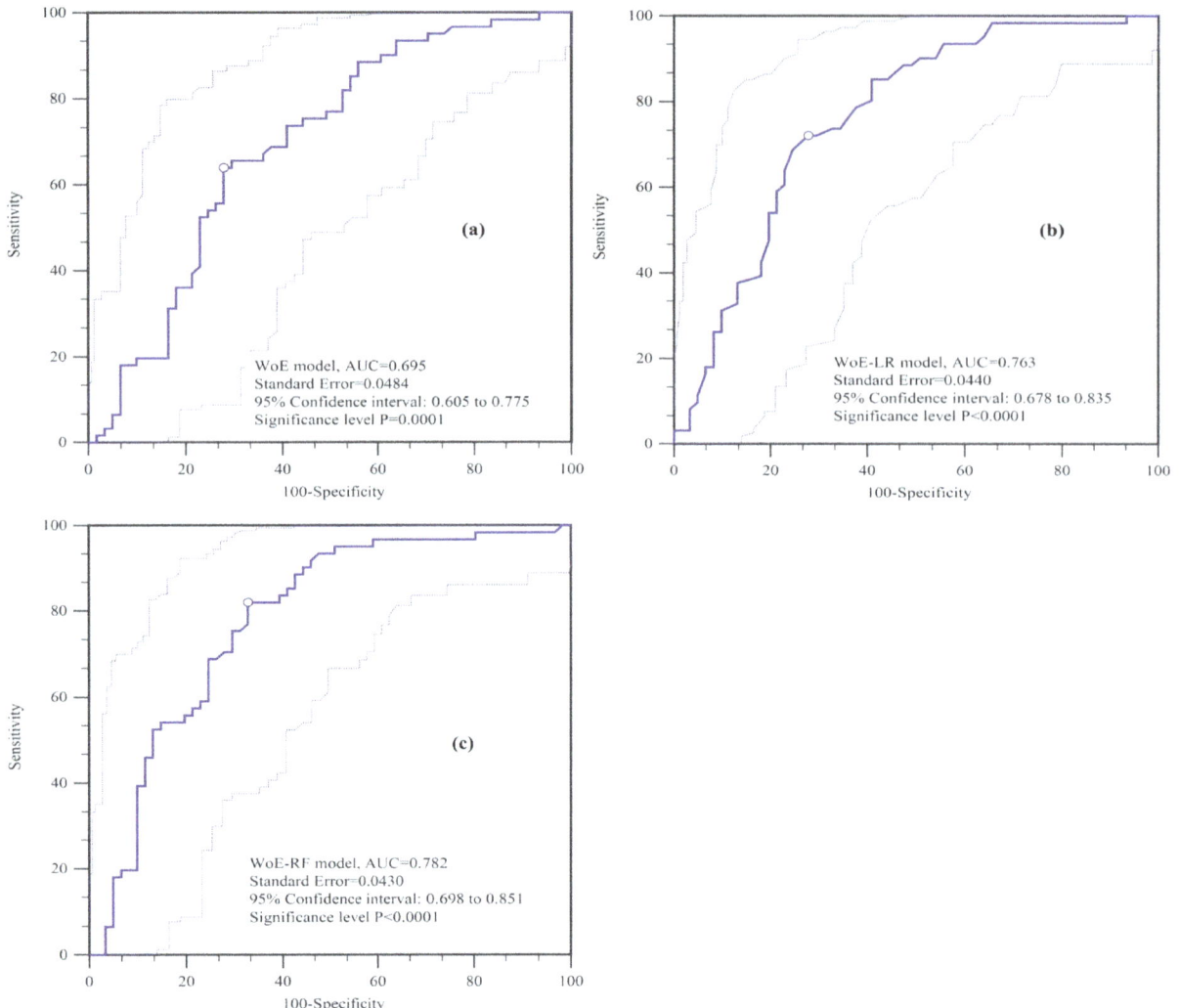

Figure 9. ROC curves using the validation dataset.

In the case of the training dataset, the WoE-RF model had the best performance with the highest AUC value of 0.802, while the AUC values of the WoE model and WoE-LR model were 0.720 and 0.773, respectively. Meanwhile, the WoE-RF model had the lowest standard error (0.0275) and a 95% confidence interval of 0.729–0.829. Thus, the WoE-RF and WoE-LR models can improve the accuracy of the traditional WoE model in this study, and the WoE-RF model showed a relatively better performance.

In the case of the validation data, it can be seen that the AUC values of the various models decreased slightly when compared with the training dataset. The AUC values were 0.695, 0.763, and 0.782 for the WoE model, WoE-LR model, and WoE-RF model, respectively. Similarly, the lowest standard error was 0.0430 for the WoE-RF model, followed by the WoE-LR model (0.0440), and the WoE model (0.0484). The detailed results demonstrated that the WoE-RF model had a prominent prediction capacity on landslide susceptibility mapping.

5. Discussions

Under the action of environmental factors and human activities, the frequency of landslide occurrence has been increasing in recent decades, which may result in catastrophic losses on lives, resources, and property [109,110]. Currently, numerous approaches have been used in landslide susceptibility mapping such as FR [23], WoE [19], IoE [111], machine learning [64,112], and ensemble learning models [53,54]. In the above-mentioned models, the probabilistic meaning and calculation procedure of the WoE model are relatively concise and specific, which makes the WoE a classical and widely used method in landslide susceptibility mapping. Nevertheless, due to the uncertainties and fuzziness in the data of the conditioning factors [113], for different datasets, the performance of the WoE models were significantly distinguished [114–116]. In the present study, the integrated ensemble WoE with LR and RF models were proposed and applied for landslide susceptibility modeling in order to improve the accuracy and generalization ability of the traditional WoE model.

Landslide inventory map is a preliminary step toward landslide susceptibility, hazard and risk assessment [59]. Generally, there are two classes of Landslide inventories: landslide-event inventories that are associated with a trigger and historical landslide inventories [59,117]. In the present study, we adopted the latter formation, which was the sum of many landslide events over a long time. However, the evidence of many smaller landslides might has been lost due to various degrees of modification by subsequent landslides, erosional processes, vegetation growth and anthropic influences [59]. Therefore, application of multi-temporal high-resolution satellites images for interpretation of smaller landslides may be an effective supplement to the current landslide inventory and efficient for improving the accuracies of landslide susceptibility maps.

According to the existing literature and multicollinearity analysis, sixteen conditioning factors were selected: slope angle, slope aspect, elevation, plan curvature, profile curvature, TWI, STI, SPI, distance to rivers, distance to roads, distance to faults, NDVI, soil, land use, lithology, and rainfall. Furthermore, based on the W_f values, the relationships between landslide occurrence and these factors were analyzed. It was demonstrated that all factors had nonlinear relationships with landslides. In addition, the RF model was employed to measure the importance of factors with two indices, the MDA and MDG. In terms of MDA, it could be observed that the most critical factor was land use, followed by soil and elevation. Slope angle had the lowest impact on landslide occurrence. However, for MDG, the importance of elevation, rainfall, and land use ranked first, second, and third, respectively, while the lowest MDG value was for profile curvature.

There are some classification techniques for a landslide susceptibility map in GIS software, such as manual, defined interval, natural break, equal interval, quantile, standard deviation, geometrical interval, and landslide percentage [118]. Generally, user-defined classification is more difficult for the reader to interpret and justify. Therefore, current automatic classification systems should be used instead of a user-defined classification [118]. Besides, when landslide susceptibility indexes have positive or negative skewness, the best classification methods are quantile or natural break [119]. In the present study, natural break method, which is the most commonly used models [120,121], is the most suitable method for modelling landslide susceptibility according to the histogram of data distribution.

In this paper, a comparison study of the WoE, WoE-LR, and WoE-RF models was implemented. LR is a widely used model for classification, particularly for binary classification problems [122]. Thus, we integrated the WoE with the LR model to acquire a better classifier. The WoE-RF model is a combination of the weight of evidence and random forest approach. It has been proven that RF is one of the most popular classification algorithms and can improve the performance of single classifiers [96,123]. Moreover, RF can decrease the dependence of the WoE model on independence among the conditioning factors. Accordingly, the results showed that both the LR model (AUC = 0.773 for training data; AUC = 0.763 for validation data) and RF model (AUC = 0.802 for training data; AUC = 0.782 for validation data) can increase the performance of the traditional WoE model (AUC = 0.720 for training data; AUC = 0.695 for validation data), and the WoE-RF model produced the best results.

Comparing the overall classification results of the three models, the results confirmed that the RF model had a better performance on improving the generalization ability of a weak classifier and raising the corresponding prediction accuracy. Therefore, the landslide susceptibility maps generated by the WoE-RF and WoE-LR models contain reference meaning for the study area to a certain extent. Furthermore, the procedure of factor selection and ensemble model construction is of some value to similar studies.

6. Conclusions

The results are indicative of the quality of the maps drawn by the hybrid approaches of traditional bivariate weights of evidence (WoE) with multivariate logistic regression (WoE-LR) and machine learning-based random forest (WoE-RF). In general, the following conclusions can be drawn:

(1) Geomorphological factors, geological factors, geo-environmental factors, and anthropogenic factors were used for the development of the landslide model. The preliminary selection of these 16 conditioning factors was based on the multicollinearity diagnosis. The TOL and VIF values of all the conditioning factors indicated no multicollinearity.

(2) According to the results of the WoE model, most occurred at slopes of 10–20° with the south aspect, elevations of 600–800 m, distance to rivers of <200 m, distance to roads of <300 m, and a farmland land cover category.

(3) WoE-RF possessed relatively good accuracy when compared to the WoE-LR and WoE models. By using the ROC curve, the AUC values of the training dataset produced by these three methods were 0.802, 0.773, and 0.720, respectively. For the validation dataset, the AUC values were 0.782, 0.763, and 0.695, respectively. It can be concluded that the proposed hybrid models are promising approaches for the spatial prediction of landslides and can also be applied in other landslide-prone areas.

Author Contributions: W.C. and Z.S. collected the field data and conducted the landslide mapping and analysis. W.C. and Z.S. wrote the manuscript. J.H. provided critical comments in planning this paper and edited the manuscript. All authors discussed the results and edited the manuscript.

Acknowledgments: This study was supported by the Opening Fund of Key Laboratory of Degraded and Unused Land Consolidation Engineering, the Ministry of Land and Resources (Grant No. SXDJ2018-04), the National Natural Science Foundation of China (Grant No. 41807192), the China Postdoctoral Science Foundation (Grant No. 2018T111084, 2017M613168), and the Project funded by the Shaanxi Province Postdoctoral Science Foundation (Grant No. 2017BSHYDZZ07).

References

1. Kim, M.S.; Onda, Y.; Kim, J.K.; Kim, S.W. Effect of topography and soil parameterisation representing soil thicknesses on shallow landslide modelling. *Quat. Int.* **2015**, *384*, 91–106. [CrossRef]
2. Liucci, L.; Melelli, L.; Suteanu, C.; Ponziani, F. The role of topography in the scaling distribution of landslide areas: A cellular automata modeling approach. *Geomorphology* **2017**, *290*, 236–249. [CrossRef]
3. Agostini, A.; Tofani, V.; Nolesini, T.; Gigli, G.; Tanteri, L.; Rosi, A.; Cardellini, S.; Casagli, N. A new appraisal of the ancona landslide based on geotechnical investigations and stability modelling. *Q. J. Eng. Geol. Hydrogeol.* **2014**, *47*, 29–43. [CrossRef]
4. Peng, D.; Xu, Q.; Liu, F.; He, Y.; Zhang, S.; Qi, X.; Zhao, K.; Zhang, X. Distribution and failure modes of the landslides in heitai terrace, china. *Eng. Geol.* **2018**, *236*, 97–110. [CrossRef]
5. Persichillo, M.G.; Bordoni, M.; Meisina, C. The role of land use changes in the distribution of shallow landslides. *Sci. Total Environ.* **2017**, *574*, 924–937. [CrossRef] [PubMed]
6. Chang, J.-M.; Chen, H.; Jou, B.J.-D.; Tsou, N.-C.; Lin, G.-W. Characteristics of rainfall intensity, duration, and kinetic energy for landslide triggering in taiwan. *Eng. Geol.* **2017**, *231*, 81–87. [CrossRef]
7. Segoni, S.; Rosi, A.; Lagomarsino, D.; Fanti, R.; Casagli, N. Brief communication: Using averaged soil moisture estimates to improve the performances of a regional-scale landslide early warning system. *Nat. Hazards Earth Syst. Sci.* **2018**, *18*, 807–812. [CrossRef]
8. Hou, X.; Vanapalli, S.K.; Li, T. Water infiltration characteristics in loess associated with irrigation activities and its influence on the slope stability in heifangtai loess highland, china. *Eng. Geol.* **2018**, *234*, 27–37. [CrossRef]
9. Wang, T.; Wu, S.R.; Shi, J.S.; Xin, P.; Wu, L.Z. Assessment of the effects of historical strong earthquakes on large-scale landslide groupings in the wei river midstream. *Eng. Geol.* **2018**, *235*, 11–19. [CrossRef]
10. Mohammadi, S.; Taiebat, H. Finite element simulation of an excavation-triggered landslide using large deformation theory. *Eng. Geol.* **2016**, *205*, 62–72. [CrossRef]

11. Alvioli, M.; Melillo, M.; Guzzetti, F.; Rossi, M.; Palazzi, E.; von Hardenberg, J.; Brunetti, M.T.; Peruccacci, S. Implications of climate change on landslide hazard in central italy. *Sci. Total Environ.* **2018**, *630*, 1528–1543. [CrossRef] [PubMed]

12. Peres, D.J.; Cancelliere, A. Modeling impacts of climate change on return period of landslide triggering. *J. Hydrol.* **2018**, *567*, 420–434. [CrossRef]

13. Gariano, S.L.; Guzzetti, F. Landslides in a changing climate. *Earth-Sci. Rev.* **2016**, *162*, 227–252. [CrossRef]

14. Fell, R.; Corominas, J.; Bonnard, C.; Cascini, L.; Leroi, E.; Savage, W.Z. Guidelines for landslide susceptibility, hazard and risk zoning for land use planning. *Eng. Geol.* **2008**, *102*, 85–98. [CrossRef]

15. Corominas, J.; van Westen, C.; Frattini, P.; Cascini, L.; Malet, J.P.; Fotopoulou, S.; Catani, F.; Van Den Eeckhaut, M.; Mavrouli, O.; Agliardi, F.; et al. Recommendations for the quantitative analysis of landslide risk. *Bull. Eng. Geol. Environ.* **2014**, *73*, 209–263. [CrossRef]

16. Pourghasemi, H.R.; Teimoori Yansari, Z.; Panagos, P.; Pradhan, B. Analysis and evaluation of landslide susceptibility: A review on articles published during 2005–2016 (periods of 2005–2012 and 2013–2016). *Arab. J. Geosci.* **2018**, *11*, 193. [CrossRef]

17. Lee, S.; Dan, N.T. Probabilistic landslide susceptibility mapping in the lai chau province of vietnam: Focus on the relationship between tectonic fractures and landslides. *Environ. Geol.* **2005**, *48*, 778–787. [CrossRef]

18. Chen, W.; Pourghasemi, H.R.; Panahi, M.; Kornejady, A.; Wang, J.; Xie, X.; Cao, S. Spatial prediction of landslide susceptibility using an adaptive neuro-fuzzy inference system combined with frequency ratio, generalized additive model, and support vector machine techniques. *Geomorphology* **2017**, *297*, 69–85. [CrossRef]

19. Xu, C.; Xu, X.; Lee, Y.H.; Tan, X.; Yu, G.; Dai, F. The 2010 yushu earthquake triggered landslide hazard mapping using gis and weight of evidence modeling. *Environ. Earth Sci.* **2012**, *66*, 1603–1616. [CrossRef]

20. Xie, Z.; Chen, G.; Meng, X.; Zhang, Y.; Qiao, L.; Tan, L. A comparative study of landslide susceptibility mapping using weight of evidence, logistic regression and support vector machine and evaluated by sbas-insar monitoring: Zhouqu to wudu segment in bailong river basin, china. *Environ. Earth Sci.* **2017**, *76*, 313. [CrossRef]

21. Mandal, S.; Mandal, K. Bivariate statistical index for landslide susceptibility mapping in the rorachu river basin of eastern sikkim himalaya, india. *Spat. Inf. Res.* **2018**, *26*, 59–75. [CrossRef]

22. Regmi, A.D.; Devkota, K.C.; Yoshida, K.; Pradhan, B.; Pourghasemi, H.R.; Kumamoto, T.; Akgun, A. Application of frequency ratio, statistical index, and weights-of-evidence models and their comparison in landslide susceptibility mapping in central nepal himalaya. *Arab. J. Geosci.* **2014**, *7*, 725–742. [CrossRef]

23. Jaafari, A.; Najafi, A.; Pourghasemi, H.R.; Rezaeian, J.; Sattarian, A. Gis-based frequency ratio and index of entropy models for landslide susceptibility assessment in the caspian forest, northern iran. *Int. J. Environ. Sci. Technol.* **2014**, *11*, 909–926. [CrossRef]

24. Tien Bui, D.; Shahabi, H.; Shirzadi, A.; Chapi, K.; Alizadeh, M.; Chen, W.; Mohammadi, A.; Ahmad, B.; Panahi, M.; Hong, H.; et al. Landslide detection and susceptibility mapping by airsar data using support vector machine and index of entropy models in cameron highlands, malaysia. *Remote Sens.* **2018**, *10*, 1527. [CrossRef]

25. Hong, H.; Chen, W.; Xu, C.; Youssef, A.M.; Pradhan, B.; Tien Bui, D. Rainfall-induced landslide susceptibility assessment at the chongren area (china) using frequency ratio, certainty factor, and index of entropy. *Geocarto Int.* **2017**, *32*, 139–154. [CrossRef]

26. Chen, W.; Li, W.; Chai, H.; Hou, E.; Li, X.; Ding, X. Gis-based landslide susceptibility mapping using analytical hierarchy process (ahp) and certainty factor (cf) models for the baozhong region of baoji city, china. *Environ. Earth Sci.* **2016**, *75*, 1–14. [CrossRef]

27. Dou, J.; Oguchi, T.; Hayakawa, Y.S.; Uchiyama, S.; Saito, H.; Paudel, U. Gis-based landslide susceptibility mapping using a certainty factor model and its validation in the chuetsu area, central japan. In *Landslide Science for a Aafer Geoenvironment*; Springer: New York, NY, USA, 2014; pp. 419–424.

28. Chen, W.; Shahabi, H.; Shirzadi, A.; Hong, H.; Akgun, A.; Tian, Y.; Liu, J.; Zhu, A.-X.; Li, S. Novel hybrid artificial intelligence approach of bivariate statistical-methods-based kernel logistic regression classifier for landslide susceptibility modeling. *Bull. Eng. Geol. Environ.* **2018**, 1–23. [CrossRef]

29. Pradhan, A.M.S.; Kim, Y.-T. Spatial data analysis and application of evidential belief functions to shallow landslide susceptibility mapping at mt. Umyeon, seoul, korea. *Bull. Eng. Geol. Environ.* **2017**, *76*, 1263–1279. [CrossRef]

30.	Ding, Q.; Chen, W.; Hong, H. Application of frequency ratio, weights of evidence and evidential belief function models in landslide susceptibility mapping. *Geocarto Int.* **2017**, *32*, 619–639. [CrossRef]

31.	Zhang, T.; Han, L.; Chen, W.; Shahabi, H. Hybrid integration approach of entropy with logistic regression and support vector machine for landslide susceptibility modeling. *Entropy* **2018**, *20*, 884. [CrossRef]

32.	Mandal, S.; Mandal, K. Modeling and mapping landslide susceptibility zones using gis based multivariate binary logistic regression (lr) model in the rorachu river basin of eastern sikkim himalaya, india. *Modeling Earth Syst. Environ.* **2018**, *4*, 69–88. [CrossRef]

33.	Chen, W.; Zhang, S.; Li, R.; Shahabi, H. Performance evaluation of the gis-based data mining techniques of best-first decision tree, random forest, and naïve bayes tree for landslide susceptibility modeling. *Sci. Total Environ.* **2018**, *644*, 1006–1018. [CrossRef]

34.	Youssef, A.M.; Pourghasemi, H.R.; Pourtaghi, Z.S.; Al-Katheeri, M.M. Landslide susceptibility mapping using random forest, boosted regression tree, classification and regression tree, and general linear models and comparison of their performance at wadi tayyah basin, asir region, saudi arabia. *Landslides* **2016**, *13*, 839–856. [CrossRef]

35.	Zhou, C.; Yin, K.; Cao, Y.; Ahmed, B.; Li, Y.; Catani, F.; Pourghasemi, H.R. Landslide susceptibility modeling applying machine learning methods: A case study from longju in the three gorges reservoir area, china. *Comput. Geosci.* **2018**, *112*, 23–37. [CrossRef]

36.	Pham, B.T.; Jaafari, A.; Prakash, I.; Bui, D.T. A novel hybrid intelligent model of support vector machines and the multiboost ensemble for landslide susceptibility modeling. *Bull. Eng. Geol. Environ.* **2018**, 1–22. [CrossRef]

37.	Hong, H.; Liu, J.; Bui, D.T.; Pradhan, B.; Acharya, T.D.; Pham, B.T.; Zhu, A.X.; Chen, W.; Ahmad, B.B. Landslide susceptibility mapping using j48 decision tree with adaboost, bagging and rotation forest ensembles in the guangchang area (china). *Catena* **2018**, *163*, 399–413. [CrossRef]

38.	Chen, W.; Shahabi, H.; Shirzadi, A.; Li, T.; Guo, C.; Hong, H.; Li, W.; Pan, D.; Hui, J.; Ma, M.; et al. A novel ensemble approach of bivariate statistical-based logistic model tree classifier for landslide susceptibility assessment. *Geocarto Int.* **2018**, *33*, 1398–1420. [CrossRef]

39.	Chen, W.; Shahabi, H.; Zhang, S.; Khosravi, K.; Shirzadi, A.; Chapi, K.; Pham, B.T.; Zhang, T.; Zhang, L.; Chai, H.; et al. Landslide susceptibility modeling based on gis and novel bagging-based kernel logistic regression. *Appl. Sci.* **2018**, *8*, 2540. [CrossRef]

40.	Truong, X.; Mitamura, M.; Kono, Y.; Raghavan, V.; Yonezawa, G.; Truong, X.; Do, T.; Tien Bui, D.; Lee, S. Enhancing prediction performance of landslide susceptibility model using hybrid machine learning approach of bagging ensemble and logistic model tree. *Appl. Sci.* **2018**, *8*. [CrossRef]

41.	Pham, B.T.; Tien Bui, D.; Prakash, I.; Dholakia, M.B. Hybrid integration of multilayer perceptron neural networks and machine learning ensembles for landslide susceptibility assessment at himalayan area (india) using gis. *CATENA* **2017**, *149*, 52–63. [CrossRef]

42.	Pham, B.T.; Shirzadi, A.; Tien Bui, D.; Prakash, I.; Dholakia, M.B. A hybrid machine learning ensemble approach based on a radial basis function neural network and rotation forest for landslide susceptibility modeling: A case study in the himalayan area, india. *Int. J. Sediment. Res.* **2017**. [CrossRef]

43.	Chen, W.; Pourghasemi, H.R.; Kornejady, A.; Zhang, N. Landslide spatial modeling: Introducing new ensembles of ann, maxent, and svm machine learning techniques. *Geoderma* **2017**, *305*, 314–327. [CrossRef]

44.	Pham, B.T.; Prakash, I.; Tien Bui, D. Spatial prediction of landslides using a hybrid machine learning approach based on random subspace and classification and regression trees. *Geomorphology* **2018**, *303*, 256–270. [CrossRef]

45.	Zabihi, M.; Pourghasemi, H.R.; Pourtaghi, Z.S.; Behzadfar, M. Gis-based multivariate adaptive regression spline and random forest models for groundwater potential mapping in iran. *Environ. Earth Sci.* **2016**, *75*, 665. [CrossRef]

46.	Lagomarsino, D.; Tofani, V.; Segoni, S.; Catani, F.; Casagli, N. A tool for classification and regression using random forest methodology: Applications to landslide susceptibility mapping and soil thickness modeling. *Environ. Modeling Assess.* **2017**, *22*, 201–214. [CrossRef]

47.	Breiman, L. Bagging predictors. *Mach. Learn.* **1996**, *24*, 123–140. [CrossRef]

48.	Tien Bui, D.; Ho, T.-C.; Pradhan, B.; Pham, B.-T.; Nhu, V.-H.; Revhaug, I. Gis-based modeling of rainfall-induced landslides using data mining-based functional trees classifier with adaboost, bagging, and multiboost ensemble frameworks. *Environ. Earth Sci.* **2016**, *75*, 1101. [CrossRef]

49. Xia, C.-k.; Su, C.-l.; Cao, J.-t.; Li, P. Multiboost with enn-based ensemble fault diagnosis method and its application in complicated chemical process. *J. Cent. South. Univ.* **2016**, *23*, 1183–1197. [CrossRef]

50. Pham, B.T.; Shirzadi, A.; Tien Bui, D.; Prakash, I.; Dholakia, M.B. A hybrid machine learning ensemble approach based on a radial basis function neural network and rotation forest for landslide susceptibility modeling: A case study in the himalayan area, india. *Int. J. Sediment. Res.* **2018**, *33*, 157–170. [CrossRef]

51. Fanos, A.M.; Pradhan, B.; Mansor, S.; Yusoff, Z.M.; Abdullah, A.F.b. A hybrid model using machine learning methods and gis for potential rockfall source identification from airborne laser scanning data. *Landslides* **2018**, *15*, 1833–1850. [CrossRef]

52. Chen, W.; Panahi, M.; Tsangaratos, P.; Shahabi, H.; Ilia, I.; Panahi, S.; Li, S.; Jaafari, A.; Ahmad, B.B. Applying population-based evolutionary algorithms and a neuro-fuzzy system for modeling landslide susceptibility. *CATENA* **2019**, *172*, 212–231. [CrossRef]

53. Naghibi, S.A.; Moghaddam, D.D.; Kalantar, B.; Pradhan, B.; Kisi, O. A comparative assessment of gis-based data mining models and a novel ensemble model in groundwater well potential mapping. *J. Hydrol.* **2017**, *548*, 471–483. [CrossRef]

54. Chen, W.; Li, H.; Hou, E.; Wang, S.; Wang, G.; Panahi, M.; Li, T.; Peng, T.; Guo, C.; Niu, C.; et al. Gis-based groundwater potential analysis using novel ensemble weights-of-evidence with logistic regression and functional tree models. *Sci. Total Environ.* **2018**, *634*, 853–867. [CrossRef] [PubMed]

55. Kadavi, P.; Lee, C.-W.; Lee, S. Application of ensemble-based machine learning models to landslide susceptibility mapping. *Remote Sens.* **2018**, *10*, 1252. [CrossRef]

56. Chen, W.; Xie, X.; Peng, J.; Shahabi, H.; Hong, H.; Bui, D.T.; Duan, Z.; Li, S.; Zhu, A.X. Gis-based landslide susceptibility evaluation using a novel hybrid integration approach of bivariate statistical based random forest method. *CATENA* **2018**, *164*, 135–149. [CrossRef]

57. Vakhshoori, V.; Pourghasemi, H.R. A novel hybrid bivariate statistical method entitled froc for landslide susceptibility assessment. *Environ. Earth Sci.* **2018**, *77*, 686. [CrossRef]

58. Fell, R.; Glastonbury, J.; Hunter, G. Rapid landslides: The importance of understanding mechanisms and rupture surface mechanics. *Q. J. Eng. Geol. Hydrogeol.* **2007**, *40*, 9–27. [CrossRef]

59. Rosi, A.; Tofani, V.; Tanteri, L.; Stefanelli, C.T.; Agostini, A.; Catani, F.; Casagli, N. The new landslide inventory of tuscany (italy) updated with ps-insar: Geomorphological features and landslide distribution. *Landslides* **2018**, *15*, 5–19. [CrossRef]

60. Hungr, O.; Leroueil, S.; Picarelli, L. The varnes classification of landslide types, an update. *Landslides* **2014**, *11*, 167–194. [CrossRef]

61. Tien Bui, D.; Tuan, T.A.; Klempe, H.; Pradhan, B.; Revhaug, I. Spatial prediction models for shallow landslide hazards: A comparative assessment of the efficacy of support vector machines, artificial neural networks, kernel logistic regression, and logistic model tree. *Landslides* **2016**, *13*, 361–378. [CrossRef]

62. Dai, F.C.; Lee, C.F.; Li, J.; Xu, Z.W. Assessment of landslide susceptibility on the natural terrain of lantau island, hong kong. *Environ. Geol.* **2001**, *40*, 381–391.

63. Nefeslioglu, H.A.; Duman, T.Y.; Durmaz, S. Landslide susceptibility mapping for a part of tectonic kelkit valley (eastern black sea region of turkey). *Geomorphology* **2008**, *94*, 401–418. [CrossRef]

64. Pham, B.T.; Khosravi, K.; Prakash, I. Application and comparison of decision tree-based machine learning methods in landside susceptibility assessment at pauri garhwal area, uttarakhand, india. *Environ. Process.* **2017**, *4*, 711–730. [CrossRef]

65. Wu, Y.; Li, W.; Liu, P.; Bai, H.; Wang, Q.; He, J.; Liu, Y.; Sun, S. Application of analytic hierarchy process model for landslide susceptibility mapping in the gangu county, gansu province, china. *Environ. Earth Sci.* **2016**, *75*, 422. [CrossRef]

66. Saadatkhah, N.; Kassim, A.; Lee, L.M. Susceptibility assessment of shallow landslides in hulu kelang area, kuala lumpur, malaysia using analytical hierarchy process and frequency ratio. *Geotech. Geol. Eng.* **2015**, *33*, 43–57. [CrossRef]

67. Aditian, A.; Kubota, T.; Shinohara, Y. Comparison of gis-based landslide susceptibility models using frequency ratio, logistic regression, and artificial neural network in a tertiary region of ambon, indonesia. *Geomorphology* **2018**, *318*, 101–111. [CrossRef]

68. Riaz Muhammad, T.; Basharat, M.; Hameed, N.; Shafique, M.; Luo, J. A data-driven approach to landslide-susceptibility mapping in mountainous terrain: Case study from the northwest himalayas, pakistan. *Nat. Hazards Rev.* **2018**, *19*, 05018007. [CrossRef]

69. Chen, W.; Li, W.; Hou, E.; Bai, H.; Chai, H.; Wang, D.; Cui, X.; Wang, Q. Application of frequency ratio, statistical index, and index of entropy models and their comparison in landslide susceptibility mapping for the baozhong region of baoji, china. *Arab. J. Geosci.* **2015**, *8*, 1829–1841. [CrossRef]

70. Moore, I.D.; Grayson, R.B.; Ladson, A.R. Digital terrain modelling: A review of hydrological, geomorphological, and biological applications. *Hydrol. Process.* **1991**, *5*, 3–30. [CrossRef]

71. Chen, W.; Peng, J.; Hong, H.; Shahabi, H.; Pradhan, B.; Liu, J.; Zhu, A.-X.; Pei, X.; Duan, Z. Landslide susceptibility modelling using gis-based machine learning techniques for chongren county, jiangxi province, china. *Sci. Total Environ.* **2018**, *626*, 1121–1135. [CrossRef]

72. Beven, K.J.; Kirkby, M.J. A physically based, variable contributing area model of basin hydrology/un modèle à base physique de zone d'appel variable de l'hydrologie du bassin versant. *Int. Assoc. Sci. Hydrol. Bull.* **1979**, *24*, 43–69. [CrossRef]

73. Ge, Y.; Chen, H.; Zhao, B.; Tang, H.; Lin, Z.; Xie, Z.; Lv, L.; Zhong, P. A comparison of five methods in landslide susceptibility assessment: A case study from the 330-kv transmission line in gansu region, china. *Environ. Earth Sci.* **2018**, *77*, 662. [CrossRef]

74. Wu, Z.; Wu, Y.; Yang, Y.; Chen, F.; Zhang, N.; Ke, Y.; Li, W. A comparative study on the landslide susceptibility mapping using logistic regression and statistical index models. *Arab. J. Geosci.* **2017**, *10*, 187. [CrossRef]

75. Yalcin, A. Gis-based landslide susceptibility mapping using analytical hierarchy process and bivariate statistics in ardesen (turkey): Comparisons of results and confirmations. *CATENA* **2008**, *72*, 1–12. [CrossRef]

76. Akgun, A. A comparison of landslide susceptibility maps produced by logistic regression, multi-criteria decision, and likelihood ratio methods: A case study at İzmir, turkey. *Landslides* **2012**, *9*, 93–106. [CrossRef]

77. Chen, W.; Yan, X.; Zhao, Z.; Hong, H.; Bui, D.T.; Pradhan, B. Spatial prediction of landslide susceptibility using data mining-based kernel logistic regression, naive bayes and rbfnetwork models for the long county area (china). *Bull. Eng. Geol. Environ.* **2018**, 1–20. [CrossRef]

78. Xu, C.; Dai, F.; Xu, X.; Lee, Y.H. Gis-based support vector machine modeling of earthquake-triggered landslide susceptibility in the jianjiang river watershed, China. *Geomorphology* **2012**, *145–146*, 70–80. [CrossRef]

79. Ohlmacher, G.C.; Davis, J.C. Using multiple logistic regression and gis technology to predict landslide hazard in northeast kansas, USA. *Eng. Geol.* **2003**, *69*, 331–343. [CrossRef]

80. Al-Abadi, A.M.; Shahid, S. A comparison between index of entropy and catastrophe theory methods for mapping groundwater potential in an arid region. *Environ. Monit. Assess.* **2015**, *187*, 576. [CrossRef]

81. Salvatici, T.; Tofani, V.; Rossi, G.; D'Ambrosio, M.; Tacconi Stefanelli, C.; Masi, E.B.; Rosi, A.; Pazzi, V.; Vannocci, P.; Petrolo, M.; et al. Application of a physically based model to forecast shallow landslides at a regional scale. *Nat. Hazards Earth Syst. Sci.* **2018**, *18*, 1919–1935. [CrossRef]

82. Peng, J.; Tong, X.; Wang, S.; Ma, P. Three-dimensional geological structures and sliding factors and modes of loess landslides. *Environ. Earth Sci.* **2018**, *77*, 675. [CrossRef]

83. Santo, A.; Di Crescenzo, G.; Forte, G.; Papa, R.; Pirone, M.; Urciuoli, G. Flow-type landslides in pyroclastic soils on flysch bedrock in southern italy: The bosco de' preti case study. *Landslides* **2018**, *15*, 63–82. [CrossRef]

84. Glade, T. Landslide occurrence as a response to land use change: A review of evidence from new zealand. *CATENA* **2003**, *51*, 297–314. [CrossRef]

85. Cui, S.-H.; Pei, X.-J.; Wu, H.-Y.; Huang, R.-Q. Centrifuge model test of an irrigation-induced loess landslide in the heifangtai loess platform, northwest china. *J. Mt. Sci.* **2018**, *15*, 130–143. [CrossRef]

86. Justice, C.O.; Townshend, J.R.G.; Holben, B.N.; Tucker, C.J. Analysis of the phenology of global vegetation using meteorological satellite data. *Int. J. Remote Sens.* **1985**, *6*, 1271–1318. [CrossRef]

87. Chen, W.; Xie, X.; Wang, J.; Pradhan, B.; Hong, H.; Tien Bui, D.; Duan, Z.; Ma, J. A comparative study of logistic model tree, random forest, and classification and regression tree models for spatial prediction of landslide susceptibility. *CATENA* **2017**, *151*, 147–160. [CrossRef]

88. Peruccacci, S.; Brunetti, M.T.; Luciani, S.; Vennari, C.; Guzzetti, F. Lithological and seasonal control on rainfall thresholds for the possible initiation of landslides in central italy. *Geomorphology* **2012**, *139–140*, 79–90. [CrossRef]

89. Tsukamoto, Y.; Ohta, T. Runoff process on a steep forested slope. *J. Hydrol.* **1988**, *102*, 165–178. [CrossRef]

90. Agterberg, F.P. Systematic approach to dealing with uncertainty of geoscience information in mineral exploration. *APCO* **1989**, *89*, 165–178.

91. Dahal, R.K.; Hasegawa, S.; Nonomura, A.; Yamanaka, M.; Masuda, T.; Nishino, K. Gis-based weights-of-evidence modelling of rainfall-induced landslides in small catchments for landslide susceptibility mapping. *Environ. Geol.* **2008**, *54*, 311–324. [CrossRef]

92. Lachenbruch, P.A.; Mccullagh, P.; Nelder, J.A. Generalized linear models. *Biometrics* **1990**, *46*, 291–303. [CrossRef]

93. Agarwal, S.; Kachroo, P.; Regentova, E. A hybrid model using logistic regression and wavelet transformation to detect traffic incidents. *IATSS Res.* **2016**, *40*, 56–63. [CrossRef]

94. Ravì, D.; Bober, M.; Farinella, G.M.; Guarnera, M.; Battiato, S. Semantic segmentation of images exploiting dct based features and random forest. *Pattern Recognit.* **2016**, *52*, 260–273. [CrossRef]

95. Masetic, Z.; Subasi, A. Congestive heart failure detection using random forest classifier. *Comput. Methods Programs Biomed.* **2016**, *130*, 54–64. [CrossRef] [PubMed]

96. Breiman, L. Random forests. *Mach. Learn.* **2001**, *45*, 5–32. [CrossRef]

97. Ozdemir, A. Landslide susceptibility mapping using bayesian approach in the sultan mountains (akşehir, turkey). *Nat. Hazards* **2011**, *59*, 1573–1607. [CrossRef]

98. Chen, W.; Xie, X.; Peng, J.; Wang, J.; Duan, Z.; Hong, H. Gis-based landslide susceptibility modelling: A comparative assessment of kernel logistic regression, naïve-bayes tree, and alternating decision tree models. *Geomat. Nat. Hazards Risk* **2017**, *8*, 950–973. [CrossRef]

99. Lin, G.-F.; Chang, M.-J.; Huang, Y.-C.; Ho, J.-Y. Assessment of susceptibility to rainfall-induced landslides using improved self-organizing linear output map, support vector machine, and logistic regression. *Eng. Geol.* **2017**, *224*, 62–74. [CrossRef]

100. Lee, J.-H.; Sameen, M.I.; Pradhan, B.; Park, H.-J. Modeling landslide susceptibility in data-scarce environments using optimized data mining and statistical methods. *Geomorphology* **2018**, *303*, 284–298. [CrossRef]

101. Yu, H.; Jiang, S.; Land, K.C. Multicollinearity in hierarchical linear models. *Soc. Sci. Res.* **2015**, *53*, 118–136. [CrossRef]

102. Ozdemir, A.; Altural, T. A comparative study of frequency ratio, weights of evidence and logistic regression methods for landslide susceptibility mapping: Sultan mountains, sw turkey. *J. Asian Earth Sci.* **2013**, *64*, 180–197. [CrossRef]

103. Ozdemir, A. Gis-based groundwater spring potential mapping in the sultan mountains (konya, turkey) using frequency ratio, weights of evidence and logistic regression methods and their comparison. *J. Hydrol.* **2011**, *411*, 290–308. [CrossRef]

104. Hong, H.; Tsangaratos, P.; Ilia, I.; Chen, W.; Xu, C. Comparing the Performance of a Logistic Regression and a Random Forest Model in Landslide Susceptibility Assessments. the Case of Wuyaun Area, China. In *Workshop on World Landslide Forum*; Mikos, M., Tiwari, B., Yin, Y., Eds.; Springer: Cham, Switzerland, 2017; pp. 1043–1050.

105. Pham, B.T.; Pradhan, B.; Tien Bui, D.; Prakash, I.; Dholakia, M.B. A comparative study of different machine learning methods for landslide susceptibility assessment: A case study of uttarakhand area (india). *Environ. Model. Softw.* **2016**, *84*, 240–250. [CrossRef]

106. Tien Bui, D.; Tuan, T.A.; Hoang, N.-D.; Thanh, N.Q.; Nguyen, D.B.; Van Liem, N.; Pradhan, B. Spatial prediction of rainfall-induced landslides for the lao cai area (vietnam) using a hybrid intelligent approach of least squares support vector machines inference model and artificial bee colony optimization. *Landslides* **2017**, *14*, 447–458. [CrossRef]

107. Chen, W.; Pourghasemi, H.R.; Naghibi, S.A. Prioritization of landslide conditioning factors and its spatial modeling in shangnan county, china using gis-based data mining algorithms. *Bull. Eng. Geol. Environ.* **2018**, *77*, 611–629. [CrossRef]

108. Chen, W.; Pourghasemi, H.R.; Naghibi, S.A. A comparative study of landslide susceptibility maps produced using support vector machine with different kernel functions and entropy data mining models in china. *Bull. Eng. Geol. Environ.* **2018**, *77*, 647–664. [CrossRef]

109. He, S.; Pan, P.; Dai, L.; Wang, H.; Liu, J. Application of kernel-based fisher discriminant analysis to map landslide susceptibility in the qinggan river delta, three gorges, china. *Geomorphology* **2012**, *171*, 30–41. [CrossRef]

110. Chen, W.; Panahi, M.; Pourghasemi, H.R. Performance evaluation of gis-based new ensemble data mining techniques of adaptive neuro-fuzzy inference system (anfis) with genetic algorithm (ga), differential evolution

(de), and particle swarm optimization (pso) for landslide spatial modelling. *CATENA* **2017**, *157*, 310–324. [CrossRef]

111. Wang, Q.; Li, W.; Wu, Y.; Pei, Y.; Xie, P. Application of statistical index and index of entropy methods to landslide susceptibility assessment in gongliu (xinjiang, china). *Environ. Earth Sci.* **2016**, *75*, 599. [CrossRef]

112. Naghibi, S.A.; Ahmadi, K.; Daneshi, A. Application of support vector machine, random forest, and genetic algorithm optimized random forest models in groundwater potential mapping. *Water Resour. Manag.* **2017**, *31*, 2761–2775. [CrossRef]

113. Chen, W.; Shirzadi, A.; Shahabi, H.; Ahmad, B.B.; Zhang, S.; Hong, H.; Zhang, N. A novel hybrid artificial intelligence approach based on the rotation forest ensemble and naïve bayes tree classifiers for a landslide susceptibility assessment in langao county, china. *Geomat. Nat. Hazards Risk* **2017**, *8*, 1955–1977. [CrossRef]

114. Ilia, I.; Tsangaratos, P. Applying weight of evidence method and sensitivity analysis to produce a landslide susceptibility map. *Landslides* **2016**, *13*, 379–397. [CrossRef]

115. Polykretis, C.; Chalkias, C. Comparison and evaluation of landslide susceptibility maps obtained from weight of evidence, logistic regression, and artificial neural network models. *Nat. Hazards* **2018**. [CrossRef]

116. Lee, S. Landslide detection and susceptibility mapping in the sagimakri area, korea using kompsat-1 and weight of evidence technique. *Environ. Earth Sci.* **2013**, *70*, 3197–3215. [CrossRef]

117. Malamud, B.D.; Turcotte, D.L.; Guzzetti, F.; Reichenbach, P. Landslide inventories and their statistical properties. *Earth Surf. Process. Landf.* **2004**, *29*, 687–711. [CrossRef]

118. Baeza, C.; Lantada, N.; Amorim, S. Statistical and spatial analysis of landslide susceptibility maps with different classification systems. *Environ. Earth Sci.* **2016**, *75*, 1318. [CrossRef]

119. Akgun, A.; Sezer, E.A.; Nefeslioglu, H.A.; Gokceoglu, C.; Pradhan, B. An easy-to-use matlab program (mamland) for the assessment of landslide susceptibility using a mamdani fuzzy algorithm. *Comput. Geosci.* **2012**, *38*, 23–34. [CrossRef]

120. Kumar, R.; Anbalagan, R. Landslide susceptibility mapping using analytical hierarchy process (ahp) in tehri reservoir rim region, uttarakhand. *J. Geol. Soc. India* **2016**, *87*, 271–286. [CrossRef]

121. Khosravi, K.; Pourghasemi, H.R.; Chapi, K.; Bahri, M. Flash flood susceptibility analysis and its mapping using different bivariate models in iran: A comparison between shannon(')s entropy, statistical index, and weighting factor models. *Environ. Monit. Assess.* **2016**, *188*, 656. [CrossRef] [PubMed]

122. Kleinbaum, D.G.; Klein, M. Introduction to logistic regression. In *Logistic Regression: A Self-Learning Text*; Kleinbaum, D.G., Ed.; Springer: New York, NY, USA, 2010.

123. Genuer, R.; Poggi, J.-M.; Tuleau-Malot, C.; Villa-Vialaneix, N. Random forests for big data. *Big Data Res.* **2017**, *9*, 28–46. [CrossRef]

5

A Single Point-Based Multilevel Features Fusion and Pyramid Neighborhood Optimization Method for ALS Point Cloud Classification

Yong Li [†], **Guofeng Tong** *, **Xiance Du** [†], **Xiang Yang, Jianjun Zhang** and **Lin Yang**

College of Information Science and Engineering, Northeastern University, Shenyang 110819, China; liyong@stumail.neu.edu.cn (Y.L.); newneulab@163.com (X.D.); yangxwkm@gmail.com (X.Y.); junneu@126.com (J.Z.); 1700922@stu.neu.edu.cn (L.Y.)

* Correspondence: tongguofeng@ise.neu.edu.cn

† These authors contributed equally to this work.

Abstract: 3D point cloud classification has wide applications in the field of scene understanding. Point cloud classification based on points can more accurately segment the boundary region between adjacent objects. In this paper, a point cloud classification algorithm based on a single point multilevel features fusion and pyramid neighborhood optimization are proposed for a Airborne Laser Scanning (ALS) point cloud. First, the proposed algorithm determines the neighborhood region of each point, after which the features of each single point are extracted. For the characteristics of the ALS point cloud, two new feature descriptors are proposed, i.e., a normal angle distribution histogram and latitude sampling histogram. Following this, multilevel features of a single point are constructed by multi-resolution of the point cloud and multi-neighborhood spaces. Next, the features are trained by the Support Vector Machine based on a Gaussian kernel function, and the points are classified by the trained model. Finally, a classification results optimization method based on a multi-scale pyramid neighborhood constructed by a multi-resolution point cloud is used. In the experiment, the algorithm is tested by a public dataset. The experimental results show that the proposed algorithm can effectively classify large-scale ALS point clouds. Compared with the existing algorithms, the proposed algorithm has a better classification performance.

Keywords: ALS point cloud; multi-scale; classification; large scene

1. Introduction

Airborne Laser Scanning (ALS) can capture large-scale point clouds of urban scenes. The point cloud classification of outdoor scenes can provide high-precision semantic maps for autonomous driving, improve the accuracy of vehicle positioning, and reconstruct a three-dimensional model of the city, which plays an important role in urban planning and dynamic management. In addition, it can improve the efficiency of resource utilization. Effectively labeling the correct class for all points in the scene is an important basis for the widespread adoption of point clouds [1–4]. However, a laser point cloud has a huge data number, high redundancy, and uneven scene distribution, which may lead to huge challenges in the point cloud classification. Therefore, it is of great significance to classify the three-dimensional point cloud in large outdoor scenes.

Currently, the number of point clouds with manual labeling in outdoor large scenes is not enough. However, machine learning can learn and classify point clouds in the case of less sample training data, and the speed is faster. At present, the point cloud classification methods can be mainly divided into two strategies: the single point-based classification and object-based classification methods.

The point-based point cloud classification is the classification of each individual point in a point cloud; this strategy uses points as the basic unit to extract features, train models, and predict class labels. There are three main steps: the neighborhood selection, feature extraction, and single point classification based on the features and classifiers.

(1) Neighborhood selection. In the neighborhood selection process, the commonly used point cloud neighborhood forms are: K nearest neighbors [5], radius neighborhoods [6], and column neighborhoods [7]. The parameter of neighborhood estimation is highly dependent on prior knowledge, and it is greatly affected by the change of the point cloud density [8,9]. For example, Hackel et al. constructed a multi-level scale pyramid, and a total of 144-dimensional features such as eigenvalues of the covariance matrix were extracted for each point in each pyramid. Subsequently, Random Forest was used for training and finally classified outdoor road scenes [10], and a better classification effect was obtained. Therefore, the selection based on multi-scale neighborhood is an important method for extracting single point effective features.

(2) Features extraction. The local feature of a point cloud is an abstract depiction of the environment around a given point. It is difficult to classify a point cloud by a single local feature. The common practice is to fuse multiple point cloud local features for classification. Normal and curvature are simple local features that clearly show some local information about the point cloud, such as the fact that the normal direction can represent the partial tangent plane of the point cloud, and the curvature can represent the smoothness of the point. For example, Fanxuan et al. [11] optimized the matching accuracy of point pairs in point clouds based on the curvature information, and the registration accuracy of point clouds was further improved. Geometric features are also common local features, also known as shape descriptors. For example, in the spin image [12], the main idea is to set up an image with the normal vector as the center; the image rotates around the axis. The number of 3D points encountered by each pixel in the point cloud is taken as its gray value. Finally, a two-dimensional array representing the local information of the three-dimensional space, that is, the rotated image feature, is obtained. The 3D Shape Context (3DSC) feature [13] is based on the specified point to construct a spherical region. In the support region, the grid is divided into three coordinate directions: the radial direction, direction angle and elevation angle. Following this, a feature histogram is constructed by entering the number of points in the grid. The 3DSC is simple in construction, strong in discrimination and insensitive to noise, but it is time-consuming. The Unique Shape Context (USC) descriptor [14] improves the 3DSC to avoid ambiguities in the classification. Point Feature Histograms (PFH) [15] are local features, which construct a feature histogram with the angles and distances of the normal vectors of any two points in the specified point neighborhood. The descriptor can accurately describe the local features of the points, but the computation is large and the real-time performance is poor. Fast Point Feature Histograms (FPFH) [16] are a simplification of PFH, which greatly reduce the time consumption while retaining most of the description performance of PFH. FPFH have an excellent performance, and are widely used in the field of point cloud classification, segmentation and registration [17]. Although these features can express the local features of the point cloud, they do not take into account the characteristics of the ALS point cloud, which has the characteristics of relative sparse, rich elevation information, as well as a horizontal and vertical distribution.

(3) Single point classification based on features and classifiers. Currently, machine learning is an important method for classification problems. The single-point classification based on machine learning takes the feature vector of the point as the input and the class label of the single point as the output. Common machine learning algorithms can accomplish this classification task, such as AdaBoost [18], Random Forest [19] and Support Vector Machine (SVM) [20]. This kind of method uses a classifier to learn the local features of each point, after which the parameters in the classifier are determined based on the training dataset. Finally, the test set is classified by the classifier. This classification strategy can more accurately segment the boundary regions between different adjacent

objects, and this method has a better performance in detail. However, due to the extremely large number of points, the calculation amount is large. Thus, the model training is slow, and there are also some misclassifications of local regions. However, there are always some errors in the final classification results. Therefore, the initial classification results are required to further optimize according to the characteristics of the point cloud.

In order to solve the above-mentioned problems, an ALS point cloud classification algorithm based on a single point multi-feature fusion is proposed. This kind of algorithm is based on the point as the basic processing unit, and the classification process assigns labels to each point in the point cloud to realize a point cloud classification. The proposed method extracts the local features of each point by constructing a multi-scale neighborhood space, along with two new features: a normal angle distribution histogram (NAD) and latitude sampling histogram (LSH) are proposed. Following this, SVM is used for training and classification. However, since each point is classified, there is a problem regarding some edge points being misclassified. In this regard, the initial classification results are further optimized according to the neighborhood classification optimization of multi-scale pyramids. Experiments prove that the classification algorithm has a higher accuracy.

The main contributions of this paper are as follows:

(1) Two local features are proposed, that is, the NAD histogram and the LSH histogram. The differences of different objects in the normal distribution, and the difference of the neighborhood points around different objects in the horizontal and vertical directions of the three-dimensional space, can be fully utilized to more effectively represent the characteristics of different objects.

(2) A multilevel single-point features fusion method based on a multi-neighborhood space and multi-resolution is proposed. The multi-scale space is constructed by changing the resolution of the point cloud and the number of the neighborhood. The features of the multi-scale are extracted from each single point, and the features are fused. Following this, the SVM classifier is used to classify the features and the better classification results have also been achieved.

(3) A fast optimization method for classification results based on a multi-scale pyramid is proposed. By changing the resolution of the point cloud, a multi-scale pyramid is constructed, and the neighbor points are further re-selected. After this, the misclassifications are eliminated according to the initial classification results of the neighbors for a post-processing optimization.

2. Method

As shown in Figure 1, the algorithm flow is given as follows. In the training part, for the point cloud scene shown in Figure 1a, multiple features of each point are first extracted. Multi-scale and multiple features are fused to a fusion feature. The SVM classifier model is then trained using the fusion features of the training set. In the test part, for the point cloud scene shown in Figure 1b, the fusion feature is first obtained. As shown in Figure 1c, the test points are initially classified using the trained SVM classification model. Following this, the point clouds of different resolutions are obtained by down-sampling the point cloud, in order to construct a multi-scale point cloud pyramid. The corresponding classification labels of the different neighboring points in the different scales are searched. Finally, the label which has the most number in the neighbors is taken as the class label of the current point. The final point cloud classification result is obtained, as shown in Figure 1d.

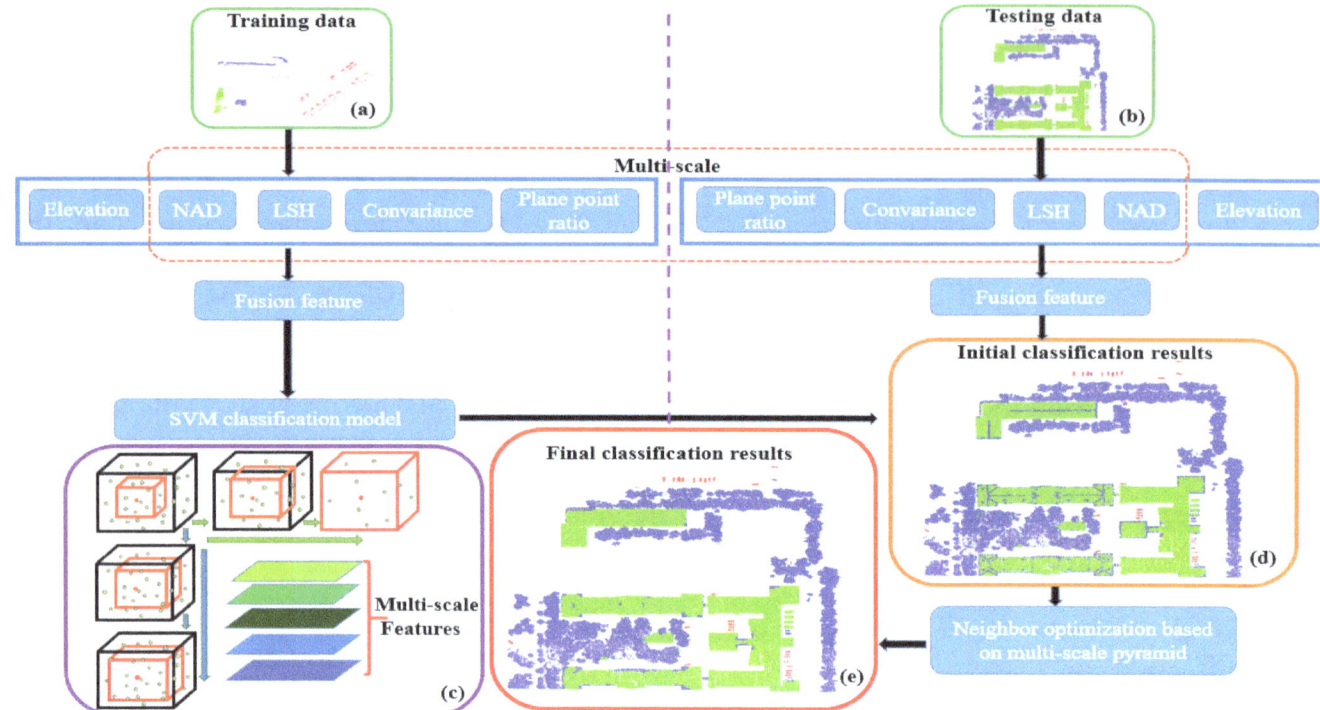

Figure 1. Flowchart of the proposed method. NAD: normal angle distribution histogram;LSH: latitude sampling histogram; SVM: Support Vector Machine.

2.1. Point Feature Extraction

In the classification task of the 3D point cloud data, the feature extraction of the point cloud plays a crucial role. It can seriously affect the final classification result. A well-behaved feature descriptor should reflect obvious differences between different types of points in the point cloud. At the same time, the descriptor should be robust and have a strong anti-interference ability. It is difficult for a single feature descriptor to have the above characteristics, so that a plurality of feature descriptor fusion methods are at present widely used. In the single point-based classification algorithm, this paper uses a variety of feature fusion methods to improve the accuracy of the classification algorithm. The specific features are as follows:

2.1.1. Feature Description

1. Elevation feature

The height is a very intuitive feature in a 3D point cloud. Generally speaking, points with large height are buildings, trees or objects with larger elevation values in the real world. When the elevation value is small, the probability is greater if the point is a vehicle point. Thus, the elevation feature is set to:

$$F_z = [Z_i, \ \frac{1}{Z_i}] \tag{1}$$

where Z_i is the distance of the i-th point from the estimated ground to the elevation value.

2. Normal angle distribution histogram

In the large scale scene, the normal direction of different objects has obvious differences. For artificial objects, such as buildings and vehicles, since the surface is relatively regular, almost all points are in the same direction, pointing in the direction of the vertical plane. However, due to the scattered distribution of the whole point cloud, the normal direction of the point cloud has a large scattered nature, and the direction does not point to a fixed direction in a uniform way. Therefore, we calculate the histogram of each point and its own normal angle distribution value in the local neighborhood point set to express the relationship between the normal of the point and the normal of the points in the neighboring region. The angle between the two normal vectors in three-dimensional space should be between $[0, \pi]$. But considering that the normal of the point on the plane can have opposite directions when the angle is larger than $\pi/2$, the corresponding angle is set to $\pi - \pi/2$. Following this, the angle of the normal vectors is defined as $[0, \pi/2]$. Considering efficiency and resolving power, we divide this interval into equal D_n parts, that is, construct a D_n dimensional histogram. After this, the number of points falling within each cell is taken as the value of the interval in the histogram. Finally, the normalization process is performed to form a histogram of the normal angle distribution, called NAD. This feature can distinguish different classes of points based on the normal angular distributions. The specific calculation formulas are as follows:

$$\Delta = \frac{\pi/2 - 0}{D_n} \tag{2}$$

$$\theta_j = a\cos(v \cdot v_j) \tag{3}$$

$$h(x_i) = \frac{n(i * \Delta \leq \theta_j \leq (i+1) * \Delta)}{N} \quad (i = 1, \ldots, D_n) \tag{4}$$

$$F_{NAD} = [h(x_1), h(x_2), \ldots, h(x_{D_n})] \tag{5}$$

where v and v_j represent the normal vectors of the current point and the j-th neighbor point, respectively. $a\cos(\cdot)$ represents the inverse cosine function. N represents the number of neighbors for the current point. $n(i * \Delta \leq \theta_j \leq (i+1) * \Delta)$ denotes the number of points for the normal angle at the range $[i * \Delta, (i+1) * \Delta]$. F_{NAD} denotes the final normal angle eigenvalue vector of the normal angle distribution. The histogram of the normal angle distributions for the randomly selected building points and tree points are shown in Figure 2.

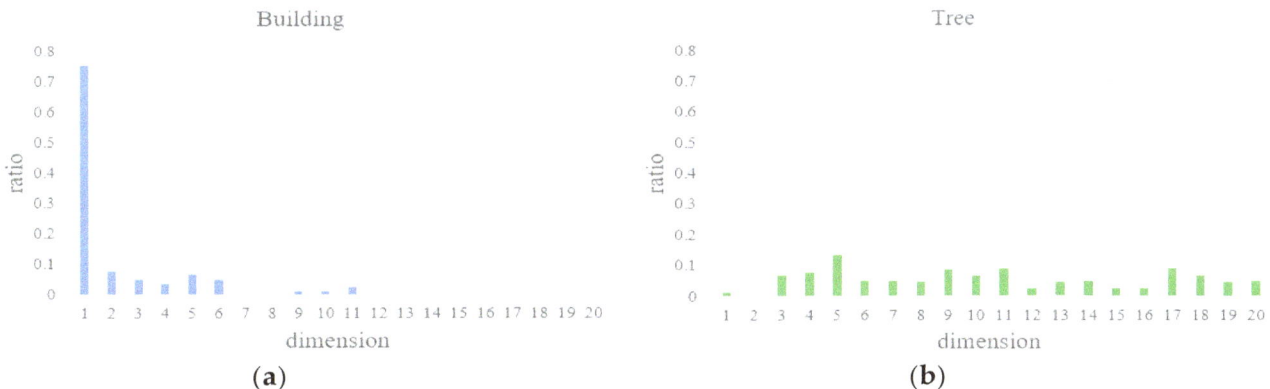

Figure 2. Normal angle distribution histogram. (**a**) Normal angle distribution histogram of a building point. (**b**) Normal angle distribution histogram of a tree point.

3.   Latitudinal sampling histogram

In the outdoor large scene environment, as for almost all points belonging to different objects, the surrounding neighborhood points have great differences in the latitudinal distribution in the three-dimensional space. For example, a building surface, which is parallel to the ground, has its neighborhood points mainly distributed near the "equator". For the points belonging to the trees, the distribution of the neighborhood points is more random and extensive, hardly concentrated in a certain latitude interval. Therefore, the selected point is regarded as the center of the sphere, and the distribution histogram of the neighborhood points in the latitudinal direction is counted. Following this, the feature of the point can be expressed. The feature is called LSH. The LSH feature can be used to distinguish different classes of points according to the distribution of neighborhood points in the latitude direction. The LSH has the advantages of anti-occlusion, without interference from the local coordinate system, as well as high efficiency. In this paper, D_l spaces are equally divided along the latitude direction. Following this, the number of points falling into each cell is counted to form a feature vector of the D_l dimension. The specific calculation formulas are:

$$\Delta = \frac{\pi - 0}{D_l} \tag{6}$$

$$\theta_j = a\cos(z \cdot (p_j - p)) \tag{7}$$

$$f(x_i) = \frac{n(i * \Delta \leq \theta_j \leq (i+1) * \Delta)}{N} \quad (i = 1, \ldots, D_l) \tag{8}$$

$$F_{LSH} = [f(x_1), f(x_2), \ldots, f(x_{D_l})] \tag{9}$$

where p and p_j represent the three-dimensional coordinates of the current point and its j-th neighbor point, respectively. $z = (0,0,1)$ represents the unit vector of the positive direction of the z axis. $n(i * \Delta \leq \theta_j \leq (i+1) * \Delta)$ represents the number of points in $[i * \Delta, (i+1) * \Delta]$ of the neighborhood points along the latitudinal direction. F_{LSH} represents the final feature vector of LSH. The LSHs of the randomly selected building points and tree points are compared, as shown in Figure 3.

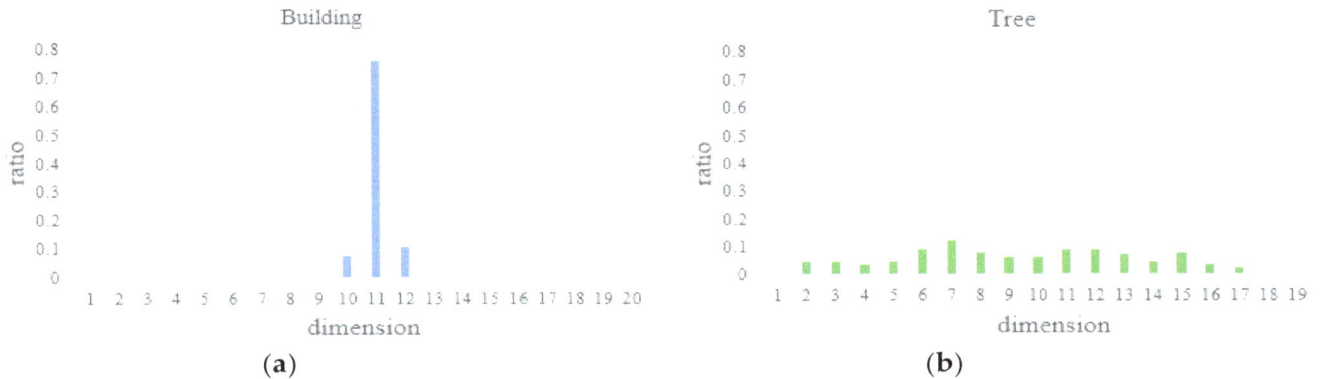

Figure 3. Latitudinal sampling histogram. **(a)** Latitudinal sampling histogram of a building point. **(b)** Latitudinal sampling histogram of a tree point.

4.   Convariance feature

First, a covariance matrix for the selected point neighborhood is constructed. After this, eigenvalues of the covariance matrix are calculated as: $\lambda_2 \geq \lambda_1 \geq \lambda_0 \geq 0$, and the corresponding eigenvectors are calculated as: v_2, v_1, v_0. Here, the covariance feature (CF) is obtained according to the relationship among the eigenvalues, as follows:

Sum of eigenvalues:
$$F_{sum} = \lambda_1 + \lambda_2 + \lambda_3 \tag{10}$$

Full variance:
$$F_{omn} = (\lambda_1 \cdot \lambda_2 \cdot \lambda_3)^{\frac{1}{3}} \tag{11}$$

Anisotropy:
$$F_{ani} = (\lambda_1 - \lambda_3)/\lambda_1 \tag{12}$$

Planarity:
$$F_{pla} = (\lambda_2 - \lambda_3)/\lambda_1 \tag{13}$$

Linearity:

$$F_{lin} = (\lambda_1 - \lambda_2)/\lambda_1 \tag{14}$$

Sphericity:

$$F_{sph} = \lambda_3/\lambda_1 \tag{15}$$

Following this, the final total covariance feature is: $F_{cov} = \left[F_{sum}, F_{omn}, F_{ani}, F_{pla}, F_{lin}, F_{sph}\right]$.

5. Plane point ratio

In outdoor large scale scenes, the classes of objects are complex and the surface shapes are also different. However, a considerable part of the surface of artificial objects exhibits planar characteristics, such as buildings, vehicles, etc. Meanwhile vegetation does not have planar characteristics, so the plane point ratio of the local point cloud can be used as a local feature to classify point clouds. The covariance feature can also reflect the planar characteristics to a certain extent, but it is greatly interfered by noise. For this reason, the Random Sample Consensus (RANSAC) [21] is employed to fit the local neighborhood of the selected point. After this, the ratio of the plane points, called PPR (Plane Point Ratio), is calculated.

RANSAC is a method used to find the subset of data that is the best match for the model from the data set with random samples that are noisy but sufficient. The points matched with the model are called the inner points, and the points unmatched with the model are called the outer points. The plane is fitted using RANSAC as follows.

(1) Select three points randomly from all the neighborhood points and calculate the current model parameters. The model is as follows:

$$ax + by + cz + d = 0 \tag{16}$$

(2) Determine whether each point is an inlier, and then determine the inlier rate ω of the current model:

$$J_i = \begin{cases} 1, & d_i \leq T_d \\ 0, & d_i > T_d \end{cases} \tag{17}$$

$$\omega = \frac{1}{N}\sum_{i=1}^{N} J_i \tag{18}$$

where d_i is the distance from the i-th point to the plane. T_d is a fixed threshold. J_i indicates whether it is an inlier or not. N is the number of neighborhood points.

(3) If the current inlier rate is larger than the previous optimal inlier rate, the optimal inlier rate is updated.

(4) To find the optimal model, repeat steps (1) to (3) k times until the probability reaches P:

$$1 - P \leq \left(1 - \omega^3\right)^k \tag{19}$$

The termination condition is:

$$k \geq \frac{\log(1 - P)}{\log(1 - \omega^3)} \tag{20}$$

When the RANSAC iteration is completed, the optimal inlier rate is the ratio of the plane points: $F_{plane} = [\omega]$.

2.1.2. Single Point Multi-Scale Multi-Feature Fusion

Since the features of the single point are dependent on the selected neighborhood space, different neighborhood spaces have different expression capabilities for different classes of point clouds.

Additionally, the structure descriptions of point clouds with different resolutions also have certain differences. The local feature description of the single scale for a point is relatively single, and there are some noise points in the point cloud, which can make the simple feature of the single scale unable to accurately describe the feature of the single point. Therefore, a multilevel features fusion method based on the multi-neighborhood space and multi-resolution is proposed. As shown in Figure 1c, the proposed method constructs the multi-scale space by changing the resolution and the number of neighborhoods of the point cloud. Following this, multi-scale features for each single point in the point cloud are extracted. Because the elevation feature is not affected by the scale changing, we select NAD, LSH, CF and PPR features to construct the multi-scale features. We extract the features of a single point in each scale by choosing μ neighborhoods with different resolutions and υ different neighborhood sizes under the original resolution. The multi-scale features of each point are expressed respectively as F'_{NAD}, F'_{LSH}, F'_{cov}, and F'_{plane}. Considering the validity of the features and the efficiency of the calculation, this generally results in $2 \leq \mu + \upsilon \leq 5$. In addition, the description of the single point feature only represents one characteristic of the point cloud. Therefore, it is necessary to fuse multiple features. After fusing the features, the multilevel features are expressed as follows:

$$F = \left[F_z, F'_{NAD}, F'_{LSH}, F'_{cov}, F'_{plane} \right] \tag{21}$$

Because we aim at an ALS point cloud, the extracted elevation features are only two-dimensional and play an important role in the point cloud classification. In addition, when the point cloud features are extracted, the values of each feature have been normalized to [0, 1]. In order to reflect the role of the non-zero feature value, the feature should be normalized again according to Formula (22) when the extracted feature F is sparse. While the extracted feature F is not sparse, there is no need to normalize the feature. Therefore, the constructed feature is $X = \left[F_z^*, F_{NAD}^*, F_{LSH}^*, F_{cov}^*, F_{plane}^* \right]$.

$$F_{i,j}^* = \frac{F_{i,j} - \min(F_j)}{\max(F_j) - \min(F_j)} \tag{22}$$

where $F_{i,j}^*$ is the value of the i-th row and the j-th column in the normalized feature matrix F^*. $F_{i,j}$ is the value of the i-th row and the j-th column in the feature matrix F. F_j is the vector of the j-th column (for all the points) in the feature matrix F.

2.2. Point Cloud Classification Based on SVM

SVM [22] is achieved by maximizing the classification interval in the feature space. For non-linear data, SVM maps them into a high-dimensional feature space by a kernel function, which make the data into linear separable data in a high-dimensional feature space. Following this, it realizes a classification by maximizing the interval. In view of the excellent generalization ability of SVM, we use SVM as a classifier for the single point classification in point cloud data. As we know, the correlation between the point cloud single point feature and neighbor points features, and the Gauss kernel function only has one parameter σ and a low model complexity. Thus, in the absence of prior knowledge, the Gauss kernel function is often better than other kernels. Therefore, we choose the Gauss kernel function as the kernel function. Here, the Gauss kernel function of the SVM classifier is defined as follows:

The fused feature space is X. The selected n d-dimensional feature samples $\{x_1, x_2, \ldots, x_n\}$:

$$\mathbf{x}_i = (x_{i1}, x_{i2}, \ldots, x_{id})' \epsilon \mathbb{R}^d \tag{23}$$

After the feature transformation, the feature space is Z. We map data in the X space to the Z space $\mathbf{z}_i = (z_{i1}, z_{i2}, \ldots, z_{id}) \epsilon \mathbb{R}^d$ via the mapping function $\phi(x)$. The function $K(x, z)$ satisfies the condition

$K(x, z) = \phi(x) \cdot \phi(z)$, and the function $K(x, z)$ is a kernel function, while $\phi(x)$ is a mapping function. The Gauss kernel function is as follows:

$$K(X, z) = \sum_i^n K(\mathbf{x}_i, z) = \sum_i^n \exp\left(-\frac{\|\mathbf{x}_i - z\|^2}{2\sigma}\right) \tag{24}$$

The corresponding decision function is:

$$f(z, \alpha^*, b^*) = sign(\sum_i^n y_i \alpha^* \exp\left(-\frac{\|\mathbf{x}_i - z\|^2}{2\sigma}\right) + b^*) \tag{25}$$

The SVM classifier is trained by the features of the training set, and the test set is classified by the trained classifier. The initial classification results for the point cloud in Figure 1b are shown in Figure 1d.

2.3. Neighborhood Optimization Based on Multi-Scale Pyramid

After the initial classification, the point clouds are basically classified correctly. Due to noise and other reasons, there are still some misclassifications in some details (such as edges). As shown in Figure 1d, most of the points in the scene have been correctly classified, and only a small part of them are misclassified. They mainly concentrate on edges and other places, and most of the points around the misclassified points are correctly classified. Therefore, it is necessary to further optimize the initial classification results to achieve a more accurate classification of the point clouds. Because local information is used as a feature to classify point clouds, the feature extraction relies heavily on a local region selection. In addition, the single point is taken as the basic unit of classification. Each point has its own characteristics, but because the two neighboring points are very close to each other and their neighborhoods are also very close, the extracted features will be very similar, which leads them to be more likely to be classified into the same class. Therefore, the neighbors of the misclassified points are also often misclassified. It is difficult to correct the misclassified points if only the points in the smaller local regions are used for the optimization. Therefore, we propose a classification results optimization method based on the multi-scale pyramid. The specific method is as follows:

First, voxel filters with different radius scales are used to down-sample the point cloud after an initial classification, as shown in Formula (24). Each minimum voxel scale is twice as large as the last down-sampling, and sparse point clouds are gradually obtained. Following this, the q-level pyramid is constructed, and the initial classes of all the points in each level are retained. According to the characteristics of the point cloud down-sampling reflecting the structure information of the shape, the scale pyramid is constructed on three scales of $q = 3$ in this paper.

Following this, the corresponding k-d tree is constructed from the point cloud in each layer of the pyramid. For each point in the original point cloud, a k-d tree is used to search for the radius of the nearest neighbors in the point cloud after the down-sampling. The class labels of the m point clouds searched within the radius of the l-th level are $L^l = \left\{L_1^l, \ldots, L_m^l\right\}, L_i^l \in \{1, \ldots, c\}, i = 1, \ldots, m, l = 1, \ldots, q$; the radius parameters are different when each layer of the point cloud chooses its nearest neighbor. The method of calculation is as follows:

$$r = k \cdot Presolution \tag{26}$$

In the formula, r represents the scale radius parameter. $Presolution$ is the resolution used by the current down-sampling point cloud. k is a fixed ratio threshold.

Finally, the initial labels of all the nearest neighbors in the q levels are counted. The discriminant function $1\left\{L_i^l = C\right\}$ represents the fact that when L_i^l belongs to class C, its value is 1; otherwise, its value is 0. This is used to count the number of the initial labels belonging to each class. As shown in Formula (27), the mode label C^* is selected as the new class label for the current point.

$$C^* = \begin{array}{c} \text{argmax} \\ C = L_i^l \in \{1,\ldots,c\} \end{array} \sum_{l=1}^{q} \sum_{i}^{m} 1\left\{ L_i^l = C \right\} \tag{27}$$

Not only do the optimized point cloud classification results avoid a situation where the nearest neighbor is also misclassified, but they also solve the problem of too many far points in a large scale, thus achieving better results. The optimized point cloud classification results are shown in Figure 1e.

3. Experimental Results and Analysis

In order to verify the performance of the point cloud classification algorithm based on single point multilevel features, we use two urban scenes' ALS data for a qualitative and quantitative comparison and analysis. This section begins by briefly introducing the experimental data, before the classification performance of the proposed method is compared with the other methods on the datasets.

3.1. Experimental Dataset

In this paper, we use two sets of dataset published in Ref [23]. The data was collected in Tianjin, China. The density of the test region point cloud is about $20 \sim 30/m^2$. The data set contains both large objects (buildings and trees) and small objects (cars). It contains different roof shapes, buildings of different heights, and dense and overlapping cars and trees. Table 1 lists the number of each class point for the two scenes. Figure 4a,b shows the training data of scene 1 and scene 2, respectively.

All the programs are run on a computer with an Intel Core i7-7700K CPU, 4.20 GHz with 24-GB RAM. The algorithm is implemented on the C++ platform based on PCL 1.8.0. Each set of data training and testing takes about 6.5 min. However, the feature extraction and optimization process can be implemented in parallel. Therefore, the efficiency of the proposed method can be further improved, and the speed of the point cloud processing will be greatly improved.

Table 1. The experimental dataset.

Scene	Training Points			Test Points		
	Tree	Building	Car	Tree	Building	Car
Scene 1	68,802	37,128	5380	213,990	200,549	7816
Scene 2	39,743	64,952	4584	73,207	156,186	7409

(a)   (b)

Figure 4. The training data of the ALS points. (**a**) scene 1, and (**b**) scene 2. (The figures are captured from the ALS points shown in Cloudcompare (http://www.cloudcompare.org/). The red points are cars, green points are buildings and blue points are trees.)

3.2. Experimental Comparison and Analysis

In order to verify the performance of the proposed algorithm, we compare it with the other seven methods shown in Table 2. Method 1 is the proposed method, which uses the features without NAD and LSH. Method 2 uses the $F_z+F_{NAD}+F_{LSH}+F_{cov}+F_{plane}$ feature fusion in single-scale without a post-processing optimization. Method 3 uses the single-scale feature and multi-scale pyramid optimization. Method 4 is the algorithm proposed in [24]. In this method, each feature uses geometry, strength, and statistics information. Following this, the JointBoost method is used to select features and classify points. Method 5 is the classification method based on the multi-scale Spin Image feature and F_{cov} feature fusion used in Ref [23]. Method 6 [25] constructs a multilevel point set using a linear transformation, and it uses Spin Image and F_{cov} features. Following this, K-means is used to construct an LDA (Latent Dirichlet Allocation) model of a multilevel point set dictionary. Method 7 [23,26,27] constructs a multilevel point set using an exponential transformation. The Spin Image and F_{cov} features are used for dictionary learning, constructing an LDA model of point sets.

Table 2. Main features of the proposed method and other comparison methods. SVM: Support Vector Machine; LDA: Latent Dirichlet Allocation; DD-SCLDA: Discriminative Dictionary based Sparse Coding and LDA.

Method	Scale	Feature Expression	Post-Processing Optimization	Classifier
Our method	Multi-scale	$F_z+F_{NAD}+F_{LSH}+F_{cov}+F_{plane}$	Multi-scale pyramid	SVM
Method 1	Multi-scale	$F_z+F_{cov}+F_{plane}$	Multi-scale pyramid	SVM
Method 2	Single scale	$F_z+F_{NAD}+F_{LSH}+F_{cov}+F_{plane}$	None	SVM
Method 3	Single scale	$F_z+F_{NAD}+F_{LSH}+F_{cov}+F_{plane}$	Multi-scale pyramid	SVM
Method 4	Multi-scale	Geometry, strength, and statistical features	Regional growth	JointBoost
Method 5	Multi-scale	Spin Image feature and F_{cov}	None	AdaBoost
Method 6	Multi-scale	LDA Model of the Spin Image feature and F_{cov} based on Multi-Level Point Sets	None	AdaBoost
Method 7	Multi-scale	DD-SCLDA Model of the Spin Image feature and F_{cov} based on Multi-Level Point Sets	None	AdaBoost

Accuracy, precision, and recall are often used to evaluate the effect of a point cloud classification [27]. The precision rate is the proportion of true positive samples in a positively predicted sample. The recall rate is the proportion of positive samples that are predicted to be successful in the actual positive samples. The accuracy rate is the ratio of all the correctly predicted samples in relation to the overall samples. In order to consider both P_a (precision rate) and R (recall rate), F_1-score values (such as Equation (28)) are generally used to represent the classification quality of the scene. In order to better evaluate the effects of each algorithm, we use the above metrics to evaluate the classification performance.

$$F_1 - score = \frac{2(R \times P_a)}{R + P_a} \tag{28}$$

The classification results of our method and of other comparison methods on scene 1 and scene 2 are shown in Table 3. Table 3 lists the precision, recall, accuracy and F_1-score statistics for the eight methods in the two scenes. It can be seen from Table 3 that the comparison between the proposed method and Method 1 shows that the accuracy of the proposed method has significantly improved. It also shows that the proposed NAD and LSH features have certain effects. By comparing Method 2 and Method 3, the proposed multi-scale pyramid optimization algorithm can effectively improve the classification accuracy. Comparing the proposed method with Method 3, the proposed multi-scale strategy has a significant effect on the improvement of the classification results.

In addition, the proposed method is compared with the methods given in other references. Method 4 and Method 5 classify the point cloud based on the single-point. Method 6 and Method 7 classify the point cloud based on the point set (object). It can be seen from the comparison of Table 3 that the proposed method has a high accuracy rate as a whole, and that it also maintains a high recall rate.

Table 3. Classification results of precision/recall, accuracy and F_1-score.

Scene 1	Tree (%)	Building (%)	Car (%)	Accuracy (%)	F_1-Score (%)
Our method	99.2/90.6	91.1/99.3	92.9/48.2	94.6	94.5/94.9/59.5
Method 1	99.2/84.9	86.8/99.3	99.9/42.7	91.9	91.5/92.7/59.8
Method 2	93.2/78.7	82.1/94.6	63.3/30.4	86.4	85.3/87.9/41.1
Method 3	96.9/81.7	84.1/97.7	98.8/23.2	89.3	88.7/90.4/37.6
Method 4	89.7/98.1	97.9/89.1	65.2/46.6	92.9	93.7/93.3/54.4
Method 5	85.7/92.9	92.0/83.8	56.9/54.7	87.9	89.2/87.7/55.8
Method 6	94.8/93.8	93.5/92.3	41.2/66.7	92.6	94.3/92.9/50.9
Method 7	93.1/96.0	95.2/92.6	73.3/62.2	93.7	94.5/93.9/67.3
Scene 2	**Tree (%)**	**Building (%)**	**Car (%)**	**Accuracy (%)**	**F_1-score (%)**
Our method	92.4/94.3	98.5/97.9	73.0/68.4	95.8	93.4/98.2/70.6
Method 1	83.2/92.9	98.5/92.8	62.6/65.7	92.0	87.8/95.6/64.1
Method 2	77.3/94.3	98.3/88.9	71.7/60.0	89.6	85.0/93.4/65.3
Method 3	91.3/92.6	96.6/96.6	63.2/55.5	93.4	91.9/96.6/59.1
Method 4	86.8/91.2	96.8/95.5	44.1/34.8	92.2	88.9/96.1/38.9
Method 5	73.9/91.2	93.6/88.2	29.5/25.4	87.2	81.6/90.8/27.3
Method 6	90.3/93.9	97.6/96.5	49.4/42.0	94.1	92.1/97.0/45.4
Method 7	94.7/94.5	98.1/97.7	53.9/60.5	95.5	94.6/97.9/57.0

For scene 1, the accuracy of the proposed method is the highest, and the value of the precision/recall is relatively high (the classification result of the proposed method is shown in Figure 1e). From the classification results evaluation of the three kinds of objects by the F_1-score, one can see that the classification effect on cars for the proposed method is not as good as for Method 7. Meanwhile, the tree and building classes can basically be classified correctly.

For scene 2, the precision/recall of trees of the proposed method are lower than for Method 7. Meanwhile, the precision/recall of buildings and cars are the maximum compared with other methods. According to the classification result of the F_1-score value, the proposed method has a better effect than the other methods, except that the tree classification performance is worse than for Method 7. Considering the accuracy, precision/recall rate and the F_1-score evaluation comprehensively, the proposed method has a better classification performance than the other methods, and the proposed method has the highest overall accuracy for both scenes. This proves that the proposed point cloud classification method based on point multilevel features is effective, and that it can accurately classify the ALS point cloud data in large scale scenes.

In order to more intuitively compare the classification performance of each method, Figure 5 shows the performance of the eight classification methods in scene 2. Figure 5 shows that the proposed method can classify most points correctly. Compared with other comparison methods, the classification accuracy of the proposed method is higher. From the comparison between Figures 5c–e and 5b, the classification effect of the proposed method is obviously better than that of the other three algorithms, especially in the buildings and trees. It can be seen from the comparison between Figure 5f,g and Figure 5b that Method 4 and Method 5 have more misclassifications for cars and buildings, and that the performance of the proposed method is significantly better than that of the other two methods. In comparing Figure 5h,i with Figure 5b, one can see that Method 6 and Method 7 have a similar classification performance to that of the proposed method. However, a certain number of architectural edge points are classified incorrectly, and the top part of the trees is also classified incorrectly. One can see from the comparison between Figures 5f–i and 5b that the compared methods have some misclassifications for the edge points and for two objects that are overlapping regions. However, the proposed method has fewer misclassifications in those regions than for the other methods. This proves that the proposed feature descriptors and post-optimization strategies can improve the classification results.

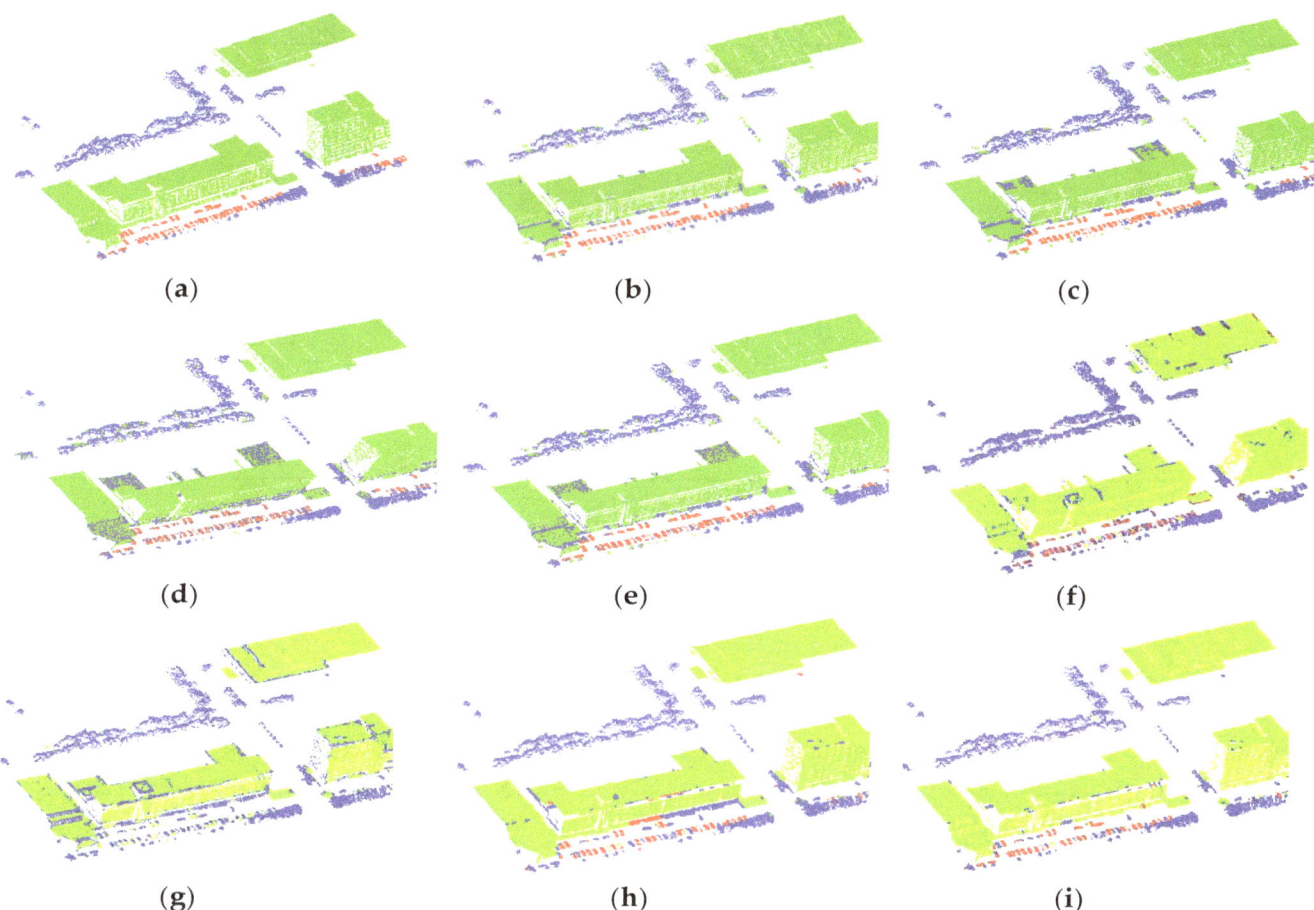

Figure 5. Scene 2 classification results. (**a**) ground truth, (**b**) proposed method, (**c**) method 1, (**d**) method 2, (**e**) method 3, (**f**) method 4, (**g**) method 5, (**h**) method 6, and (**i**) method 7. ((**f–i**) are taken from Ref [23]. The red points are cars, green points are buildings and blue points are trees.)

3.3. Sensitivity of the Parameters

In this part, we focus on the D_m in NAD, D_l in LSH and the number of neighborhood scales ($\mu + \upsilon$). Here, we select the parameters shown in Table 4; the data of scene 1 is used to compare the influence of different parameters on the proposed method. In order to evaluate the classification effect of the three kinds of objects as a whole, we average the F_1–score values of the three object classification results to obtain the mean value mF_1, which is used as the overall classification effect evaluation metric. As shown in Table 4, considering the results of mF_1 and the accuracy in combination, in comparing the parameters of the first three rows one can see that when D_m is 15, the classification effect is better; however, the value of D_m is not sensitive to the classification effect. According to the results of rows 2, 4 and 5, the classification effect of D_l is obviously improved at 15. However, when the value of D_l is too large, the classification effect is reduced. Therefore, the value of D_l is relatively sensitive to the classification result. According to the results of rows 4, 6, 7 and 8–10, the classification effect is improved when the value of the scale $\mu + \upsilon$ is increased. However, when the scale exceeds 4, the classification effect will be reduced to some extent. Therefore, the value of the scale is sensitive to the results of the classification and needs to be within a reasonable range. One can see from Table 4 that the tree and building classes are relatively less affected by the changes of the D_m and D_l, and that the results are susceptible to the size of the scale. The increasing values of D_m and D_l would likely cause a change in the car classification effect. Considering the overall effect of the classification and the factors of accuracy and feature dimension, we select $D_m = 15$, $D_l = 15$ and ($\mu + \upsilon$) = 3 as the optimal parameter values.

Table 4. Parameters comparison based on mF_1 and Accuracy.

	D_m	D_l	$\mu+\upsilon$	Tree (%)	Building (%)	Car (%)	Accuracy (%)	mF_1 (%)
1	10	10	3	94.0	94.4	60.7	93.9	82.9
2	15	10	3	94.0	94.5	60.7	94.0	83.1
3	20	10	3	93.9	94.5	60.3	94.0	82.9
4	15	15	3	94.7	95.0	63.5	94.6	84.4
5	15	20	3	94.3	94.7	60.2	94.3	83.1
6	15	15	4	94.7	95.0	62.6	94.6	84.1
7	15	15	5	94.7	95.0	62.6	94.6	83.8
8	20	20	2	93.5	94.1	54.5	93.5	80.7
9	20	20	3	94.5	94.9	59.5	94.4	83.0
10	20	20	5	94.5	94.9	59.0	94.4	82.8

4. Conclusions

The classification of the ALS point cloud is an important technology for urban planning, digital city and intelligent transportation. We propose a multilevel features fusion and pyramid neighborhood optimization ALS point cloud classification method based on a single point. The proposed method presents two local features, i.e., the NAD and LSH. They are fused with the covariance and elevation features. Following this, the multilevel features are constructed by changing the point cloud resolution and the neighborhood size. The fused features are used to train a classification model based on the Gaussian kernel function SVM for an initial classification. Finally, the point cloud classification is optimized based on the initial classification result using a multi-scale pyramid. The optimized classification results have a higher accuracy. The experimental results prove the effectiveness of the proposed method via the experiments on the two sets of public ALS datasets.

Author Contributions: Conceptualization, Y.L. and X.D.; methodology, Y.L. and X.D.; software, Y.L. and X.D.; validation, G.T., Y.L. and X.D.; data curation, G.T., J.Z. and X.D.; writing—original draft preparation, Y.L. and X.Y.; writing—review and editing, Y.L., L.Y. and X.Y.; visualization, L.Y. and X.Y.; supervision, G.T.; project administration, G.T.; funding acquisition, G.T.

Acknowledgments: The authors would like to thank Lihao Cao in Northeastern University for helping to check the grammar and spelling of the paper.   ·

References

1.   Maturana, D.; Chou, P.-W.; Uenoyama, M.; Scherer, S. Real-time semantic mapping for autonomous off-road navigation. In *Field and Service Robotics*; Hutter, M., Siegwart, R., Eds.; Springer: Cham, Germany, 2018; pp. 335–350. ISBN 978-3-319-67361-5.

2.   Li, Y.; Tong, G.F.; Yang, J.C.; Zhang, L.Q.; Peng, H.; Gao, H.S. A Summary of Key Technologies for 3D Point Cloud Scene Data Acquisition and Scene Understanding. *Laser Optoelectron. Prog.* **2019**, *56*, 040002.

3.   Yuan, L.W.; Yu, Z.Y.; Luo, W.; Hu, Y.; Feng, L.Y.; Zhu, A.-X. A hierarchical tensor-based approach to compressing, updating and querying geospatial data. *IEEE Trans. Knowl. Data Eng.* **2015**, *27*, 312–325. [CrossRef]

4.   Chen, D.; Zhang, L.Q.; Mathiopoulos, P.T.; Huang, X.F. A methodology for automated segmentation and reconstruction of urban 3-D buildings from ALS point clouds. *IEEE J. Sel. Top. Appl. Earth Observ. Remote Sens.* **2014**, *7*, 4199–4217. [CrossRef]

5.   Linsen, L.; Prautzsch, H. Local versus global triangulations. In Proceedings of the 22th Annual Conference of the European Association for Computer Graphics, EUROGRAPHICS 2001, Manchester, UK, 5–7 September 2001.

6.   Lee, I.; Schenk, T. Perceptual organization of 3D surface points. *Int. Arch. Photogramm. Remote Sens. Spat. Inf. Sci.* **2002**, *34*, 193–198.

7. Filins, S.; Pfeifer, N. Neighborhood systems for airborne laser data. *Photgramm. Eng. Remote Sens.* **2005**, *71*, 743–755. [CrossRef]
8. He, E.; Chen, Q.; Wang, H.; Liu, X. A curvature based adaptive neighborhood for individual point cloud classification. *Int. Arch. Photogramm. Remote Sens. Spat. Inf. Sci.* **2017**, *42*, 219–225. [CrossRef]
9. Weinmann, M.; Jutzi, B.; Hinz, S.; Mallet, C. Semantic point cloud interpretation based on optimal neighborhoods, relevant features and efficient classifiers. *ISPRS J. Photogramm. Remote Sens.* **2015**, *105*, 286–304. [CrossRef]
10. Hackel, T.; Wegner, J.D.; Schindler, K. Fast Semantic Segmentation of 3D Point Clouds with Strongly Varying Density. *ISPRS Ann. Photogramm. Remote Sens. Spat. Inf.* **2016**, *III*, 177–184. [CrossRef]
11. Zeng, F.X.; Li, L.; Diao, X.P. Iterative closest point algorithm registration based on curvature features. *Laser Optoelectron. Prog.* **2017**, *54*. [CrossRef]
12. Johnson, A.E.; Hebert, M. Using spin images for efficient object recognition in cluttered 3D scenes. *IEEE Trans. Pattern Anal. Mach. Intell.* **1999**, *21*, 433–449. [CrossRef]
13. Frome, A.; Huber, D.; Kolluri, R.; Bülow, T.; Malik, J. Recognizing objects in range data using regional point descriptors. *Eur. Conf. Comput. Vis.* **2004**, *3023*, 224–237. [CrossRef]
14. Tombari, F.; Salti, S.; Di Stefano, L. Unique shape context for 3D data description. In Proceedings of the International Workshop on 3D Object Retrieval (3DOR 10)—In Conjunction with ACM Multimedia, Firenze, Italy, 25 October 2010; pp. 57–62.
15. Rusu, R.B.; Blodow, N.; Marton, Z.C.; Beetz, M. Aligning point cloud views using persistent feature histograms. In Proceedings of the 21st IEEE/RSJ International Conference on Intelligent Robots and Systems (IROS 2008), Nice, France, 22–26 September 2008.
16. Rusu, R.B.; Blodow, N.; Beetz, M. Fast point feature histograms (FPFH) for 3D registration. In Proceedings of the IEEE International Conference on Robotics and Automation (ICRA), Kobe, Japan, 12–17 May 2009; pp. 3212–3217. [CrossRef]
17. Jeong, J.; Lee, I. Classification of LIDAR Data for Generating a High-Precision Roadway Map. *Int. Arch. Photogramm. Remote Sens. Spat. Inf. Sci.* **2016**, *XLI-B3*, 251–254. [CrossRef]
18. Lodha, S.K.; Fitzpatrick, D.M.; Helmbold, D.P. Aerial lidar data classification using AdaBoost. In Proceedings of the International Conference on 3-D Digital Imaging and Modeling (3DIM), Montreal, QC, Canada, 21–23 August 2007; pp. 435–442. [CrossRef]
19. Babahajiani, P.; Fan, L.; KÄMÄRÄINEN, J.K.; Gabbouj, M. Urban 3D segmentation and modelling from street view images and LiDAR point clouds. *Mach. Vis. Appl.* **2017**, *28*, 679–694. [CrossRef]
20. Zhang, J.X.; Lin, X.G.; Ning, X.G. SVM-based classification of segmented airborne lidar point clouds in urban areas. *Remote Sens.* **2012**, *5*, 3749–3775. [CrossRef]
21. Fischler, M.A.; Bolles, R.C. Random Sample Consensus: A Paradigm for Model Fitting with Applications to Image Analysis and Automated Cartography. *Commun. ACM.* **1981**, *24*, 381–395. [CrossRef]
22. Chang, C.-C.; Lin, C.-J. LIBSVM: A library for support vector machines. *ACM Trans. Intell. Syst. Technol.* **2011**, *2*, 27. [CrossRef]
23. Zhang, Z.X.; Zhang, L.Q.; Tong, X.H.; Wang, Z.; Guo, B.; Huang, X.F.; Wang, Y.B. A Multi-Level Point Cluster-based Discriminative Feature for ALS Point Cloud Classification. *IEEE Trans. Geosci. Remote Sens.* **2016**, *54*, 4714–4726.
24. Guo, B.; Huang, X.; Zhang, F.; Sohn, G. Classification of airborne laser scanning data using JointBoost. *ISPRS J. Photogramm. Remote Sens.* **2015**, *100*, 71–83. [CrossRef]
25. Wang, Z.; Zhang, L.; Fang, T.; Mathiopoulos, P.T.; Tong, X.; Qu, H.; Xiao, Z.; Li, F.; Chen, D. A Multiscale and Hierarchical Feature Extraction Method for Terrestrial Laser Scanning Point Cloud Classification. *IEEE Trans. Geosci. Remote Sens.* **2015**, *53*, 2409–2425. [CrossRef]
26. Zhang, Z.; Zhang, L.; Tong, X.; Guo, B.; Zhang, L.; Xing, X. Discriminative dictionary learning-based multi-level point-cluster features for ALS point cloud classification. *IEEE Trans. Geosci. Remote Sens.* **2016**, *54*, 7309–7322. [CrossRef]
27. Zhang, Z.X. ALS Point Cloud Classification Based on Multilevel Point Cluster Features. Ph.D. Thesis, Beijing Normal University, Beijing, China, 2017.

6

Enhancing Prediction Performance of Landslide Susceptibility Model using Hybrid Machine Learning Approach of Bagging Ensemble and Logistic Model Tree

Xuan Luan Truong [1,*]**, Muneki Mitamura** [2]**, Yasuyuki Kono** [3]**, Venkatesh Raghavan** [2]**,
Go Yonezawa** [2]**, Xuan Quang Truong** [4]**, Thi Hang Do** [1,2]**, Dieu Tien Bui** [5,*] **and Saro Lee** [6,7,*]

[1] Faculty of Information Technology, Hanoi University of Mining and Geology, No.14 Vien Street, Bac Tu Liem, Hanoi 10000, Vietnam; dohanghumg@gmail.com

[2] Graduate School for Creative Cities, Osaka City University, Osaka 558-8585, Japan; mitamura@sci.osaka-cu.ac.jp (M.M.); raghavan@media.osaka-cu.ac.jp (V.R.); yonezawa@media.osaka-cu.ac.jp (G.Y.)

[3] Center for Southeast Asian Studies, Kyoto University, Kyoto 606-8502, Japan; kono@cseas.kyoto-u.ac.jp

[4] Faculty of Information Technology, Hanoi University of Natural Resources and Environment, No. 14 Phu Dien, Bac Tu Liem, Hanoi 10000, Vietnam; txquang@hunre.edu.vn

[5] Geographic Information System Group, Department of Business and IT, University College of Southeast Norway, Gulbringvegen 36, N-3800 Bø i Telemark, Norway

[6] Geological Research Division, Korea Institute of Geoscience and Mineral Resources (KIGAM), 124, Gwahak-ro, Yuseong-gu, Daejeon 34132, Korea

[7] Department of Geophysical Exploration, Korea University of Science and Technology, 217 Gajeong-ro Yuseong-gu, Daejeon 305-350, Korea

[*] Correspondence: truongxuanluan@humg.edu.vn (X.-L.T.); Dieu.T.Bui@usn.no (D.T.B.); leesaro@kigam.re.kr (S.L.)

Abstract: The objective of this research is introduce a new machine learning ensemble approach that is a hybridization of Bagging ensemble (BE) and Logistic Model Trees (LMTree), named as BE-LMtree, for improving the performance of the landslide susceptibility model. The LMTree is a relatively new machine learning algorithm that was rarely explored for landslide study, whereas BE is an ensemble framework that has proven highly efficient for landslide modeling. Upper Reaches Area of Red River Basin (URRB) in Northwest region of Viet Nam was employed as a case study. For this work, a GIS database for the URRB area has been established, which contains a total of 255 landslide polygons and eight predisposing factors i.e., slope, aspect, elevation, land cover, soil type, lithology, distance to fault, and distance to river. The database was then used to construct and validate the proposed BE-LMTree model. Quality of the final BE-LMTree model was checked using confusion matrix and a set of statistical measures. The result showed that the performance of the proposed BE-LMTree model is high with the classification accuracy is 93.81% on the training dataset and the prediction capability is 83.4% on the on the validation dataset. When compared to the support vector machine model and the LMTree model, the proposed BE-LMTree model performs better; therefore, we concluded that the BE-LMTree could prove to be a new efficient tool that should be used for landslide modeling. This research could provide useful results for landslide modeling in landslide prone areas.

Keywords: landslide; bagging ensemble; Logistic Model Trees; GIS; Vietnam

1. Introduction

The problem of rainfall-induced landslides, which are triggered by high intense and long lasting precipitation, seems to be more serious in recent years in many regions around the world due to the effects of climate changes i.e., extreme rainfall events [1–8]. The rainfall-triggered landslide is especially exacerbated in countries that are located in storm centers of the world, such as Vietnam [9], Philippines [10], and China [11]. For example, the tropical typhoon of Rasmussen caused various floods and landslides with the total damages were estimated at $7 billion [12]. It anticipates that the number of landslides in the future will continue to rise due to effects of extreme rainfall events and changes of hydrological cycles [13]. Thus, landslide has become one of the hottest subject of the research community, however, accurately prediction of landslide still is a challenging real-world problem [14]. Therefore, more researches on landslide are still urgently required for deriving better detailed knowledge of slope failure and its mechanisms for designing remedial measures.

The development of a hazard map that provides detailed dimensional information of spatial distributions, temporal predictions, and destructive power of landslide is considered as an efficient tool for designing mitigation measures and management policies. However, the hazard map at the regional scale requires very detailed temporal landslide inventories that are hardly available, especially in developing countries [15]. For this context, a landslide susceptibility map (LS-map) could be alternatively employed since it helps to identify areas with high landslide probability. According to Ciampalini, et al. [16], LS-map is a valuable decision-support tool that assists local authorities in land use infrastructural planning and management

To produce susceptibility map, a variety of studying approaches has been introduced because the accuracy of the susceptibility map at regional analysis scale is controlled not only by the quality of the input maps, but also the algorithms and techniques that are employed [17]. These approaches vary from expert weighting methods to deterministic and statistical models. Evaluation of these approaches has been well presented i.e., in Chacon et al. [18] and Van Westen, et al. [19]. In recent years, new approaches that are based on advanced statistical and machine learning methods have been proposed i.e., fuzzy k-Nearest Neighbor [17]; fuzzy rule based models [20–23]; neural networks [24–30]; support vector machines [31–38]; Random Forests; metaheuristic optimized least squares support vector machines [39,40]; Cuckoo optimized relevance vector machines [41]; Chi-squared automatic interaction detection (CHAID) [42]; tree-based algorithms [43–47]; and, gene expression programming [48]. The main advantage of these methods is that they are capable of involving several to a large number of variables for reliable results, and overall, these methods are able to provide better performance models when compared to those of conventional methods [43,49,50].

In the last years, the integration of advanced machine learning algorithms and homogeneous ensemble frameworks has been explored for landslide susceptibility modeling with promising results. For example, Tien Bui, et al. [51] show that the landslide model based on a combination of functional trees with Bagging performs better than the neural network models. Pham et al. [23] concluded that the hybridization of Fuzzy Unordered Rules Induction Algorithm and Rotation forest ensemble has increased the prediction performance of the landslide model when compared to the benchmark of support vector machines model. Pham et al. [26] reported that the landslide model derived from a combinations of MultiBoost and Dagging with neural networks has significantly improved the prediction power of the landslide model using only the neural network. Thus, it could be concluded that homogeneous ensembles of machine learning are promising and should be further investigated aiming to improve the prediction capability of landslide susceptibility model.

Based on the mentioned motivation, this research aim is to expand the body knowledge of landslide modeling through introducing a new machine learning ensemble approach that combines the Logistic Model Trees (LMTree) algorithm [52] and Bagging Ensemble (BE) [53], named as BE-LMtree, for enhancing the performance of the landslide model. LMTree is a relative new and promising machine learning algorithm that was rarely explored for the landslide study, whereas Bagging ensemble is an framework that has proven efficient in landslide modeling [51,54]. Consequently, a combination of

BE and LMTree has resulted in a new powerful prediction method, and to the best of our knowledge, this is the first time that the BE-LMTree is studied for landslide susceptibility.

2. Theoretical Background of the Methods

2.1. Logistic Model Tree

Logistic Model Trees (LMTree), which is a relatively new machine learning algorithm, is developed based on the integration of tree induction algorithm and additive logistic regression [52]. The difference of LMTree when compared to the other decision tree algorithms is that the tree growing process is carried out using the LogitBoost algorithm [52,55] and the tree pruning is performed using Classification And Regression Tree (CART) [56].

Given a training dataset $T = (x_i, y_i)_{i=1}^{ds}$ with $x_i \in R^D$ is the input vector, ds is the number of data samples, D is the dimension of the training dataset, and $y_i \in (1, 0)$ is the label class. In this research context, the input vector consists of eight variables (slope, aspect, elevation, land cover, soil type, lithology, distance to fault, and distance to river), whereas the label class contains two classes, landslide (LS) and non-landslide (Non-LS). The landslide class is coded as "1" and the non-landslide is coded as "0". The objective of LMTree is to construct a tree-like structure model that is capable of classifying the training dataset into the two above classes in term of probability. The predicted numeric value to the landslide class of sample is used as susceptibility index.

Structurally, the LMTree model consists of a root node, a set of inner nodes, and a set of leaves. The aim of the training phase that includes the tree growing and the tree pruning processes is to determine the best tree structure with numbers of inner nodes and leaves. Accordingly, first, a logistic regression model Equation (1) is built at the root note using the binary LogitBoost algorithm [57] and the training dataset. In the next step, the training dataset at the root is split using the C4.5 splitting rule [58] in order to sort appropriate sub-datasets for the inner nodes, and then, logistic regression models Equation (1) for these inner nodes is built using their associated sorted datasets and the binary LogitBoost. The tree continues growing in the same procedure until it meets the stopping criterion of less than 15 samples at nodes. Finally, to prevent the LMTree model from over-fitting, the tree pruning is performed using the CART algorithm that is based on a combination of the model error and the model complexity [52].

In the LMTree building process, the binary LogitBoost algorithm [57] is used to generate logistic regression models Equation (1) for all of the inner nodes and leaves, as follows.

$$f_{LS,Non-LS}(x) = \sum_{i=1}^{D} \beta_i x_i + \beta_0 \qquad (1)$$

where D is the total number of landslide input factors and β_i is the logistic coefficient.

The membership probability [52] of the landslide class at the leaves of the LMTree model is posterior probabilities derived using Equation (2) and is used as landslide susceptibility index.

$$p((LS, Non - LS)|x) = \frac{\exp f_{LS,Non-LS}(x)}{\exp f_{LS}(x) + \exp f_{Non-LS}(x)} \qquad (2)$$

The complexity of the LMTree model could be estimated using the following equation [52]:

$$MC = O(dept * ds * \log n + ds * D^2 * dept + nt^2) \qquad (3)$$

where MC is the model complexity; $dept$ is the depth of the initial unpruned tree; nt is the number of nodes in the LMTree; ds is the number of training samples; and, D is the number of landslide predisposing factors.

2.2. Bagging Ensemble

Ensemble learning is a machine learning paradigm where multiple classifiers are trained and combined to enhance the prediction capability of a model. Different from popular machine learning approaches where one model is built from the training data, ensemble frameworks try to generate a set of sub-datasets from the training data, and then, each sub-dataset is used to construct a classifier, which is also called a based learner. At last, all of the based learners are combined to form the final prediction model using combination techniques i.e., averaging or majority voting [59].

Different ensemble techniques have been successfully proposed i.e., Bagging, AdaBoost, Multiboost, Stacking, and Rotation forest [60]; however, in landslide modeling, Bagging ensemble has proven robust and better than other ensembles [26,51,54], therefore, it is selected for this study.

Bagging also called Bootstrap aggregating in the full name is one of the earliest procedure for generating sub-datasets and combining based learners proposed by Breiman [53]. Using the training dataset, this technique generates bootstrap samples in which some of the samples are replicated and some samples are omitted. These bootstrap samples, which are called bootstrapped sub-datasets, are used to construct based learners using the same classification algorithm i.e., the LMTree in this work. These based learners are then combined using the majority voting strategy.

3. The Study Area and Spatial Datasets

3.1. Description of the Upper Reaches Area of Red River Basin

The study area is the Upper reaches area of the Red River Basin (URRB) (103°33′36″–104°30′50″ E, 22°05′40″–22°47′52″ N) that belongs to the Lao Cai, a north-western mountainous province in Vietnam (Figure 1). The URRB covers an area of 3273.5 km^2 with complex topography, steep slopes, and narrow valleys. The topography is highly fragmented with high mountains ranges, wide valleys, and deep streams, which result in high relief amplitudes [40]. The altitude varies from 48.1 m to 2812.6 m above sea level, with the mean and the standard deviation of 528.6 m and 484.9 m, respectively. Topographically, 61.8% of the URRB is occupied by slope angles that are higher than 15°, whereas areas with slopes less than 5° cover approximately 7.3% the total area of the URRB. The remaining 30.9% are areas located in the slope group 5–15°.

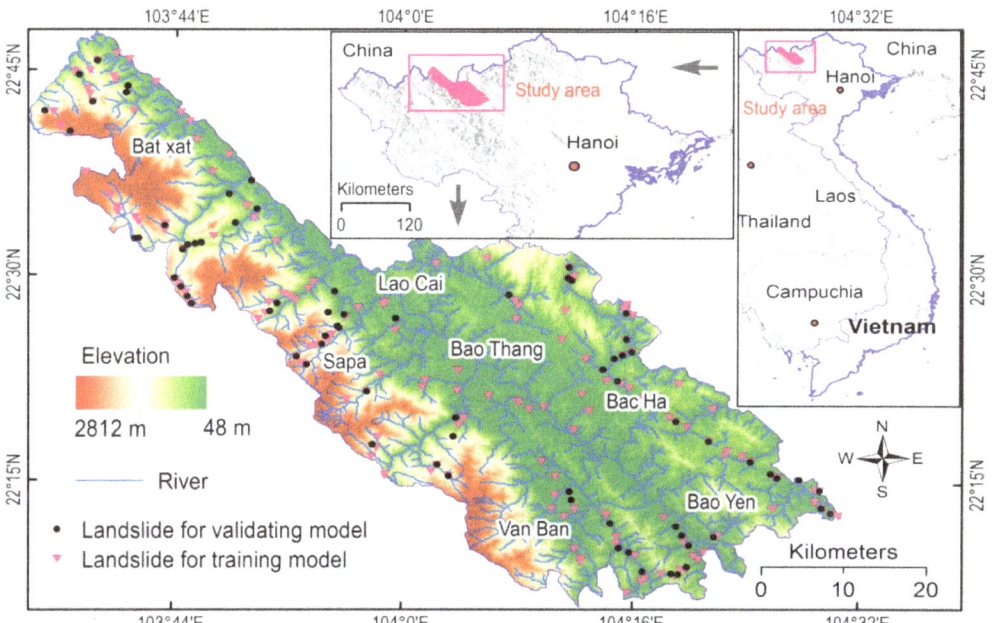

Figure 1. Location of the Upper Reaches Area of Red River Basin (Vietnam).

Hydrologically, due to the fragmentation of the terrain, the river system in the study area is dense and evenly distributed (Figure 1). These rivers are characterized by being narrow and steep, which are favorable conditions for the occurrence of flash flood and landslides. The Red River, which is the second largest river in Vietnam, is the major channel system of the URRB. This river originates from Yunnan province (China) and flows south-eastward to the study area [61].

The climate of URRB is divided into two seasons: the rainy season begins from April to October and the dry season lasts from November to March next year. The average temperature ranges range from 23 °C t o2 9 °C [62] and the average annual rainfall is from 1400 mm to 1900 mm [63].

The URRB is located in an active tectonic region with the relatively fast movement of the Red River fault zone that results in continuously landslide occurrences over the years [40]. It should be noted that the Red River fault zone is one of the four main tectonic features in north Vietnam that begins from Tibetan plateau (China) and extends to the Red River area of Vietnam [64,65]. Twenty seven geological formations outcrop in the basin with varied area and space distribution (Figure 2). Quaternary deposits, which consist of mainly granule, grit, breccia, pebble, boulder, and sand, cover 7.04% of the total area of the basin. Whereas, 86.68% of the basin is covered by nine geological formations, Suoi Chieng (23.62%), Ha Giang (10.96%), Nui Con Voi (10.54%), Sinh Quyen (10.43%), Ngoi Chi (8.44%), Cam Duong (8.29%), Ye Yen Sun (6.23%), Po Sen (5.96%), and Muong Hum (2.21%). The main lithologies are biotite schist, garnet-biotite gneiss, coaly shale, marble cherty shale, quartz-plagioclase-biotite schist, and two-mica schist. Detailed distribution of the lithological formations in the basin is shown in Figure 2.

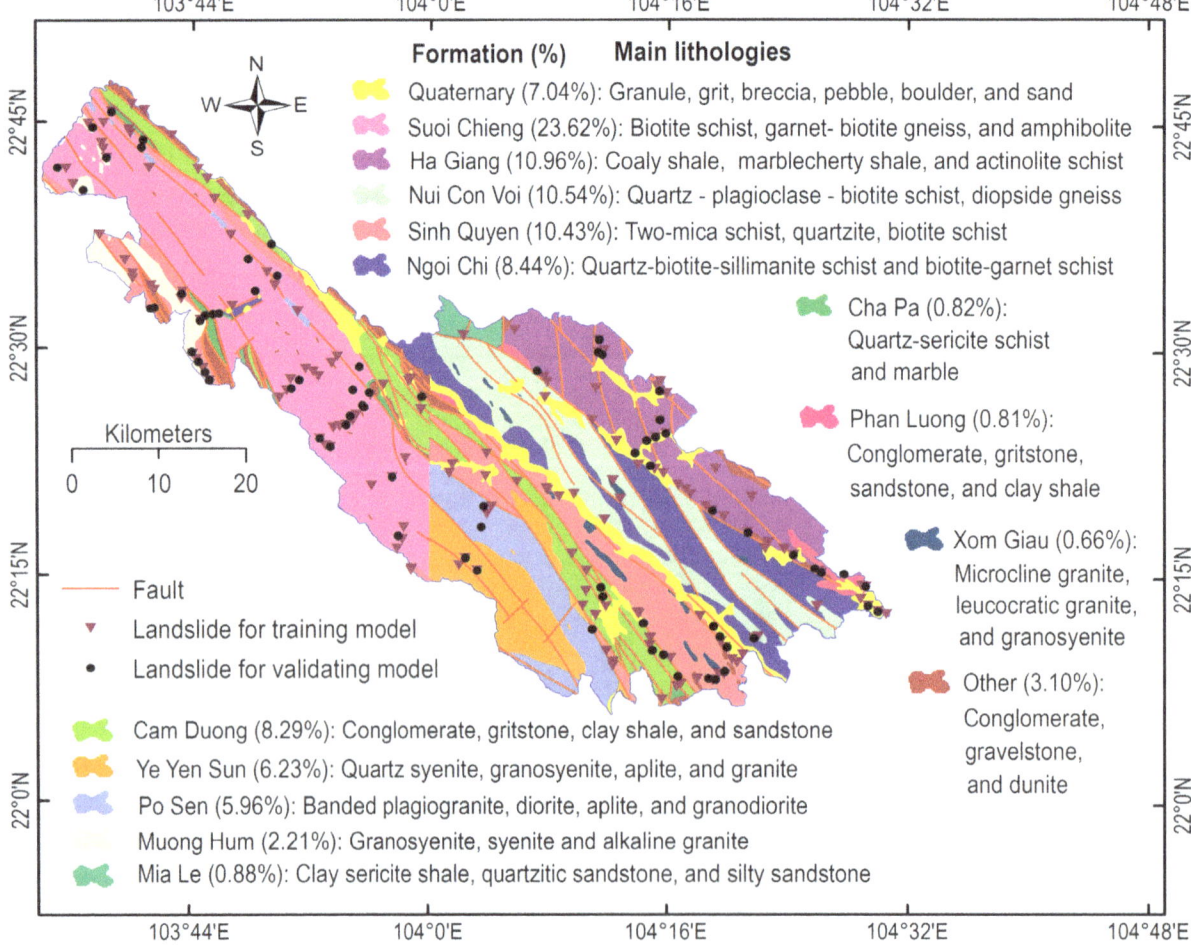

Figure 2. Geological map of the study area.

3.2. Geospatial Data

Landslide inventory map for the URRB was constructed from two main sources: (i) historic landslides from the project VAST05.02/14-15 in 2015, which was prepared by Tien Bui et al. [40]; and, (ii) landslide polygons from the State-Funded Landslide Project (SFLP) 2016 [9], a national landslide program that is carrying out in Vietnam. The SFLP project has systematically investigated and collected historic landslides for all northwest mountainous provinces in Vietnam, including the study area. Accordingly, these landslides were mainly interpreted and mapped using aerial photos and field investigations. Detailed descriptions of methods and techniques for obtaining these historic landslides in the SFLP project are present in [9].

As result, a total of 255 historic soil-mixed-boulder slides that occurred during the last two decades were registered for the landslide inventory map (Figure 1). It is noted that many rock falls were excluded out of this research because their falling mechanism are very different when compared to that of the soil-mixed-boulder slides. Analysis of the landslide inventory map showed that these slides occurred due to rainfall during tropical rainstorms [40]. Our statistical analysis of these slides showed that the largest and the smallest landslides are 116627.9 m^2 and 6.2 m^2, respectively, with the mean is 3742.5 m^2 and the standard deviation is 11467.3 m^2. Approximately 9.1% of the landslide inventories are large landslides (>10,000 m^2), whereas 9.1% of the landslide inventories are medium landslides (1000–10,000 m^2), and the remaining are landslides less than 1000 m^2. Two examples of landslide photos in the study area are shown in Figure 3.

Figure 3. Two photos of landslides in the study area: (**a**) Landslide at the Mong Sen area and (**b**) Landslide at Km 7 Lao Cai. The two photos were taken by Xuan-Luan Truong in August 2014.

Because the rainfall-trigged landslides in this study area occurred due to interactions of various geo-environmental factors, including topography, land cover, lithology, soil type, and river network [9,40,66,67], these factors were selected for this analysis. Digital elevation model (DEM) with resolution of 25 × 25 m for the URRB area was constructed using digital topographic maps 1:50,000 scale provided by the Ministry of Natural Resource and Environment of Vietnam. Using this DEM, three morphometric factors, slope, elevation, and aspect, were generated. To build the slope map (Figure 4a), seven categories were used. For the elevation map (Figure 4b), eight categories were considered. These categories were determined using Jenks natural break available in ArcGIS. For the aspect map, nice facing slopes were used (Figure 4c).

Land cover map (Figure 4d) at scale of 1:50,000 with nine classes for the URRB area was derived from the project No.02/2012/ HD-HTSP funded by Ministry of Education and Training of Vietnam. The nine classes were obtained through the classification of Landsat 8 OLI imagery in 2013 using ENVI software. Soil type map (Figure 4e) at 1:100,000 scale with 13 soil types for the URRB area was provided by Department of Agriculture and Rural Development of the Lao Cai province.

Lithological map for the URRB area was constructed based on National Geological and Mineral Resources Maps at scale of 1: 200,000, as provided by the Ministry of Natural Resource and Environment of Vietnam. Our analysis showed that more than 15 formations outcrop in the URRB area (see Figure 2). For this research, the lithological map with seven categories was constructed (Figure 4f) and these categories were separated based on clay composition, weathering characteristics, and material strength [24,68,69]. Detailed characteristics of the seven categories could be found in Tien Bui, et al. [70]. Fault is an popular factor for landslide susceptibility that was used various works i.e., in [71–73], and especially, it is an important factor for landslide modeling in areas that are affected by tectonic activities [74]. In this research, distance to fault map (Figure 4g) with seven classes [40] for the URRB area was constructed by buffering the fault lines extracted from the National Geological and Mineral Resources Maps above.

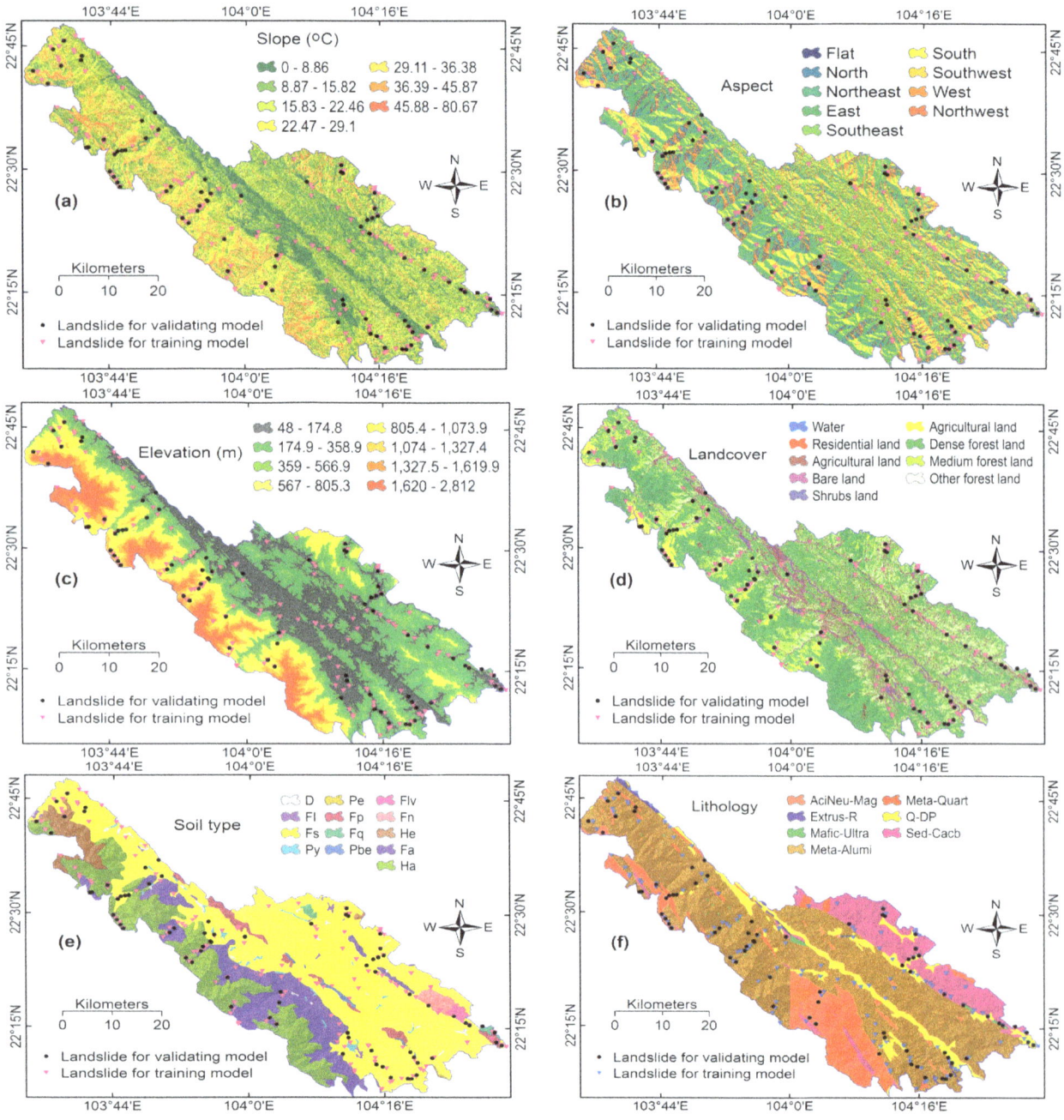

Figure 4. *Cont.*

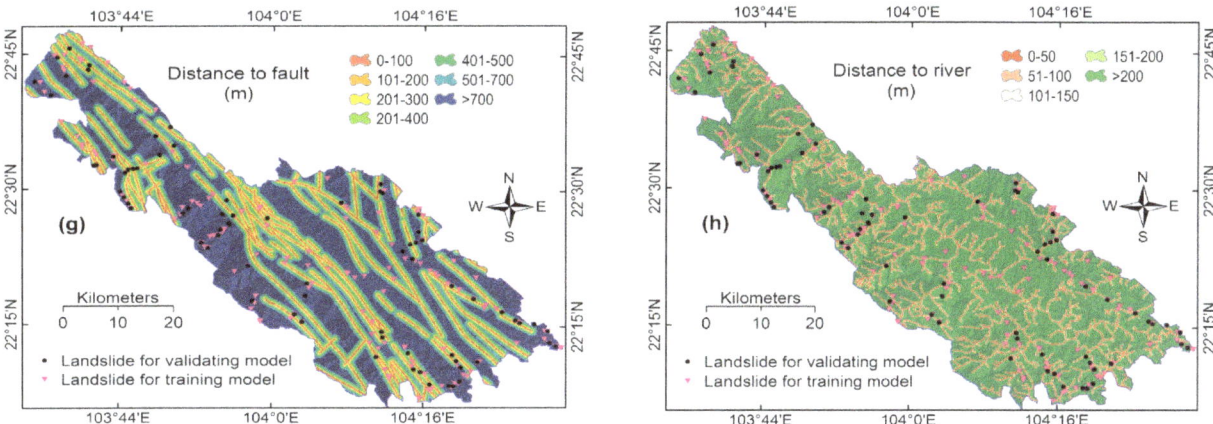

Figure 4. Landslide predisposing factors used in this study: (**a**) Slope; (b) Aspect; (**c**) Elevation; (**d**) Land cover; (**e**) Soil type ; (**f**) Lithology; (**g**) Distance to fault; and, (**h**) Distance to river.

Soil type (e) legend: **D**: Sloping soil; **Fl**: Cultivated rice yellowish red soil; **Fs**: Yellowish red soil on claystone and metamorphic rocks; **Py**: Alluvial soil deposited by river; **Pe**: neutral-less acidic and light texture alluvial soil; **Fp**: Brown-yellowish soil on old alluvium; **Fq**: Light yellowish soil on sandstone; **Pbe**: Neutral and less acidic alluvial soil; **Flv**: Red soil on limestone; **Fn**: Brown-yellowish soil on limestone; **He**: Humus yellow red soil on claystone and metamorphic rocks; **Fa**: Yellowish red soil on acid magmatic rock; and **Ha**: Humus yellow red soil on acid igneous rock. **Lithology (f) legend**: **AciNeu-Mag**: Acid-neutral magmatic rocks; **Extrus-R**: Extrusive rocks; **Mafic-ultra**: Mafic-ultramafic rocks; **Meta-Alumi**: Metamorphic rock with aluminosilicate components; **Meta-Quart**: Metamorphic rock with rich quarts components; **Q-DP**: Quaternary deposits; and, **Sed-Cacb**: Sedimentary carbonate rocks.

4. Proposed a Hybrid Machine Learning Approach of Bagging Ensemble (BE) and Logistic Model Tree (LMTree)

In this section, the proposed hybrid machine learning approach for Landslide Susceptibility Modeling at Upper Reaches Area of Red River Basin (Viet Nam) is described and presented in the first time. Methodological concept of the proposed BE-LMT model used in this study is shown in Figure 5.

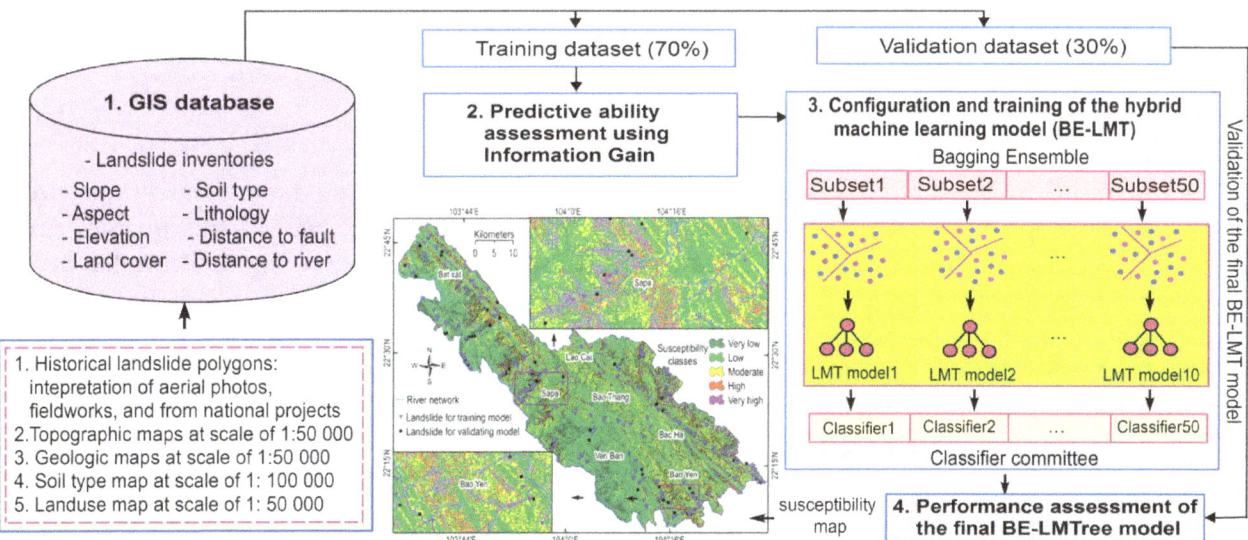

Figure 5. Methodological concept of the proposed Bagging ensemble (BE)-Logistic Model Trees (LMTree) model used in this study.

The proposed approach is a hybridization of LMTree and BE and is named as BE-LMTree. It should be noted that the data processing and coding were conducted using IDRISI Selva 17.0 (Clark University, Worcester, MA, USA, 2017) and ArcGIS 10.4 (ESRI Inc., Redlands, CA, USA 2017). The BE code is from Kuncheva [59] whereas the Logistic Model Tree algorithm is available at Weka's API [75]. The proposed BE-LMTree model was programmed by the authors in the Matlab environment.

4.1. Establishment of GIS Database, the Training Dataset and the Validation Dataset

In the first step, a GIS database for this project was designed and established using ArcCatalog software. Accordingly, the File Geodatabase format was used due to the ability to host and process very large geographic datasets with their different data types in a only one file system [76]. Accordingly, the GIS database consists of 255 landslide polygons and eight predisposing factors (slope, aspect, elevation, land cover, soil type, lithology, distance to fault, and distance to river). These landslide polygons and factors were converted to raster format with a resolution of 25 m. In this research, the categories of the eight predisposing factors were coded and normalized, as suggested in [24,77], to avoid the imbalance of categorical magnitudes [78].

In landslide modeling, cross validation [79] that has proven efficient for evaluating the model performance should be used. Accordingly, in this research, 179 landslide polygons (70%, 1006 pixels) were randomly extracted [80] and used for training the landslide models, whereas the other 76 landslides (30%, 441 pixels) were used for assessing the prediction capability of the models. Because the proposed approach in this study employs "on-off" classification, the equal amount of non-landslide pixels were also randomly sampled in the not-yet landslide areas of the basin, area with slope angles less than $5°$, as suggested in [32]. Detailed discussions on sampling strategies can be found at [81]. In the next step, values of the eight predisposing factors for all of the aforementioned pixels were extracted to build the training dataset and the validation dataset. Finally, the coding process that was proposed in [17] was performed, in which the landslide pixels were assigned "1" and the non-landslide pixels were assigned "0".

Because the aforementioned partition of the landslide dataset into the training and validation datasets was randomly generated only once; therefore, a further cross validation was additionally used to ensure that the modeling result is the objective. Accordingly, 10-fold cross validation was employed in the training phase with the training dataset to build landslide models. Thus, the training dataset was randomly partitioned into 10 equally sized subsets; nine subsets were used for building the landslide model, whereas the remaining subset was used for testing the landslide model. This procedure was repeated 10 times where each subset was being used once as the testing dataset. Once the model was successfully trained using the training dataset with the 10-fold cross validation procedure, the model was again validated using the validation dataset.

4.2. Merit Evaluation of Factor

Identification of relevant features is an essential task when employing machine learning techniques for landslide susceptibility [82]. This is because landslide is a typical real-world problem that is influenced by various factors, but the contribution of these factors to the prediction model is different. If non-contribution factors are included in the model, then they may cause noises that reduce the prediction power of the final model; therefore, these factors should be excluded.

To detect non-relevant factors in this study, Pearson technique was employed to quantify the predictive power of all landslide predisposing factors. Accordingly, the meritof these features were estimated using Pearson correlation values [83] of the predisposing factors and the output using the following equation:

$$Merit_i = \frac{covr(IF_i, y)}{\sqrt{varr(IF_i) * varr(y)}} \tag{4}$$

where $Merit_i$ is the correlation value of landslide predisposing factor *IFi* and the label class y; $covr(.)$ is the covariance; and, $varr(.)$ is the variance.

4.3. Configuration and Training of the BE-LMTree Model

Configuration of the BE-LMTree model consists of two steps: (i) Determining the minimum number of samples (NS) that are used for growing the LMTree; and, (ii) Determining the number of bootstrap subsets (BS) used for BE. Because at least five samples are required to build a logistic regression model at a tree node [52], we varied NS from 5 to 100 with a step size of 1, and then, estimating the classification rate of the corresponding LMTree model on both the training dataset and the validation dataset. As a result, minimum of 10 samples is the best for the data at hand; therefore, NS of 10 was selected. For the case of determining the number of the bootstrap subsets, since no thumb rule is available, an empirical test was carried out by varying BS from 2 to 100, and then, compute their classification rates of the LMTree model both on the training dataset and the validation dataset. The test result revealed that the BE-LMTree with 50 tree-based classifiers provided the highest classification accuracy for the data at hand; therefore, BS of 50 is selected. Once the BE-LMTree model had been configured, the training process was carried out to derive the final BE-LMTree model.

4.4. Performance Assessment of the Final BE-LMTree Model

Because the landslide modeling in this research is considered to be a binary form of pattern recognition, therefore the performance of the final BE-LMTree model could be assessed using confusion matrix (Figure 6) [40], both on the training dataset and the validation dataset. Based on the matrix, several model measures are further derived i.e., sensitivity (SEN), specificity (SPE), positive predictive power (PP2), and negative predictive power (NP2), Kappa statistics, and classification accuracy (CLA) for the assessment, as suggested in [50]. It should be noted that a perfect landslide model will have 100% for SEN, SPE, PP2, NP2, and CLA.

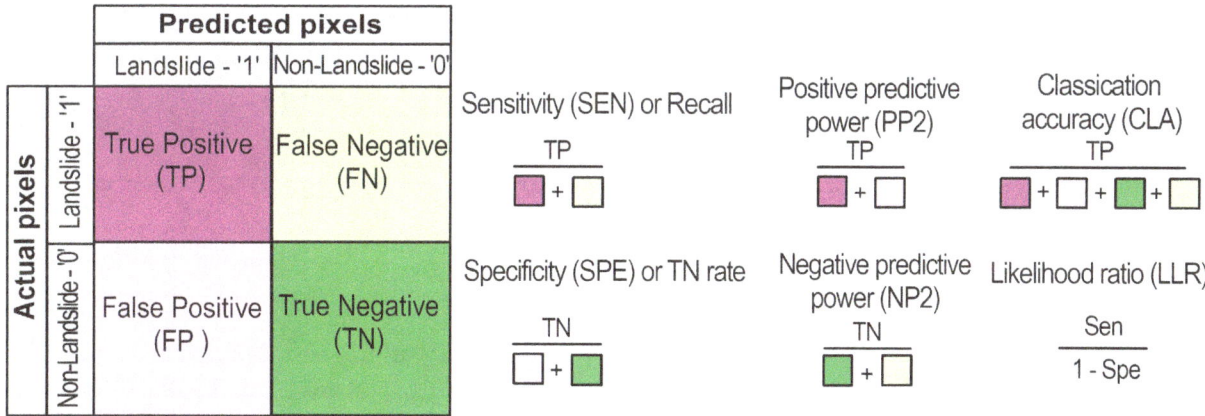

Figure 6. Confusion matrix and model measures used in this research.

For the case of CLA, although CLA provides the overall performance of the landslide model, however, a landslide model with a high CLA value may not classify the landslide pixels well. Therefore, the likelihood ratio (LLR) is additionally used [84]. LLR is a metric that assesses the trade-off of both SEN and SPE of landslide models. The higher the LLR value, the better the landslide model.

Global performance of the BE-LMTree model is summarized and assessed using the Receiver Operating Characteristic (ROC) Curve and Area Under the curve (AUC) [40,41,85]. In general, the closer the curve to the upper left corner, the better performance of the landslide model. Once the ROC curve is constructed, AUC for the model is computed and used to quantify the quality of the model. Accordingly, the performance of the model is excellent (AUC belong to 0.9–1), good (AUC belong to 0.8–0.9), fair (AUC belong to 0.7–0.8), and poor (AUC is less than 0.7) [86].

4.5. Computing Landslide Susceptibility Index

When the final BE-LMTree model is satisfied in the performance assessment check, the model is used to compute susceptibility index for all the pixels of the study area. These susceptibility indices are then converted to the ASCII raster format in ArcGIS using a Python application that was developed by the authors. Finally, the landslide susceptibility map is classified by five susceptibility classes: very high, high, moderate, low, and very low [87].

5. Results and Discussion

5.1. Predictive Ability Assessment

Result of the predictive ability evaluation of the eight predisposing factors is shown in Table 1. It is noted that the 10-fold cross validation was used to ensure the stable assessment result, as suggested in [88]. It could be seen that slope the highest predictive with the average merit (AM) is 0.225, followed by distance to river (AM of 0.171), lithology (AM of 0.148), aspect (AM of 0.129), and elevation (AM of 0.102). In contrast, soil type (AM of 0.038), distance to fault (AM of 0.055), and land cover (AM of 0.077) have low predictive ability values (Table 1).

The findings are reasonable because slope is widely recognized as the most important factor for landslide in various projects [89,90]. From the above results, it could be seen that all predisposing factors revealed predictive values to landslide model; therefore, we concluded that they are all relevant factors and are included in this analysis.

Table 1. Predictive ability of eight landslide predisposing factors using Pearson technique and 10-fold cross validation techniques.

No.	Predisposing Factors	Average Merit	Standard Deviation
1	Slope	0.225	0.008
2	Distance to river	0.171	0.008
3	Lithology	0.148	0.008
4	Aspect	0.129	0.008
5	Elevation	0.102	0.006
6	Land cover	0.077	0.008
7	Distance to fault	0.055	0.005
8	Soil type	0.038	0.005

5.2. Model Training and Evaluation

Using the eight predisposing factors, the BE-LMTree model was trained using the training dataset with the 10-fold cross validation technique. The training result is shown in Figure 7. It could be seen that the CLA of the BE-LMTree model is 93.81%, indicating a high degree of fit of the model with the dataset. Kappa statistics of 0.876 indicates the high agreement of the model and the training dataset. SEN of the BE-LMTree model is 93.02%, indicating that the proportion of the landslide pixels is correctly classified to the landslide class is 93.02%. Whereas, SPE is 94.63%, indicating that the proportion of the non-landslide pixels is correctly classified to the non-landslide class is 94.63%. PP2 is 94.72%, indicating that the probability that the BE-LMTree model correctly classifies pixels to the landslide class is 94.72%. NP2 is 92.89% indicating that the probability the BE-LMTree model correctly classifies pixels to the non-landslide class is 92.89%. Overall, these above measures have demonstrated that the BE-LMTree model performed very well with the training dataset.

To assess the contribution of landslide factors to the BE-LMT model, each factor was removed, and then, the classification accuracy (CLA) was estimated. The reduction of CLA of the BE-LMT model when one or more factors were removed indicates the contribution of these factors to the model. The result is shown in Table 2. It could be seen that when Distance to Fault and Soil type were removed from the LMT model, the CLA was reduced 2.12%. Therefore, although the average merit of Distance to fault (0.055) and Soil type (0.038) are small (see Table 1), the two factors contributed to 2.12% increasing classification accuracy of the BE-LMT model. An even larger accuracy decrease (4.3%, see Table 2) occurred when the four most significant variables (Slope, Distance to river, Lithology, and Aspect) are used into the BE-LMT model. Overall, it is reasonable of the to keep all factors in this research.

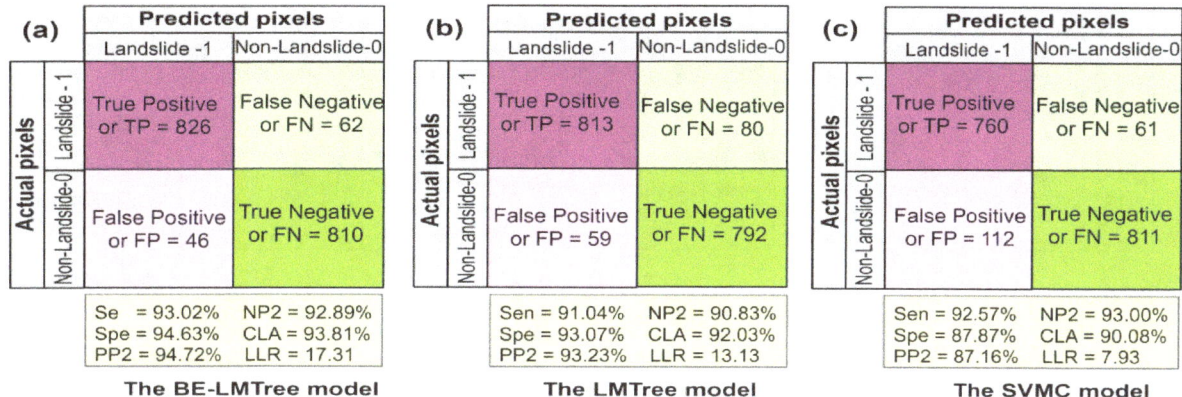

Figure 7. Confusion matrices and performance measures of the three landslide models using the training dataset: (**a**) the BE-LMTree model; (**b**) the LMTree model; and (**c**) the SVMC model.

Table 2. Contribution of the landslide predisposing factors to the BE-LMT model.

No.	Removing Factor	Classification Accuracy-CLA (%)
1	Slope	91.74
2	Aspect	92.31
3	Elevation	92.49
4	Land cover	93.60
5	Soil type	93.59
6	Lithology	91.97
7	Distance to fault	92.83
8	Distance to river	93.35
9	Distance to Fault and Soil type	91.69
10	Elevation, Land cover, Distance to fault and Soil type	89.51

The prediction performance of the BE-LMTree model is assessed using the validation dataset and the result is shown in Figure 8. It could be observed that the CLA is 87.89%, indicating a high prediction result. Kappa statistics of 0.759 indicates that the prediction performance of the model is 75.9% better than random. SEN of the BE-LMTree model is 92.25%, indicating that the proportion of the landslide pixels, which is accurately predicted, is 92.25%. SPE of the BE-LMTree model is 84.35%, indicating that the proportion of the non-landslide pixels is accurately predicted is 84.35%. PP2 of the model is 82.73%, indicating that the probability that the BE-LMTree model accurately predicts pixels to the landslide class is 82.73%. NP2 is 93.05%, indicating that the probability that the BE-LMTree model accurately predicts pixels to the non-landslide class is 93.05%.

Figure 9 shows 72 mispredicted landslide pixels (false positive) and 29 mispredicted non-landslide pixels (false negative) for the study area. We see that the 76.4% and 20.8% of the mispredicted landslide pixels were located in areas with slope angles <8.86° or slope angles from 36.39° to 5.87°, respectively. The mispredicted landslide pixels were also mainly located in elevation 174.78–358.94 m (76.4%), the lithology of sedimentary carbonate rocks (73.6%), the yellowish red soil on claystone and metamorphic rocks (87.5%), distance to fault >700 m (76.4%), and distance to river >200 m (79.2%). Distribution of the mispredicted landslide pixels in the classes in the other factors was more even. Regarding the mispredicted non-landslide pixels, they were mainly located in the distance to river >200 m (79.3%), the dense forest land (69.0%), and the yellowish red soil on claystone and metamorphic rocks (62.1%). For the other factors, the distribution of the mispredicted non-landslide pixels in their classes was quite even.

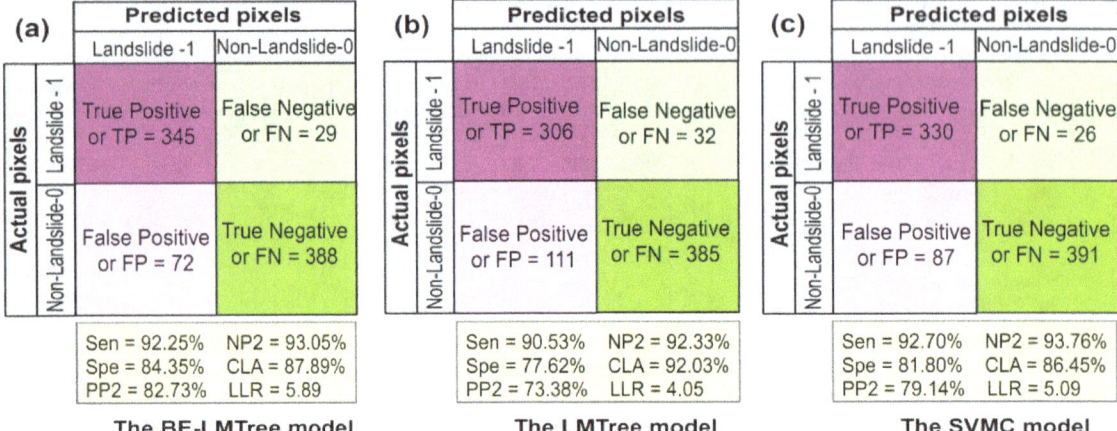

Figure 8. Confusion matrices and prediction measures of the three landslide models using the validation dataset: (**a**) the BE-LMTree model; (**b**) the LMTree model; and (**c**) the SVMC model.

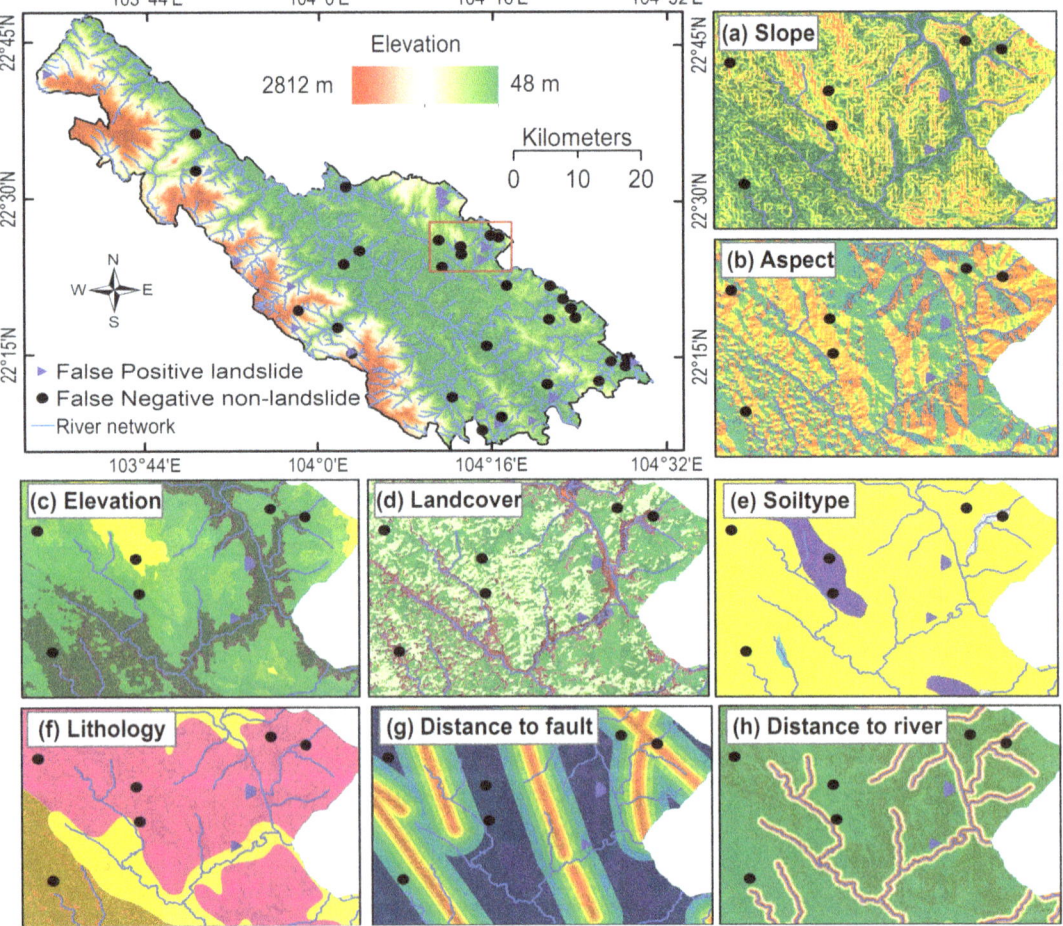

Figure 9. Mispredicted landslide pixels (false positive) and mispredicted non-landslide pixels in the validation dataset versus the eight landslide predisposing factors (legend for the eight factors was the same as in Figure 4). (**a**) Slope; (**b**) Aspect; (**c**) Elevation; (**d**) Landcover; (**e**) Soil type; (**f**) Lithology; (**g**) Distance to fault; and (**h**) Distance to river.

The global prediction capability of the BE-LMTree model is summarized and presented using the ROC curve and AUC (Figure 10). It can be seen that AUC is 0.834, indicating that the prediction capability of the proposed model is 83.4%, which is a high prediction capability.

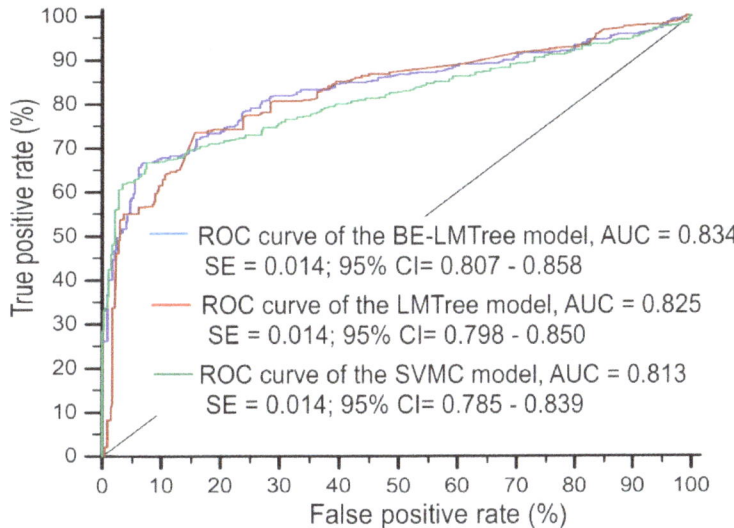

Figure 10. Receiver Operating Characteristic (ROC) curve and Area Under the curve (AUC) of the BE-LMTree model, the LMTree model, and the SVMC model using the validation dataset. SE: Standard Error; CI: Confidence Interval.

5.3. Comparison of the BE-LMTree Model with Benchmark

Because this is the first time that the BE-LMTree model is investigated for landslide modeling, the validity of the proposed model therefore was evaluated and compared with the benchmark. We select support vector machine (SVMC) as a benchmark because SVMC has proven efficient and outperforms other conventional methods [38,91]. For constructing the SVMC model, the radial basic function (RBF) kernel [41,92,93] was selected and the grid-search method [94–96] was used to derive the best the regularization (C = 9) and kernel width (γ = 0.245). In addition, the performance of the LMTree model was also included to present the merit of the proposed BE-LMTree model that is an integration of the Bagging ensemble and the LMTree.

The result is shown in Figures 7, 8, and 10. Using the training dataset, the CLA of the SVMC model (90.08%) and the LMTree model (92.03%) is slightly lower than CLA (93.81%) of the BE-LMTree model. Regarding LLR, the SVMC model (7.93) and the LMTree model (13.13) have lower values when compared to that of the BE-LMTree model (17.31). The other detailed metrics of the two models are shown in Figure 7. Overall, the BE-LMTree model performs better than the SVMC model and the LMTree model in the training dataset.

Using the validation dataset, the prediction performance of the SVMC model and the LMTree model is evaluated (Figure 8). It could be seen that the proposed BE-LMTree model (CLA = 87.98, LLR = 5.89) has a higher prediction performance when compared to those of the SVMC model (CLA = 86.45%, LLR = 5.09) and the LMTree model (CLA = 82.85%, LLR = 4.05). The global prediction capabilities of the three landslide models are assessed using the ROC curve and AUC (Figure 10). It could be been that the proposed BE-LMTree model (AUC = 0.834) is slightly higher than those of the SVMC model (AUC = 0.825) and the LMTree model (AUC = 0.813). Other detailed prediction performances of the three models are presented in Figure 8. Based on the aforementioned analysis, it could be concluded that the proposed BE-LMTree model is capable of producing the best landslide susceptibility result for this study area.

5.4. The Landslide Susceptibility Map

The final BE-LMTree model derived from the training step above was then used to compute landslide susceptibility indices for the Upper Reaches Area of Red River Basin (URRB), Vietnam. Accordingly, all of the predisposing factors in the raster maps were converted into ASCII format, and then fed to the BE-LMTree model to generate susceptibility indices. Distribution of these susceptibility indices is shown in Figure 11.

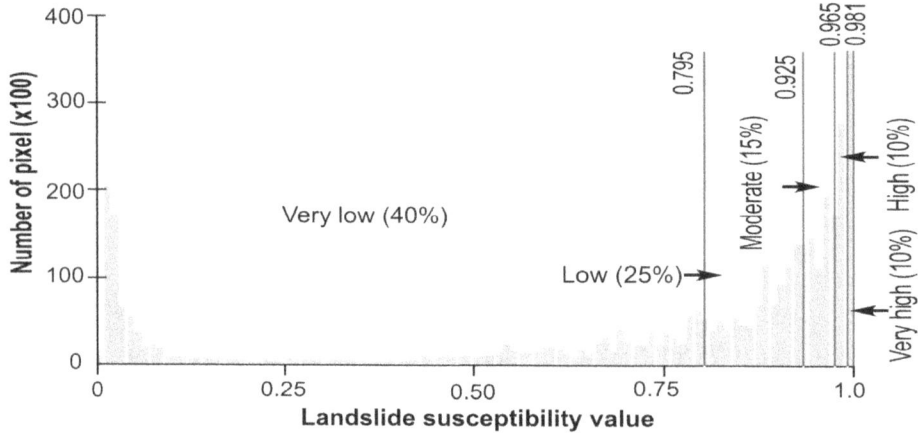

Figure 11. Distribution of these susceptibility indices versus of the five susceptibility classes.

These landslide susceptibility indices were then transformed to the raster format to manage in ArcGIS software using a python application that was programmed by the authors. Finally, the landslide susceptibility map (Figure 12) for the URRB was cartographically presented by five classes: very high (10%), high (10%), moderate (15%), low (25%), and very low (40%). To determine the thresholds for these classes, the extensively used graphic curve method has been considered to be the most suitable; a detailed explanation of it is available in [87,97,98]. The thresholds for these classes were determined based on an analysis of the susceptibility index map and the landslide inventory map, and then, the percentage of the landslide pixel versus the percentage of the susceptibility indices was calculated. At last, the four thresholds for the five classes were obtained.

Characteristics of the five landslide susceptibility classes that were derived from the BE-LMTree model the study area are shown in Table 3. Accordingly, the overall landslide frequency (OLF) proposed in [99] for the five classes was derived, and theoretically, the overall frequency should gradually grow from the very low class to the very high class [87]. It can be seen that the very high occupied only 10% of the study area, but it has the highest OLF value (4.40), followed by the high class (OLF = 1.59), the moderate class (OLF = 0.86), the low class (OLF = 0.43), and the very low class (OLF = 0.41). These confirm that the BE-LMTree model performed well with the URRB area.

Table 3. Characteristics of the landslide susceptibility classes derived from the BE-LMTree model the study area.

No.	Index Interval	Landslide Susceptibility (%)	Expression	Overall Landslide Frequency (OLF)	Areas (km²)
1	1.000–0.981	90–100	Very high	4.40	327.4
2	0.965–0.980	80–90	High	1.59	327.4
3	0.925–0.964	65–80	Moderate	0.86	491.0
4	0.795–0.924	40–65	Low	0.43	818.4
5	0.000–0. 794	0–50	Very low	0.41	1309.4

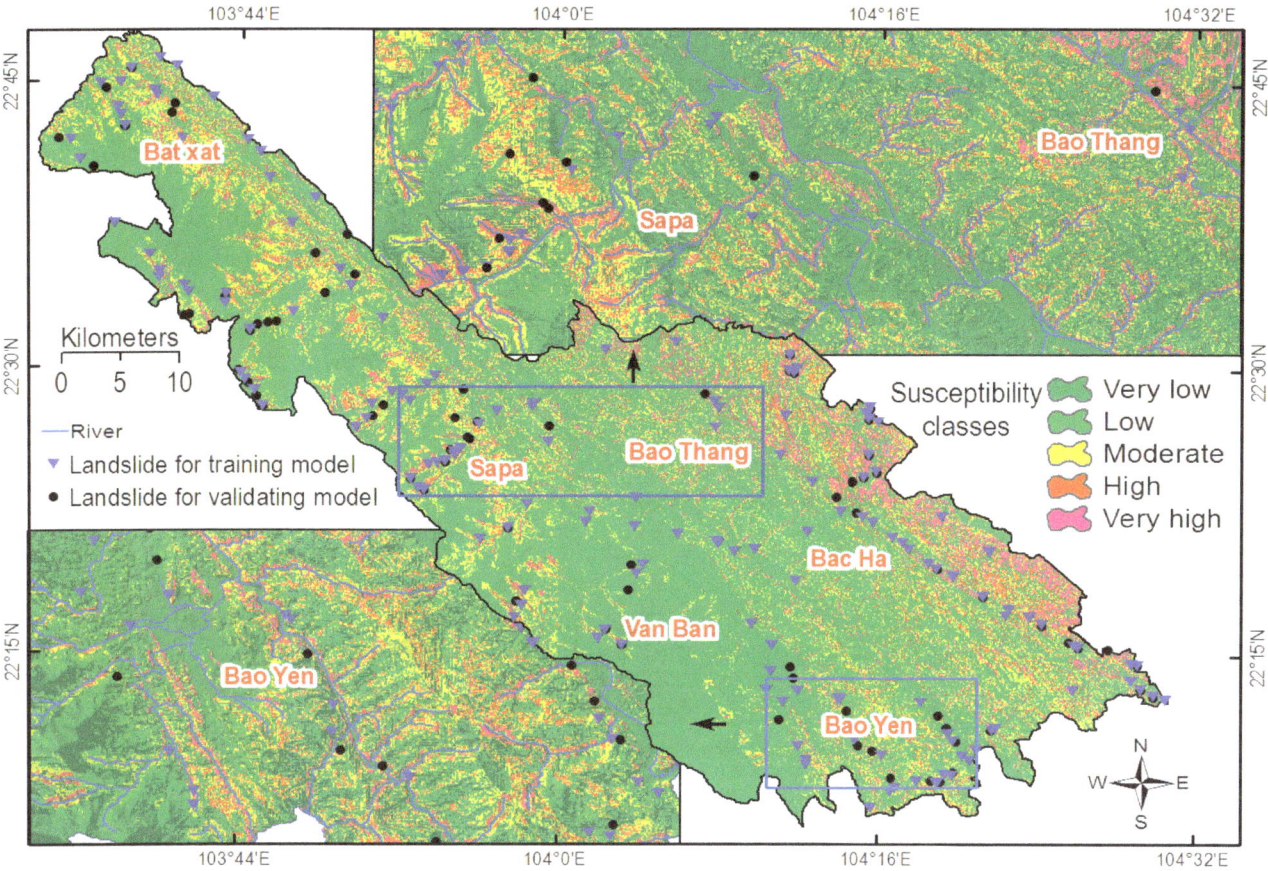

Figure 12. Landslide susceptibility map for the study area using the proposed BE-LMTree model.

Visual interpretation of the map (Figure 12) shows that the high probability of landslide is for areas i.e., Sapa, Bat Xat, and Bao Yen, therefore these areas should receive more attention in the development of remedial measures for the landslide prevention. Inversely, the low probability of landslide is for the Van Ban area. In fact, this area belongs to the Hoang Lien National Park, which is covered by the protected and dense tropical forest [100], therefore, having a low probability of landslide.

6. Concluding Remarks

This paper proposes a new modeling approach that is a hybrid intelligence of BE-LMTree for landslide susceptibility mapping with a case study at URRB. According to current literature, the BE-LMTree model has not been used for landslide modeling. For this purpose, the GIS database for the URRB area has been established, which contains a total of 255 historic soil-mixed-boulder slides and eight geo-environmental factors. These factors checked their merits to landslide using the Pearson correlation. The GIS database was then used to construct and verify the BE-LMTree model. Quality of the final BE-LMTree model was checked using confusion matrices and several model measures.

The results in this study point out that the new approach of the BE-LMTree could help to model landslide susceptibility with desirable prediction capability. When compared to the support vector machines (SVMC), a recognized benchmark in landslide modeling, the proposed BE-LMTree model presents a better performance. Therefore, the BE-LMTree is a new promising tool that could be used to enhance the quality of landslide susceptibility mapping.

For the case of the LMTree, this technique has been recently investigated for landslide susceptibility mapping with promising results i.e., in [50], the performance of the LMTree model in this research is lower than that of the SVMC model and the BE-LMTree model (Figures 4 and 5). Therefore, it could be concluded that the integration of the BE and the LMTree has significantly improved the quality of the LMTree model. This is due to the stability and robustness of the BE procedures itself with the ability to reduce variances [101]. This finding agrees with [51], who concluded that the performance of the landslide model is enhanced with the use of ensemble frameworks.

The main disadvantage of the proposed approach is that the quality of the BE-LMTree model is heavily controlled by the minimum number of samples (NS) that is used for growing the LMTree and the number of bootstrap subsets (BS) used in the BE. In this research, NS and BS were determined using an empirical test. Although the NS and the BS found results in the high performance BE-LMTree model, however these do not warrant them being the optimal parameters. Therefore, the performance of the BE-LMTree model may be further enhanced if optimization algorithms are considered to integrate in the model. In addition, the BE-LMTree may create a complex forest trees i.e., 50 trees in this research. Therefore, the interpretation of the BE-LMTree model may be complicated. Despite the aforementioned limitations, the BE-LMTree can be considered as a new and valid tool for landslide susceptibility modeling.

Author Contributions: X.L.T., M.M., Y.K., V.R., G.Y., X.Q.T., and T.H.D collected data and processed input data. X.L.T., X.Q.T., D.T.B., and S.L. carried out the modeling process and wrote the paper.

Acknowledgments: We would like to thank three anonymous reviewers for their valuable and constructive comments on the earlier version of the manuscript.

References

1. Huggel, C.; Clague, J.J.; Korup, O. Is climate change responsible for changing landslide activity in high mountains? *Earth Surf. Process. Landf.* **2012**, *37*, 77–91. [CrossRef]

2. Uchida, T.; Sakurai, W.; Okamoto, A. Historical Patterns of Heavy Rainfall Event and Deep-Seated Rapid Landslide Occurrence in Japan: Insight for Effects of Climate Change on Landslide Occurrence. In *Advancing Culture of Living with Landslides, Proceedings of the World Landslide Forum WLF 2017, Ljubljana, Slovenia, 29 May–2 June 2017*; Springer: Cham, Switzerland, 2017; pp. 251–257.

3. Ciervo, F.; Rianna, G.; Mercogliano, P.; Papa, M. Effects of climate change on shallow landslides in a small coastal catchment in southern Italy. *Landslides* **2017**, *14*, 1043–1055. [CrossRef]

4. Sewell, R.; Parry, S.; Millis, S.; Wang, N.; Rieser, U.; DeWitt, R. Dating of debris flow fan complexes from Lantau Island, Hong Kong, China: The potential relationship between landslide activity and climate change. *Geomorphology* **2015**, *248*, 205–227. [CrossRef]

5. Gallina, V.; Torresan, S.; Critto, A.; Sperotto, A.; Glade, T.; Marcomini, A. A review of multi-risk methodologies for natural hazards: Consequences and challenges for a climate change impact assessment. *J. Environ. Manag.* **2016**, *168*, 123–132. [CrossRef] [PubMed]

6. Montz, B.E.; Tobin, G.A.; Hagelman, R.R., III. *Natural Hazards: Explanation and Integration*; Guilford Publications: New York, NY, USA, 2017.

7. Maes, J.; Kervyn, M.; de Hontheim, A.; Dewitte, O.; Jacobs, L.; Mertens, K.; Vanmaercke, M.; Vranken, L.; Poesen, J. Landslide risk reduction measures: A review of practices and challenges for the tropics. *Prog. Phys. Geogr.* **2017**, *41*, 191–221. [CrossRef]

8. Gian, Q.A.; Tran, D.-T.; Nguyen, D.C.; Nhu, V.H.; Tien Bui, D. Design and implementation of site-specific rainfall-induced landslide early warning and monitoring system: a case study at Nam Dan landslide (Vietnam). *Geomat. Nat. Hazards Risk* **2017**, *8*, 1978–1996. [CrossRef]

9. Hung, L.Q.; Van, N.T.H.; Son, P.V.; Ninh, N.H.; Tam, N.; Huyen, N.T. Landslide Inventory Mapping in the Fourteen Northern Provinces of Vietnam: Achievements and Difficulties. In *Advancing Culture of Living with Landslides: Volume 1 ISDR-ICL Sendai Partnerships 2015–2025*; Sassa, K., Mikoš, M., Yin, Y., Eds.; Springer International Publishing: Cham, Switzerland, 2017; pp. 501–510.

10. Acosta, L.A.; Eugenio, E.A.; Macandog, P.B.M.; Magcale-Macandog, D.B.; Lin, E.K.-H.; Abucay, E.R.; Cura, A.L.; Primavera, M.G. Loss and damage from typhoon-induced floods and landslides in the Philippines: Community perceptions on climate impacts and adaptation options. *Int. J. Glob. Warm.* **2016**, *9*, 33–65. [CrossRef]

11. Shan, W.; Hu, Z.; Guo, Y.; Zhang, C.; Wang, C.; Jiang, H.; Liu, Y.; Xiao, J. The impact of climate change on landslides in southeastern of high-latitude permafrost regions of China. *Front. Earth Sci.* **2015**, *3*, 7. [CrossRef]

12. LeComte, D. International weather highlights 2014: Winter storms, typhoons, hurricanes, and flooding. *Weatherwise* **2015**, *68*, 20–26. [CrossRef]

13. Jiménez-Perálvarez, J.; El Hamdouni, R.; Palenzuela, J.; Irigaray, C.; Chacón, J. Landslide-hazard mapping through multi-technique activity assessment: An example from the Betic Cordillera (southern Spain). *Landslides* **2017**, *4*, 1975–1991. [CrossRef]

14. Pham, B.; Tien Bui, D.; Pourghasemi, H.; Indra, P.; Dholakia, M.B. Landslide susceptibility assesssment in the Uttarakhand area (India) using GIS: A comparison study of prediction capability of naïve bayes, multilayer perceptron neural networks, and functional trees methods. *Theor. Appl. Climatol.* **2015**, *128*, 255–273. [CrossRef]

15. Corominas, J.; van Westen, C.; Frattini, P.; Cascini, L.; Malet, J.P.; Fotopoulou, S.; Catani, F.; Van Den Eeckhaut, M.; Mavrouli, O.; Agliardi, F.; et al. . Recommendations for the quantitative analysis of landslide risk. *Bull. Eng. Geol. Environ.* **2014**, *73*, 209–263. [CrossRef]

16. Ciampalini, A.; Raspini, F.; Lagomarsino, D.; Catani, F.; Casagli, N. Landslide susceptibility map refinement using PSInSAR data. *Remote Sens. Environ.* **2016**, *184*, 302–315. [CrossRef]

17. Tien Bui, D.; Nguyen, Q.-P.; Hoang, N.-D.; Klempe, H. A Novel Fuzzy K-Nearest Neighbor Inference model with Differential Evolution for Spatial Prediction of Rainfall-Induced Shallow Landslides in a Tropical Hilly Area using GIS. *Landslides* **2017**, *14*, 1–17. [CrossRef]

18. Chacon, J.; Irigaray, C.; Fernandez, T.; El Hamdouni, R. Engineering geology maps: Landslides and geographical information systems. *Bull. Eng. Geol. Environ.* **2006**, *65*, 341–411. [CrossRef]

19. Van Westen, C.J.; Van Asch, T.W.J.; Soeters, R. Landslide hazard and risk zonation—Why is it still so difficult? *Bull. Eng. Geol. Environ.* **2006**, *65*, 167–184. [CrossRef]

20. Akgun, A.; Sezer, E.A.; Nefeslioglu, H.A.; Gokceoglu, C.; Pradhan, B. An easy-to-use MATLAB program (MamLand) for the assessment of landslide susceptibility using a Mamdani fuzzy algorithm. *Comput. Geosci.* **2012**, *38*, 23–34. [CrossRef]

21. Meng, Q.; Miao, F.; Zhen, J.; Wang, X.; Wang, A.; Peng, Y.; Fan, Q. GIS-based landslide susceptibility mapping with logistic regression, analytical hierarchy process, and combined fuzzy and support vector machine methods: A case study from Wolong Giant Panda Natural Reserve, China. *Bull. Eng. Geol. Environ.* **2016**, *75*, 923–944. [CrossRef]

22. Gheshlaghi, H.A.; Feizizadeh, B. An integrated approach of analytical network process and fuzzy based spatial decision making systems applied to landslide risk mapping. *J. Afr. Earth Sci.* **2017**, *133*, 15–24. [CrossRef]

23. Pham, B.T.; Tien Bui, D.; Prakash, I.; Dholakia, M.B. Rotation forest fuzzy rule-based classifier ensemble for spatial prediction of landslides using GIS. *Natl. Hazards* **2016**, *83*, 97–127. [CrossRef]

24. Tien Bui, D.; Pradhan, B.; Lofman, O.; Revhaug, I.; Dick, O.B. Landslide susceptibility assessment in the Hoa Binh province of Vietnam: A comparison of the Levenberg-Marquardt and Bayesian regularized neural networks. *Geomorphology* **2012**, *171–172*, 12–29. [CrossRef]

25. Yilmaz, I. Landslide susceptibility mapping using frequency ratio, logistic regression, artificial neural networks and their comparison: A case study from Kat landslides (Tokat-Turkey). *Comput. Geosci.* **2009**, *35*, 1125–1138. [CrossRef]

26. Pham, B.T.; Tien Bui, D.; Prakash, I.; Dholakia, M.B. Hybrid integration of Multilayer Perceptron Neural Networks and machine learning ensembles for landslide susceptibility assessment at Himalayan area (India) using GIS. *Catena* **2017**, *149 Pt 1*, 52–63. [CrossRef]

27. Gorsevski, P.V.; Brown, M.K.; Panter, K.; Onasch, C.M.; Simic, A.; Snyder, J. Landslide detection and susceptibility mapping using LiDAR and an artificial neural network approach: A case study in the Cuyahoga Valley National Park, Ohio. *Landslides* **2016**, *13*, 467–484. [CrossRef]

28. Oh, H.-J.; Lee, S. Shallow Landslide Susceptibility Modeling Using the Data Mining Models Artificial Neural Network and Boosted Tree. *Appl. Sci.* **2017**, *7*, 1000. [CrossRef]

29. Conforti, M.; Pascale, S.; Robustelli, G.; Sdao, F. Evaluation of prediction capability of the artificial neural networks for mapping landslide susceptibility in the Turbolo River catchment (northern Calabria, Italy). *Catena* **2014**, *113*, 236–250. [CrossRef]

30. Pascale, S.; Parisi, S.; Mancini, A.; Schiattarella, M.; Conforti, M.; Sole, A.; Murgante, B.; Sdao, F. Landslide susceptibility mapping using artificial neural network in the Urban area of Senise and San Costantino Albanese (Basilicata, Southern Italy). In *International Conference on Computational Science and Its Applications*; Springer: Berlin, Germany, 2013; pp. 473–488.

31. Yao, X.; Tham, L.G.; Dai, F.C. Landslide susceptibility mapping based on Support Vector Machine: A case study on natural slopes of Hong Kong, China. *Geomorphology* **2008**, *101*, 572–582. [CrossRef]

32. Kavzoglu, T.; Sahin, E.; Colkesen, I. Landslide susceptibility mapping using GIS-based multi-criteria decision analysis, support vector machines, and logistic regression. *Landslides* **2014**, *11*, 425–439. [CrossRef]

33. Kumar, D.; Thakur, M.; Dubey, C.S.; Shukla, D.P. Landslide susceptibility mapping & prediction using Support Vector Machine for Mandakini River Basin, Garhwal Himalaya, India. *Geomorphology* **2017**, *295*, 115–125.

34. Colkesen, I.; Sahin, E.K.; Kavzoglu, T. Susceptibility mapping of shallow landslides using kernel-based Gaussian process, support vector machines and logistic regression. *J. Afr. Earth Sci.* **2016**, *118*, 53–64. [CrossRef]

35. Pham, B.T.; Bui, D.T.; Prakash, I.; Nguyen, L.H.; Dholakia, M. A comparative study of sequential minimal optimization-based support vector machines, vote feature intervals, and logistic regression in landslide susceptibility assessment using GIS. *Environ. Earth Sci.* **2017**, *76*, 371. [CrossRef]

36. Hong, H.; Pradhan, B.; Bui, D.T.; Xu, C.; Youssef, A.M.; Chen, W. Comparison of four kernel functions used in support vector machines for landslide susceptibility mapping: A case study at Suichuan area (China). *Geomat. Natl. Hazards Risk* **2016**, *8*, 544–569. [CrossRef]

37. Pham, B.T.; Jaafari, A.; Prakash, I.; Bui, D.T. A novel hybrid intelligent model of support vector machines and the MultiBoost ensemble for landslide susceptibility modeling. *Bull. Eng. Geol. Environ.* **2018**. [CrossRef]

38. Pham, B.T.; Tien Bui, D.; Prakash, I. Bagging based Support Vector Machines for spatial prediction of landslides. *Environ. Earth Sci.* **2018**, *77*, 146. [CrossRef]

39. Tien Bui, D.; Pham, T.B.; Nguyen, Q.-P.; Hoang, N.-D. Spatial Prediction of Rainfall-Induced Shallow Landslides Using Hybrid Integration Approach of Least Squares Support Vector Machines and Differential Evolution Optimization: A Case Study in Central Vietnam. *Int. J. Dig. Earth* **2016**, *9*, 1077–1097. [CrossRef]

40. Tien Bui, D.; Anh Tuan, T.; Hoang, N.-D.; Quoc Thanh, N.; Nguyen, B.D.; Van Liem, N.; Pradhan, B. Spatial Prediction of Rainfall-induced Landslides for the Lao Cai area (Vietnam) Using a Novel hybrid Intelligent Approach of Least Squares Support Vector Machines Inference Model and Artificial Bee Colony Optimization. *Landslides* **2017**, *14*, 447–458. [CrossRef]

41. Hoang, N.-D.; Tien Bui, D. A Novel Relevance Vector Machine Classifier with Cuckoo Search Optimization for Spatial Prediction of Landslides. *J. Comput. Civ. Eng.* **2016**, *30*, 1–10. [CrossRef]

42. Althuwaynee, O.F.; Pradhan, B.; Lee, S. A novel integrated model for assessing landslide susceptibility mapping using CHAID and AHP pair-wise comparison. *Int. J. Remote Sens.* **2016**, *37*, 1190–1209. [CrossRef]

43. Youssef, A.M.; Pourghasemi, H.R.; Pourtaghi, Z.S.; Al-Katheeri, M.M. Landslide susceptibility mapping using random forest, boosted regression tree, classification and regression tree, and general linear models and comparison of their performance at Wadi Tayyah Basin, Asir Region, Saudi Arabia. *Landslides* **2015**, *13*, 839–856. [CrossRef]

44. Lagomarsino, D.; Tofani, V.; Segoni, S.; Catani, F.; Casagli, N. A Tool for Classification and Regression Using Random Forest Methodology: Applications to Landslide Susceptibility Mapping and Soil Thickness Modeling. *Environ. Model. Assess.* **2017**, *22*, 201–214. [CrossRef]

45. Tsangaratos, P.; Ilia, I. Landslide susceptibility mapping using a modified decision tree classifier in the Xanthi Perfection, Greece. *Landslides* **2015**, *13*, 305–320. [CrossRef]

46. Kim, J.-C.; Lee, S.; Jung, H.-S.; Lee, S. Landslide susceptibility mapping using random forest and boosted tree models in Pyeong-Chang, Korea. *Geocarto Int.* **2017**, *33*, 1000–1015. [CrossRef]

47. Hong, H.; Liu, J.; Bui, D.T.; Pradhan, B.; Acharya, T.D.; Pham, B.T.; Zhu, A.X.; Chen, W.; Ahmad, B.B. Landslide susceptibility mapping using J48 Decision Tree with AdaBoost, Bagging and Rotation Forest ensembles in the Guangchang area (China). *CATENA* **2018**, *163*, 399–413. [CrossRef]

48. Hoang, N.-D.; Tien Bui, D. Spatial prediction of rainfall-induced shallow landslides using gene expression programming integrated with GIS: A case study in Vietnam. *Natl. Hazards* **2018**, *92*, 1871–1887. [CrossRef]

49. Pradhan, B. A comparative study on the predictive ability of the decision tree, support vector machine and neuro-fuzzy models in landslide susceptibility mapping using GIS. *Comput. Geosci.* **2013**, *51*, 350–365. [CrossRef]

50. Tien Bui, D.; Tuan, T.A.; Klempe, H.; Pradhan, B.; Revhaug, I. Spatial prediction models for shallow landslide hazards: A comparative assessment of the efficacy of support vector machines, artificial neural networks, kernel logistic regression, and logistic model tree. *Landslides* **2016**, *13*, 361–378. [CrossRef]

51. Tien Bui, D.; Ho, T.-C.; Pradhan, B.; Pham, B.-T.; Nhu, V.-H.; Revhaug, I. GIS-Based Modeling of Rainfall-Induced Landslides Using Data Mining Based Functional Trees Classifier with AdaBoost, Bagging, and MultiBoost Ensemble Frameworks. *Environ. Earth Sci.* **2016**, *75*, 1101–1123. [CrossRef]

52. Landwehr, N.; Hall, M.; Frank, E. Logistic Model Trees. *Mach. Learn.* **2005**, *59*, 161–205. [CrossRef]

53. Breiman, L. Bagging Predictors. *Mach. Learn.* **1996**, *24*, 123–140. [CrossRef]

54. Tien Bui, D.; Ho, T.C.; Revhaug, I.; Pradhan, B.; Nguyen, D. Landslide Susceptibility Mapping Along the National Road 32 of Vietnam Using GIS-Based J48 Decision Tree Classifier and Its Ensembles. In *Cartography from Pole to Pole*; Buchroithner, M., Prechtel, N., Burghardt, D., Eds.; Springer: Berlin/Heidelberg, Germany, 2013; pp. 303–317.

55. Pham, T.D.; Bui, D.T.; Yoshino, K.; Le, N.N. Optimized rule-based logistic model tree algorithm for mapping mangrove species using ALOS PALSAR imagery and GIS in the tropical region. *Environ. Earth Sci.* **2018**, *77*, 159. [CrossRef]

56. Breiman, L.; Friedman, J.H.; Olshen, R.A.; Stone, C.J. *Classification and Regression Trees*; Chapman and Hall/CRC: New York, NY, USA, 1984.

57. Doetsch, P.; Buck, C.; Golik, P.; Hoppe, N.; Kramp, M.; Laudenberg, J.; Oberdörfer, C.; Steingrube, P.; Forster, J.; Mauser, A. Logistic Model Trees with AUC Split Criterion for the KDD Cup 2009 Small Challenge. In Proceedings of the 2009 International Conference on KDD-Cup 2009, Paris, France, 28 June–1 July 2009; pp. 77–88.

58. Quinlan, J.R. *C4.5: Programs for Machine Learning*; Morgan Kaufmann: San Mateo, CA, USA, 1993.

59. Kuncheva, L.I. *Combining Pattern Classifiers: Methods and Algorithms*, 2nd ed.; John Wiley & Sons: Hoboken, NJ, USA, 2014.

60. Kotsiantis, S. Combining bagging, boosting, rotation forest and random subspace methods. *Artif. Intell. Rev.* **2011**, *35*, 223–240. [CrossRef]

61. Lu, X.X.; Oeurng, C.; Le, T.P.Q.; Thuy, D.T. Sediment budget as affected by construction of a sequence of dams in the lower Red River, Viet Nam. *Geomorphology* **2015**, *248*, 125–133. [CrossRef]

62. Do, T.; Nguyen, C.; Phung, T. *Assessment of Natural Disasters in Vietnam's Northern Mountains*; Munich University Library: Munich, Germany, 2013; p. 57.

63. Tran, T. Climate change adaptation from small and medium scale hydropower plants: A case study for Lao Cai province. *VNU J. Sci. Earth Environ. Sci.* **2011**, *27*, 32–38.

64. Jolivet, L.; Beyssac, O.; Goffe, B.; Avigad, D.; Lepvrier, C.; Maluski, H.; Thang, T.T. Oligo-Miocene midcrustal subhorizontal shear zone in Indochina. *Tectonics* **2001**, *20*, 46–57. [CrossRef]

65. Duan, B.V. The relation between fault movement potential and seismic activity of major faults in Northwestern Vietnam. *Vietnam J. Earth Sci.* **2017**, *39*, 240–255. [CrossRef]

66. Hue, T.T.; Duong, T.V.; Toan, D.V.; Nghinh, L.T.; Minh, V.C.; Pho, N.V.; Xuan, P.T.; Hoan, L.T.; Huyen, N.X.; Pha, P.D.; et al. *Investigation and Assessment of the Types of Geological Hazard in the Territory of Vietnam and Recommendation of Remedial Measures. Phase II: A Study of the Northern Mountainous Province of Vietnam*; Institute of Geological Sciences, Vietnam Academy of Science and Technology: Hanoi, Vietnam, 2004; p. 361.

67. Yem, N.T.; Thanh, N.Q.; Anh, P.L.; Chi, C.T.; Du, C.D.; Dung, N.P.; Dung, P.D.; Hai, N.P.; Hien, T.T.; Hoang, N.V.; et al. *Assessment of Landslides and Debris Flows at Some Prone Mountainouns Areas Vietnam and Recommendation of Remedial Measures. Phase I: A Study of the East Side of the Hoang Lien Son Mountainous Area of Vietnam*; Institute of Geological Sciences, Vietnam Academy of Science and Technology: Hanoi, Vietnam, 2006; p. 361.

68. Van, T.T.; Tuy, P.K.; Giap, N.X.; Ke, T.D.; Thai, T.N.; Giang, N.T.; Tho, H.M.; Tuat, L.T.; San, D.N.; Hung, L.Q.; et al. *Assessment and Prediction of Geological Hazards in the 8 Coastal Provinces of Central Vietnam from Quang Binh to Phu Yen—Current Status, Causes, Prediction and Recommendation of Remedial Measures*; Vietnam Institude of Geosciences and Mineral Resourses: Hanoi, Vietnam, 2002; p. 215.

69. Van, T.T.; Anh, D.T.; Hieu, H.H.; Giap, N.X.; Ke, T.D.; Nam, T.D.; Ngoc, D.; Ngoc, D.T.Y.; Thai, T.N.; Thang, D.V.; et al. *Investigation and Assessment of the Current Status and Potential of Landslides in Some Sections of the Ho Chi Minh Road, National Road 1A and Proposed Remedial Measures to Prevent Landslides from Threat of Safety of People, Property, and Infrastructure*; Vietnam Institute of Geosciences and Mineral Resources: Hanoi, Vietnam, 2006; p. 249.

70. Tien Bui, D.; Pradhan, B.; Lofman, O.; Revhaug, I.; Dick, O.B. Landslide susceptibility mapping at Hoa Binh province (Vietnam) using an adaptive neuro-fuzzy inference system and GIS. *Comput. Geosci.* **2012**, *45*, 199–211. [CrossRef]

71. Cevik, E.; Topal, T. GIS-based landslide susceptibility mapping for a problematic segment of the natural gas pipeline, Hendek (Turkey). *Environ. Geol.* **2003**, *44*, 949–962. [CrossRef]

72. Conforti, M.; Pascale, S.; Pepe, M.; Sdao, F.; Sole, A. Denudation processes and landforms map of the Camastra River catchment (Basilicata–South Italy). *J. Maps* **2013**, *9*, 444–455. [CrossRef]

73. Yilmaz, I. A case study from Koyulhisar (Sivas-Turkey) for landslide susceptibility mapping by artificial neural networks. *Bull. Eng. Geol. Environ.* **2009**, *68*, 297–306. [CrossRef]

74. Pachauri, A.; Pant, M. Landslide hazard mapping based on geological attributes. *Eng. Geol.* **1992**, *32*, 81–100. [CrossRef]

75. Witten, I.H.; Frank, E.; Hall, M.A.; Pal, C.J. *Data Mining: Practical Machine Learning Tools and Techniques*; Morgan Kaufmann: San Mateo, CA, USA, 2016.

76. Zeiller, M. *Modeling Our World: The ESRI Guide to Geodatabase Concepts*; ESRI Press: Redlands, CA, USA, 2010.

77. Tien Bui, D.; Hoang, N.-D. A Bayesian framework based on a Gaussian mixture model and radial-basis-function Fisher discriminant analysis (BayGmmKda V1. 1) for spatial prediction of floods. *Geosci. Model Dev.* **2017**, *10*, 3391. [CrossRef]

78. Dang, V.-H.; Dieu, T.B.; Tran, X.-L.; Hoang, N.-D. Enhancing the accuracy of rainfall-induced landslide prediction along mountain roads with a GIS-based random forest classifier. *Bull. Eng. Geol. Environ.* **2018**. [CrossRef]

79. Goetz, J.N.; Brenning, A.; Petschko, H.; Leopold, P. Evaluating machine learning and statistical prediction techniques for landslide susceptibility modeling. *Comput. Geosci.* **2015**, *81*, 1–11. [CrossRef]

80. Micheletti, N.; Foresti, L.; Robert, S.; Leuenberger, M.; Pedrazzini, A.; Jaboyedoff, M.; Kanevski, M. Machine learning feature selection methods for landslide susceptibility mapping. *Math. Geosci.* **2014**, *46*, 33–57. [CrossRef]

81. Erener, A.; Sivas, A.A.; Selcuk-Kestel, A.S.; Düzgün, H.S. Analysis of training sample selection strategies for regression-based quantitative landslide susceptibility mapping methods. *Comput. Geosci.* **2017**, *104*, 62–74. [CrossRef]

82. Nguyen, Q.-K.; Tien Bui, D.; Hoang, N.-D.; Trinh, P.T.; Nguyen, V.-H.; Yilmaz, I. A Novel Hybrid Approach Based on Instance Based Learning Classifier and Rotation Forest Ensemble for Spatial Prediction of Rainfall-Induced Shallow Landslides using GIS. *Sustainability* **2017**, *9*, 813. [CrossRef]

83. Guyon, I.; Elisseeff, A. An introduction to variable and feature selection. *J. Mach. Learn. Res.* **2003**, *3*, 1157–1182.

84. Lagomarsino, D.; Segoni, S.; Rosi, A.; Rossi, G.; Battistini, A.; Catani, F.; Casagli, N. Quantitative comparison between two different methodologies to define rainfall thresholds for landslide forecasting. *Natl. Hazards Earth Syst. Sci.* **2015**, *15*, 2413–2423. [CrossRef]

85. Lucà, F.; Conforti, M.; Robustelli, G. Comparison of GIS-based gullying susceptibility mapping using bivariate and multivariate statistics: Northern Calabria, South Italy. *Geomorphology* **2011**, *134*, 297–308. [CrossRef]

86. Cantor, S.B.; Kattan, M.W. Determining the area under the ROC curve for a binary diagnostic test. *Med. Decis. Mak.* **2000**, *20*, 468–470. [CrossRef] [PubMed]

87. Pradhan, B.; Lee, S. Landslide susceptibility assessment and factor effect analysis: Backpropagation artificial neural networks and their comparison with frequency ratio and bivariate logistic regression modelling. *Environ. Model. Softw.* **2010**, *25*, 747–759. [CrossRef]

88. Fushiki, T. Estimation of prediction error by using K-fold cross-validation. *Stat. Comput.* **2011**, *21*, 137–146. [CrossRef]

89. Van Den Eeckhaut, M.; Vanwalleghem, T.; Poesen, J.; Govers, G.; Verstraeten, G.; Vandekerckhove, L. Prediction of landslide susceptibility using rare events logistic regression: A case-study in the Flemish Ardennes (Belgium). *Geomorphology* **2006**, *76*, 392–410. [CrossRef]

90. Costanzo, D.; Rotigliano, E.; Irigaray, C.; Jiménez-Perálvarez, J.D.; Chacón, J. Factors selection in landslide susceptibility modelling on large scale following the gis matrix method: Application to the river Beiro basin (Spain). *Natl. Hazards Earth Syst. Sci.* **2012**, *12*, 327–340. [CrossRef]

91. Tien Bui, D.; Pradhan, B.; Lofman, O.; Revhaug, I.; Dick, O. Regional prediction of landslide hazard using probability analysis of intense rainfall in the Hoa Binh province, Vietnam. *Natl. Hazards* **2013**, *66*, 707–730. [CrossRef]

92. Hoang, N.-D.; Tien Bui, D. Predicting earthquake-induced soil liquefaction based on a hybridization of kernel Fisher discriminant analysis and a least squares support vector machine: A multi-dataset study. *Bull. Eng. Geol. Environ.* **2018**, *77*, 191–204. [CrossRef]

93. Hoang, N.-D.; Tien Bui, D.; Liao, K.-W. Groutability estimation of grouting processes with cement grouts using Differential Flower Pollination Optimized Support Vector Machine. *Appl. Soft Comput.* **2016**, *45*, 173–186. [CrossRef]

94. Ngoc-Thach, N.; Ngo, D.B.-T.; Xuan-Canh, P.; Hong-Thi, N.; Thi, B.H.; NhatDuc, H.; Dieu, T.B. Spatial pattern assessment of tropical forest fire danger at Thuan Chau area (Vietnam) using GIS-based advanced machine learning algorithms: A comparative study. *Ecol. Inform.* **2018**, *46*, 74–85. [CrossRef]

95. Vafaei, S.; Soosani, J.; Adeli, K.; Fadaei, H.; Naghavi, H.; Pham, T.D.; Tien Bui, D. Improving Accuracy Estimation of Forest Aboveground Biomass Based on Incorporation of ALOS-2 PALSAR-2 and Sentinel-2A Imagery and Machine Learning: A Case Study of the Hyrcanian Forest Area (Iran). *Remote Sens.* **2018**, *10*, 172. [CrossRef]

96. Tien Bui, D.; Pradhan, B.; Lofman, O.; Revhaug, I.; Dick, O.B. Application of support vector machines in landslide susceptibility assessment for the Hoa Binh province (Vietnam) with kernel functions analysis. In *iEMSs 2012—Managing Resources of a Limited Planet, Proceedings of the 6th Biennial Meeting of the International Environmental Modelling and Software Society, Leipzig, Germany, 1 July 2012*; Brigham Young University: Provo, UT, USA, 2012; pp. 382–389.

97. Chung, C.-J.; Fabbri, A.G. Predicting landslides for risk analysis—Spatial models tested by a cross-validation technique. *Geomorphology* **2008**, *94*, 438–452. [CrossRef]

98. Tien Bui, D.; Pradhan, B.; Lofman, O.; Revhaug, I.; Dick, O.B. Spatial prediction of landslide hazards in Hoa Binh province (Vietnam): A comparative assessment of the efficacy of evidential belief functions and fuzzy logic models. *Catena* **2012**, *96*, 28–40. [CrossRef]

99. Sarkar, S.; Kanungo, D.P. An integrated approach for landslide susceptibility mapping using remote sensing and GIS. *Photogramm. Eng. Remote Sens.* **2004**, *70*, 617–625. [CrossRef]

100. Kieu, Q.L.; Nguyen, T.T. Study on the distribution characteristics of the vegetation in high levations in Hoang Lien National park of Vietnam. *J. Vietnam. Environ.* **2015**, *6*, 84–88.

101. Mert, A.; Kılıç, N.; Akan, A. Evaluation of bagging ensemble method with time-domain feature extraction for diagnosing of arrhythmia beats. *Neural Comput. Appl.* **2014**, *24*, 317–326. [CrossRef]

7

Deep Fusion Feature based Object Detection Method for High Resolution Optical Remote Sensing Images

Eric Ke Wang [1], Yueping Li [2,*], Zhe Nie [2], Juntao Yu [1], Zuodong Liang [1], Xun Zhang [1] and Siu Ming Yiu [3]

[1]   Harbin Institute of Technology, Shenzhen 518055, China; wk_hit@hit.edu.cn (E.K.W.);
      yujuntao@stu.hit.edu.cn (J.Y.); liangzuodong@stu.hit.edu.cn (Z.L.); zhangxun@stu.hit.edu.cn (X.Z.)
[2]   School of Computer Engineering, Shenzhen Polytechnic, Shenzhen 518055, China; niezhe@szpt.edu.cn
[3]   Department of Computer Science, University of Hong Kong, Pokfulam Road, Hong Kong, China;
      smyiu@cs.hku.hk
*   Correspondence: liyueping@szpt.edu.cn

Abstract: With the rapid growth of high-resolution remote sensing image-based applications, one of the fundamental problems in managing the increasing number of remote sensing images is automatic object detection. In this paper, we present a fusion feature-based deep learning approach to detect objects in high-resolution remote sensing images. It employs fine-tuning from ImageNet as a pre-training model to address the challenge of it lacking a large amount of training datasets in remote sensing. Besides, we improve the binarized normed gradients algorithm by multiple weak feature scoring models for candidate window selection and design a deep fusion feature extraction method with the context feature and object feature. Experiments are performed on different sizes of high-resolution optical remote sensing images. The results show that our model is better than regular models, and the average detection accuracy is 8.86% higher than objNet.

Keywords: high-resolution; optical remote sensing; object detection; deep learning; transfer learning

1. Introduction

Object detection for remote sensing images is an important research field. With the development of remote sensing technology, information carried by remote sensing images is more abundant than before. The applications of object detection in remote sensing images are more and more popular, such as city planning and environmental exploration.

However, object detection for remote sensing images is a more difficult job since remote sensing images are quite different from regular images. Objects in regular images have some properties: many objects rarely appear in one image; the main objects are regularly located in the image center, occupying the main parts and significantly different from the background.

However, one high-resolution optical remote sensing image contains more objects with more shapes and texture information than a regular image, and the objects may be scattered in the whole image. Besides, the object to be detected is relatively small and close to the background. If we zoom out from a remote sensing image to a small size for a global view, we would lose many details, and the objects may almost be invisible. Therefore, object detection for remote sensing images is harder work than for regular images to some extent.

At present, object detection methods in remote sensing images are mainly based on traditional image processing technology with machine learning, which requires rich experience and complete prior knowledge. Furthermore, most of them are only effective in a specific environment, so they have poor scalability. With the advent of deep learning technology, we introduce deep learning into the field of object detection for remote sensing images. Nevertheless, deep learning is still in its infancy for remote sensing images. One of its biggest problems is the decency on labeled datasets.

However, with the fast improvement of deep learning-based object detection on regular images, many labeled regular image datasets have appeared in recent years. Therefore, in this paper, we present a novel transfer deep learning approach to detect objects in high-resolution remote sensing images. It employs transfer learning to supply the gap that deep learning on remote sensing images lacks labeled training datasets. Besides, we improved the candidate window selection process and designed a deep feature extraction method with context scene feature fusion and detection. Finally, the proposed approach is validated on different scales of high-resolution optical remote sensing images.

This paper is organized as follows: Section 2 introduces the related works. Section 3 describes the framework of our algorithm. Our size-scalable region proposal algorithm is given in Section 4. Our deep feature extraction method with context scene feature fusion and detection is proposed in Section 5. Section 6 concludes the paper.

2. Related Works

The development of object detection in high-resolution remote sensing images has been in three stages: template matching-based, knowledge testing-based, and machine learning-based.

Weber et al. [1] performed template matching by extending the hit-or-miss transformation in a multivariate fashion to detect storage tanks and the shoreline. Sirmacek et al. [2] demonstrated the detection of urban buildings by using the SIFT feature to represent a two-building template. Knowledge testing-based object detection transforms detection into the hypothesis testing problem by establishing a set of knowledge and rules. Yokoya et al. [3] used buildings and shadows to detect buildings of arbitrary shape automatically. Machine learning-based methods have been the main direction of object detection in the remote sensing field. For example, Tao et al. [4] described the airplane objects by a set of key point SIFT feature and a graph model to detect them. Sun et al. [5] constructed a bag-of-words model by clustering SIFT feature points to represent targets and classified them by SVM. Gan et al. [6] performed ship detection on remote sensing images by extracting the HOG feature of the sliding window and extracted the continuous window feature by rotating the sliding window to achieve certain rotation invariance. Mostafa et al. [7] clustered the urban railroad point clouds into three classes of rail track, contact cable, and catenary cable by a template matching approach. Aytekin et al. [8] automatically selected a representative subset of texture features by the AdaBoost algorithm and used them to identify airport runways and then detect the airport. Felzenszwalb et al. [9] obtained good results in general object detection by combining the pyramid HOG feature with the partial deformation model and training an SVM with hidden variables. Some scholars have applied weak-supervised object detection algorithms to remote sensing images [10].

However, these methods mostly relied on a variety of well-designed prior knowledge or shallow features, etc. [11,12]. That is, they require a wealth of experience and a tedious trial and error process, and they have limitations.

In the field of object detection on regular images, Krizhevsky et al. [13] proposed an image classification algorithm based on deep learning, which automatically extracts the higher level features of the images by a convolutional neural network (CNN), which made the image classification task more concise and the classification accuracy significantly better.

Thanks to the development of deep learning and the development of region proposal algorithms, the object detection field has also made great breakthroughs. One of the main challenges of object detection in remote sensing images is how to reduce the computational complexity. In view of the large scale of remotely-sensed images, there will be a large number of candidates if the conventional multi-scale scanning window exhaustive strategy is used to obtain the region of interest, which makes the subsequent feature extraction and classification cost too much to achieve fast detection. Therefore, how to reduce the search space in this field is a key problem. In recent years, many region proposal algorithms have been proposed [14–17]. These algorithms can be divided into two categories:

(1) segmentation and combination methods: divide the input image into fragments and then combine these fragments by some bottom-up strategy to generate regions of interest; (2) the window scoring method: define the scoring criteria of the probability of the candidate window containing the object, scoring each possible window by the sliding window method, and selecting the candidate with a high score. There are two more regularly-used region proposal algorithms: the selective search algorithm [14] and the binarized normed gradients (BING) algorithm [15].

In 2014, Girshick et al. [18] proposed the R-CNN (region-based convolutional network) framework for object detection, in which a region proposal algorithm was designed to obtain the candidate window instead of the sliding window strategy to improve detection efficiency, and then, the CNN was employed to extract high-level features before an SVM classification was used. The proposed R-CNN framework greatly improved the detection accuracy and brought much inspiration to the object detection field, and many object detection algorithms based on the deep learning have been proposed [19–21]. There are more research works on object detection using deep learning. For example, Kong et al. [22] proposed the HyperNet network structure, which combines the multi-level features of the deep network and merges them to select the region and detect the object, which resulted in a more accurate localization. Ouyang et al. [23] proposed to combine CNN with the deformation model, which made the process of objection detection more sensitive through multiple models, multi-stage cascade, and other integrated approached. Redmon et al. [24] considered the objection detection as a regression prediction problem. They designed the YOLO network structure, of which the network input is the whole map. This original map was divided into 7×7 grids. This structure greatly improves the detection speed and real-time detection. However, their method is not effective with respect to objects that are located close to one another, and the objects have an irregular aspect ratio. Meanwhile, the deep learning method in the field of object detection in remote sensing images is still in a relatively nascent stage.

3. The Overall Idea of Our Method

In object detection in remote sensing images, template matching-based methods are simple and easy to implement, but the template design becomes more and more complicated when directions and shapes of objects vary greatly. Knowledge-based object detection methods can gain better detection performance through abundant a priori knowledge, but how to define the a priori knowledge and rules is a hard problem, which usually requires much experience. While machine learning-based object detection methods are based on shallow feature extraction methods, such as HOG, SIFT, and other classic features presented in object detection on regular images, and have a better detection effect for some specific scenes, when the remote sensing background is complex and the objects are diversified, the scalability of these methods is poor. Deep learning has a great advantage in automatically learning deep-level features, but it is still at a relatively early stage for object detection in remote sensing images. Meanwhile deep learning requires a large amount of labeled data, which are less available in remote sensing images. Aiming at these shortcomings, we propose to employ abundant labeled regular image datasets to assist the object detection in the remote sensing images through the transfer learning method and explore the validity of the transfer learning. As shown in Figure 1, from the detection process, the framework of the proposed approach can be divided into three steps: rapid candidate region proposal, deep feature extraction of the candidate window, context scene feature fusion, classification, and post-processing. The training process is mainly to train the models used in the steps of the detection stage: (a) train the candidate region proposal model; (b) combine transfer learning to train the deep feature extraction network; (c) combine transfer learning to train the context scene feature extraction network; (d) train the classification model.

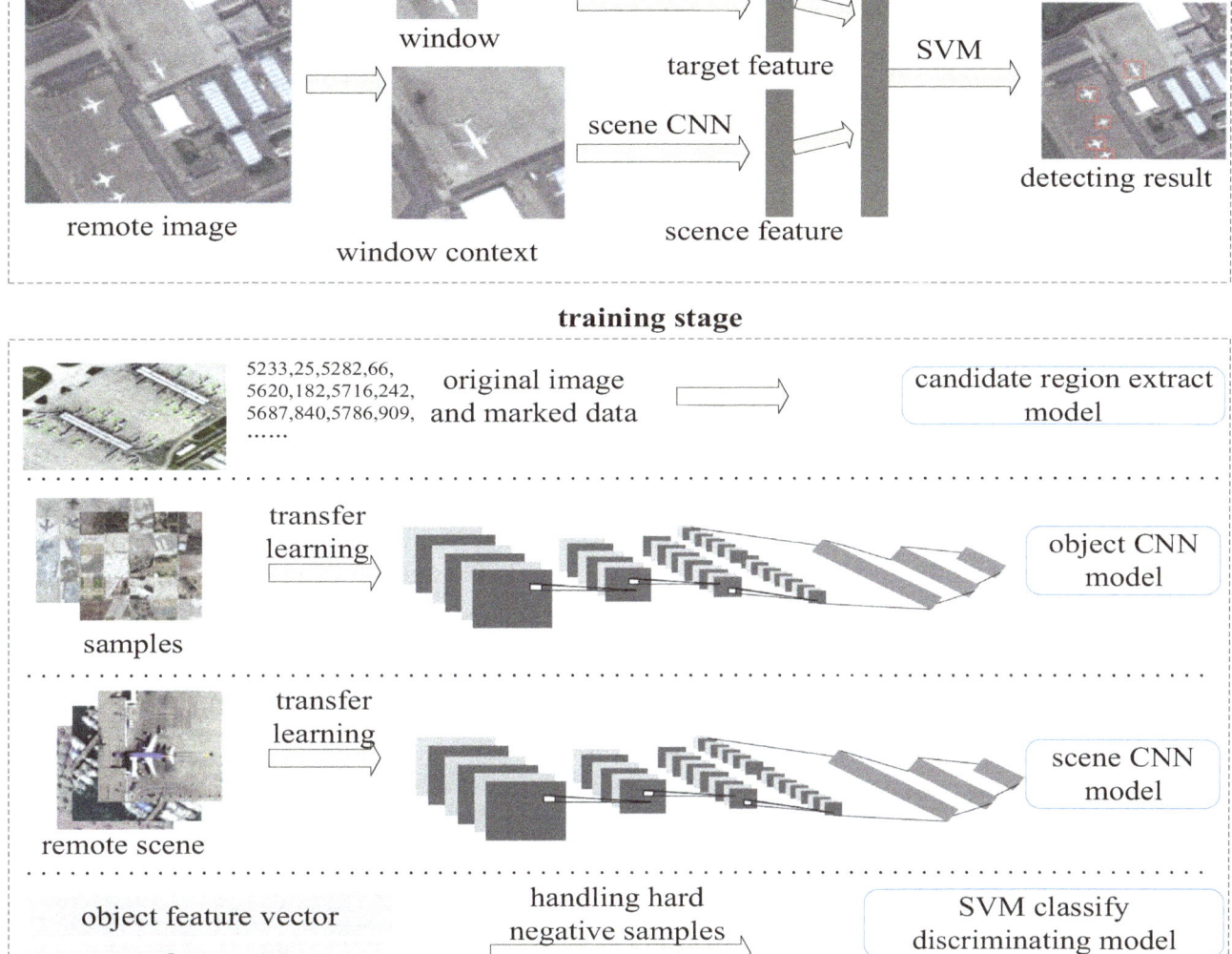

Figure 1. Proposed object detection framework.

In particular, we improve the stages of candidate region proposal and feature extraction:

1 For the stage of candidate region proposal, by analyzing the particularity of the remote sensing image object detection task, we select the BING algorithm to improve it by integrating multiple weak feature scoring to extend to large-scale images. The experiment shows that the improved algorithm achieves a better detection rate and more accurate object coverage when obtaining the same number of candidate windows.

2 For the stage of feature extraction, we employ CNN to extract deep features of the candidate windows and the windows' context scene, respectively, and then fuse the two kinds of features for detection, which improves the detection performance. In addition, we solve the problem of the insufficient annotations on remote sensing images by transfer learning, which reduces the risk of over-fitting and improves the network's ability of feature expression on remote sensing objects and scenes.

In the stage of classification and post-processing, we employ the faster linear SVM with the hard negative mining method to reduce the impact of overfull negative samples. Finally, we filter out duplicates by the non-maximum suppression algorithm to further optimize the test results.

In the following sections, we mainly discuss the two stages: candidate region proposal and feature extraction, and analyze the results of the experiments respectively.

4. Size-Scalable Candidate Region Proposal Algorithm

In the field of object detection, the traditional methods mostly employ the multi-scale sliding window method, which is an exhaustive search strategy. In order to guarantee the detection speed, only the simple feature of the candidate window can be extracted to be classified, which may lead to higher false detections. Therefore, a good region proposal algorithm plays an important role in the whole object detection process. In addition, since the size of remote sensing images is usually large, the region proposal algorithm needs to be scalable to the large image size. However, the segmentation-based selective search algorithm used in the classical deep learning-based object detection framework R-CNN needs to build a graph model: each pixel of the image acts as a node, and it includes a large number of similarity calculations and adjacent regions. This means that the algorithm needs to maintain many intermediate results, which has a heavy cost in time and in memory. Therefore, it becomes one of the bottlenecks of the whole framework. Besides, when the image size is very large, the problem is particularly critical, and memory is more likely to be insufficient in that case. The BING algorithm is based on the sliding window scoring mechanism with an accelerating optimization, and it is considered to be the fastest region proposal algorithm [25]. Furthermore, when increasing the image size, the speed and the memory usage of the algorithm increase linearly at most, and this is acceptable. In addition, because the candidate windows obtained by the BING algorithm contain probabilistic scores, we can select the appropriate number of candidate windows when necessary.

4.1. BING Algorithm

The binarized normed gradients (BING) algorithm is a very simple, but highly efficient region proposal algorithm, which is essentially a two-stage cascade classifier. The first stage uses a multi-scale sliding window to scan the image, and each window is scaled to a uniform size of 8×8, while a linear scoring model is used:

$$s_l = \langle w, g_l \rangle \tag{1}$$

$$l = (i, x, y) \tag{2}$$

where s_l is the filter score, g_l is the window feature, and $l, i, (x, y)$ are the location size and top left corner coordinates of the window, respectively.

The windows with a higher score for each size are selected as the possible candidate windows. During this period, the number of true objects in each size is counted, and the sliding window score is only applied to the size with the objects' number exceeding a certain threshold. The first-stage linear model w is trained by a linear SVM, and the NG features (norm of the gradients) of the real object windows and random sampling background windows are taken as positive and negative samples, respectively.

In the second stage, considering the different possibilities for different window sizes containing a object, such as the square window of 64×64 is more likely to contain an object than the 5×128 one, a score calibrator is trained for each size. We update the score for each candidate window:

$$o_l = v_i \times s_l + t_i \tag{3}$$

where $v_i, t_i \in R$ are the calibration coefficients that are learned for different sizes. This step is necessary only if you need to reorder the candidate windows obtained in one stage. The learning of the parameters v_i and t_i is also performed by the linear SVM.

The biggest contribution of the BING algorithm is the improvement of detection speed. In order to speed up the feature extraction and scoring process, the algorithm uses the idea of model binary approximation [26,27]. The linear model w is approximated by a set of binary basis vectors α_j:

$$w \approx \sum_{j=1}^{N_w} \beta_j \alpha_j \tag{4}$$

where N_w denotes the number of basis vectors, $\alpha_j \in \{-1,1\}^{64}$ denotes a base vector, and β_j denotes the corresponding coefficients.

Further, representing each α_j by using a binary vector and its complement: $\alpha_j = \alpha_j^+ - \overline{\alpha_j^+}$, where $\alpha_j^+ \in \{0,1\}^{64}$. These transforms allow subsequent scoring calculations for a binarized feature just using fast BITWISE andand bit countoperations, as shown in Equation (5):

$$w, b \approx \sum_{j=1}^{N_w} \beta_j \left(2a_j^+, b - |b| \right) \tag{5}$$

However, the NG features are real numbers, and how to binarize the NG features to speed up the calculation is one of the difficulties of the algorithm. This algorithm approximates the NG feature values (each saved as a BYTEvalue) using the top N_g binary bits of the BYTE values. Thus, a 64-dimensional NG feature g_l can be approximated by N_g binarized normed gradient (BING) features:

$$g_l \approx \sum_{k=1}^{N_g} 2^{8-k} b_{k,l} \tag{6}$$

Therefore, the score for whether or not an image window contains an object can be approximated as:

$$s_l = w, g_l \approx \sum_{j=1}^{N_w} \beta_j \sum_{k=1}^{N_g} 2^{8-k} \left(2a_j^+, b - |b_{k,l}| \right) \tag{7}$$

where $2^{8-k} \left(2a_j^+, b - |b_{k,l}| \right)$ can be computed using fast CPU atomic operation: BITWISE and POPCNTSSEoperators, which speeds up the calculation.

4.2. pBING Algorithm with Multiple Weak Feature Scoring

In practice, we found that the ambiguous objects were missed by the original BING algorithm. The possible reason for this phenomenon is that only a simple NG feature is used in the algorithm, and in order to speed up the calculation, the NG feature serves as the BING feature. Thus, some information would be lost in the process, which further reduces the distinguishing degree of the feature. However, in order to preserve the time and space complexity advantages of the BING algorithm, we are not able to use too complex features such as SIFT and HOG features due to the parallel optimization strategy of the algorithm. Thanks to the AdaBoost algorithm [28] that inspired us: integrating multiple weak classifiers is used to obtain a strong classifier. Therefore, we improved the BING algorithm by integrating multiple weak feature scoring models.

We improved the first stage of the candidate window scoring model, and the second stage of score correction was same as the original algorithm. For convenience, we name the improved BING algorithm as the pBING algorithm. Figure 2 shows the flowchart of the pBING algorithm, and the shaded part is our improvement. For each training sample, we extract multiple weak feature channels and train a scoring model for each weak feature channel; in the detection process, we integrate the score of multiple weak feature scoring models as the final score of the candidate window, in which a simple linear weighting method is adopted, and the weights are determined by the accuracy of each model. Each scoring model still uses an efficient linear SVM algorithm. The score of each candidate window in each scoring model is: $s_{kl} = \langle w_k, g_{kl} \rangle$, where s_{kl} denotes the score of model k for window l, w_k denotes the parameter of model k, and g_{kl} denotes the feature k of window l. The final window score for the first stage is as follows:

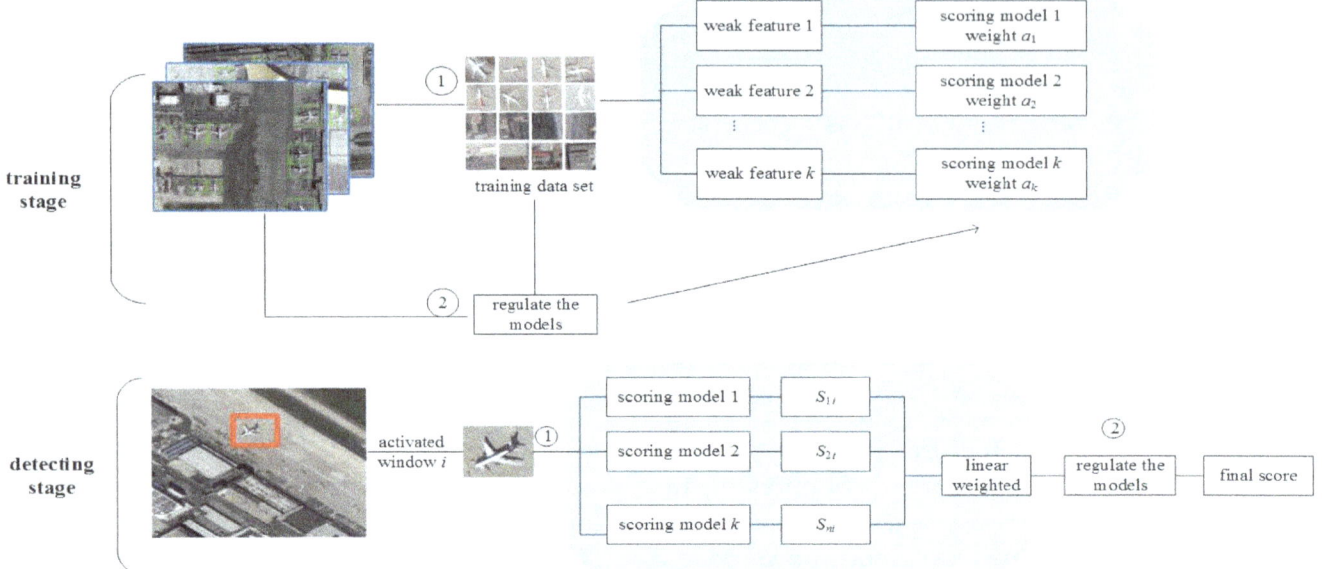

Figure 2. pBING algorithm flowchart.

The NG feature map used in the BING algorithm captures the edge intensity information of the original image by computing the gradient magnitude of each pixel, but this feature is simple and susceptible to noise. In this paper, the NG feature map is replaced by the Sobel feature map with better edge information capture. At the same time, the local binary pattern (LBP) feature map and the difference of Gaussians (DoG) feature map are introduced.

The Sobel feature map uses 3×3 Sobel operators to convolute the original image to obtain the horizontal and vertical direction of the approximate gradient, as shown in Formula (8). Then, it computes the gradient amplitude through Formula (9).

$$G_x = \begin{bmatrix} -1 & 0 & +1 \\ -2 & 0 & +2 \\ -1 & 0 & +1 \end{bmatrix} \otimes A$$

$$G_y = \begin{bmatrix} +1 & +2 & +1 \\ 0 & 0 & 0 \\ -1 & -2 & -1 \end{bmatrix} \otimes A \tag{8}$$

$$G = \sqrt{G_x^2 + G_y^2} \tag{9}$$

where A represents the original image matrix, G_x and G_y represent the horizontal and vertical gradient of the image, respectively, and G represents the gradient magnitude matrix, which is the obtained feature map. $\otimes$ represents the convolution operation.

LBP can be used to describe the local texture features, often used for face classification, pedestrian detection [29], and so on. We can get the LBP feature map by computing the LBP code for each pixel point. In order to simplify the calculation, we use the simplest 3×3 LBP operator to calculate in the gray image.

The DoG feature map is obtained by subtracting two different degrees of blurred images from the original image. The blurred image is obtained by the Gaussian kernel convolution of different standard deviation parameters on the gray image. Two Gaussian blurred image subtractions can increase the visibility of edges and other details, and the DoG algorithm does not enhance noise because Gaussian blur suppresses high-frequency noise. The two-dimensional Gaussian kernel function is defined as follows:

$$G_{\sigma_1}(x,y) = \frac{1}{\sqrt{2\pi\sigma_1^2}} exp\left(-\frac{x^2+y^2}{2\sigma_1^2}\right) \tag{10}$$

Then, the Gaussian filtering of the two blurred images is expressed as:

$$\begin{aligned} g_1(x,y) &= G_{\sigma_1}(x,y) * f(x,y) \\ g_2(x,y) &= G_{\sigma_2}(x,y) * f(x,y) \end{aligned} \tag{11}$$

Therefore, the DoG feature map is obtained by subtracting the two images $g_1(x,y)$ and $g_2(x,y)$, where σ_1 and σ_2 are two Gaussian kernel parameters, respectively. When the DoG is used for different purposes, the ratio of the two Gaussian kernel parameters is different. When used for image enhancement, usually $\sigma_2{:}\sigma_1$ is set to 4:1 or 5:1. In this paper, $\sigma_2 = 2.0$ and $\sigma_1 = 0.5$.

4.3. Experiments

4.3.1. Dataset and Evaluation

Two different scales of remote sensing image datasets are used in this paper:

(1) SMALL-FIELD-RSIs: Each image was cropped from the Google Earth software, and all airplane objects were manual marked. As shown in Figure 3, for each airplane object, a minimum enclosing rectangle was used. Each bounding box was represented by its top-left and bottom-right coordinates: (x_1, y_1, x_2, y_2). The dataset contained 980 high-resolution optical remote sensing images, and the image size was about 1300 × 800 pixels, with the spatial resolution of the image being about 0.6 m. A total of 7452 airplane objects were marked.

(2) LARGE-FIELD-RSIs: Each image was downloaded from the commercial professional remote sensing software, and the map level was 19, the scale 1:2257, the data source being from QuickBird satellite, and the spatial resolution 0.6 m. Compared to Google Earth software, the software can download any size of high-resolution optical remote sensing images with no watermark or other labels, and the images are relatively clearer. There were 110 images, and the average size of all images was about 5000 × 5000 pixels. A total of 3380 airplane objects were manually marked. As shown in Figure 4, the left is a high-resolution remote sensing image of an airport (International Airport, Shenzhen, China), and the right is an enlarged view of the red area on the left. As can be seen, for the entire image, the airplane object was very small. When zooming out to a relatively small size, the airplane objects were almost invisible. Table 1 shows the statistical details of the two datasets.

Figure 3. The annotation method of the airplane object.

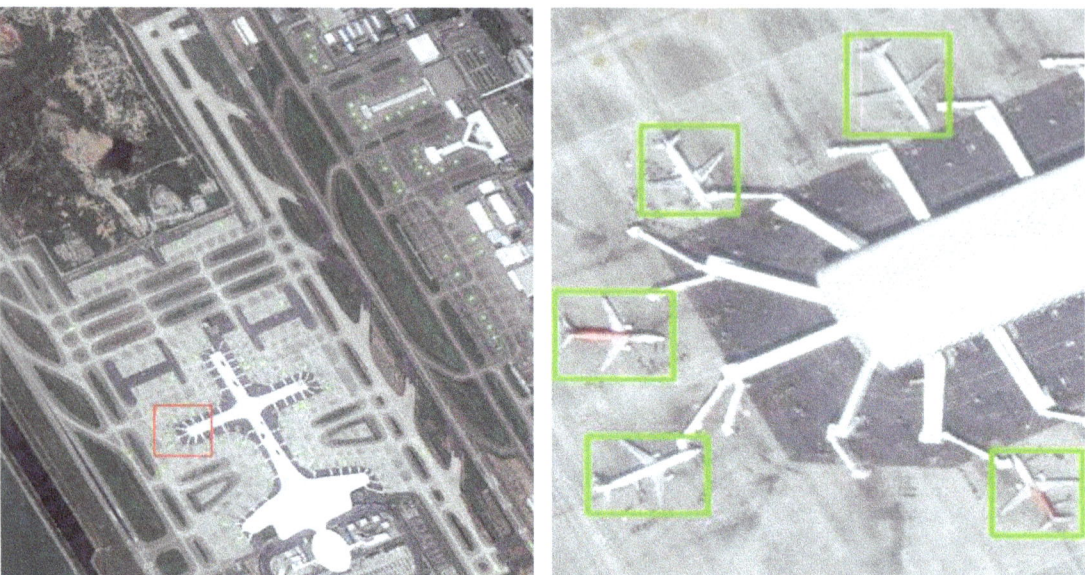

Figure 4. The high-resolution remote sensing image of Baoan Airport.

Table 1. The experimental dataset.

Dataset Name	Spatial Resolution	Image Size (Pixels)	Number of Images	Number of Objects
SMALL-FIELD-RSIs	0.6 m	1300 × 800	980	7452
LARGE-FIELD-RSIs	0.6 m	5000 × 5000	109	3380

In the region proposal algorithm, we evaluated the improved algorithm by the DR-#WIN curve and MABO-#WIN curve, where DR refers to the detection rate and MABO refers to mean average best overlap, while #WIN refers to the number of candidate windows proposed.

4.3.2. Results and Analysis

The experiments in this section were divided into two parts: (1) we briefly analyzed the applicability of the selective search algorithm for candidate region proposal in remote sensing images; (2) we evaluated the result of the BING algorithm and pBING algorithm. In the experiment, each dataset was divided into a training set and a test set, where the test set was 20%. The DR and MABO below were the average of all the test image detection results. Table 2 shows the average number of candidate windows and running times of the selective search algorithm on different sizes of remote sensing images. This experiment was performed in fast mode using only two color spaces and two similarity functions. As in the remote sensing image, when the spatial resolution of the image is determined, the size range of the object in the image can be determined. Using this prior knowledge, we filtered out the irrational size of the candidate windows, and the filtered results are shown in the third row of Table 2. It can be found that the average number of candidate windows generated by the algorithm increased significantly, and the running time of the algorithm increased sharply with the increase of the image size. In addition, in the experimental process, we found that when the image size increased to 2000 × 2000 pixels, the machine was out of memory, and the MATLAB compiler was in a stuck situation. Therefore, this algorithm has a poor scalability for image size.

Table 2. The performance analysis for the selective search algorithm.

remote image scale (pixel)	500 ×500	800 ×800	1000 ×1000	1500 ×1500
average number of candidate windows without constraints	3161	7123	9060	13,810
average number of candidate windows after filtering	3039	6270	8469	12,782
average computing time (s)	2.7	7.6	14.3	29.8

In summary, the selective search algorithm had higher complexity, and the number of candidate windows was larger, which was not scalable for the image size. The following are the performances of BING algorithm and pBING algorithm when extracting candidate regions in two high-resolution remote sensing images.

As shown in Figure 5, when the number of candidate windows was 1000, the DR of the pBING algorithm was 97.21% on the SMALL-FIELD-RSIs, which was higher than the original algorithm (95.74%). The MABO score increased from 63.43–65.30%, which indicates that the improved algorithm had a better quality of the candidate region proposal on the remote sensing images. In addition, it shows that when the number of candidate windows was 2000, the detection rate of pBING algorithm in this dataset was as high as 98.9%. Therefore, as the input of the second stage, the number of candidate windows in this stage can be 2000, that is, for each image, we output the first 2000 candidate windows of the pBING detection results to further classify.

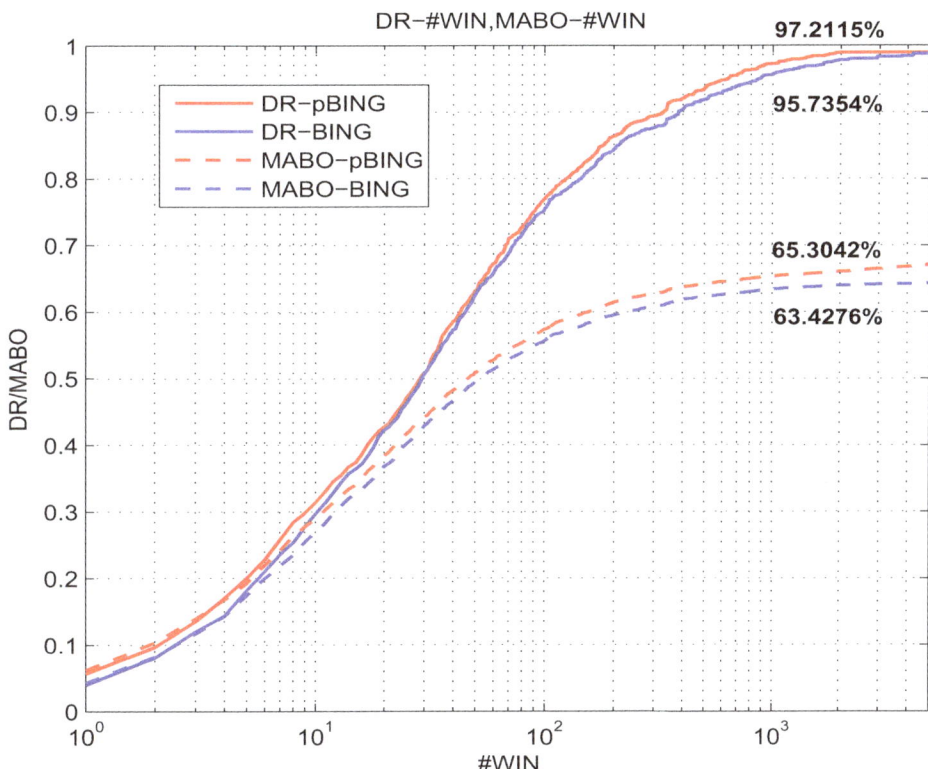

Figure 5. Tradeoff between the number of windows (#WIN) and DR/MABO on the dataset SMALL-FIELD-RSIs.

Figure 6 shows the performance of the two algorithms on dataset LARGE-FIELD-RSIs. Since the image of the dataset is relatively large, we mark DR and MABO of the algorithm when the given number of candidate windows is 8000. Figure 7 shows the details. It appears that DR and MABO are increasing as the number of candidate regions increases. Obviously, that is because that a remote sensing image with large size usually has more objects, and search space of objects' possible locations significantly increases, and then more candidate windows are needed in order to achieve a certain detection rate. As shown in Figure 7, in order to ensure that the follow-up classification to achieve a certain rate of recall, in this stage we select the first 9000 candidates as the input of the second stage. However, since the image size is not fixed in this dataset: the average size of images varies from 2000 pixels to 9000 pixels. In order to further reduce the number of candidate windows, we select the number of output candidate according to the image size.

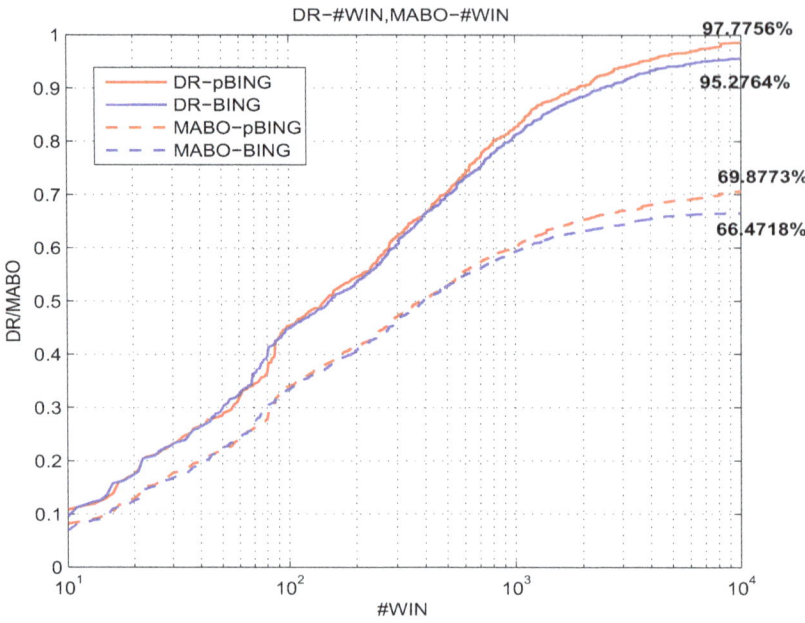

Figure 6. Tradeoff between #WIN and the detection rate (DR)/mean average best overlap (MABO) on the dataset LARGE-FIELD-RSIs.

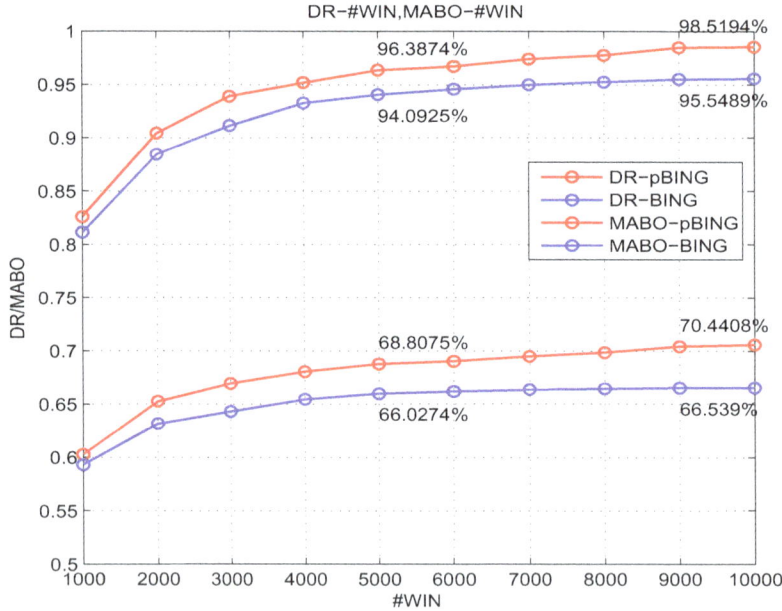

Figure 7. Tradeoff between #WIN and DR/MABO on the dataset LARGE-FIELD-RSIs when #WIN varied from 1000–10,000.

Figure 8 shows the average running time of the BING and pBING algorithms on two datasets to obtain the candidate region proposal. It shows that when the size of the remote sensing image was large, the time required to obtain the candidate region was greatly increased, but the test time was still less than 1 s, indicating that the time complexity of the algorithm did not increase sharply with the image size expanding. In addition, it reveals that the pBING algorithm took more time to acquire candidates, about three-times slower. However, the algorithm itself is very fast; even if the time expansion of the original was three-times, it is still within the acceptable range.

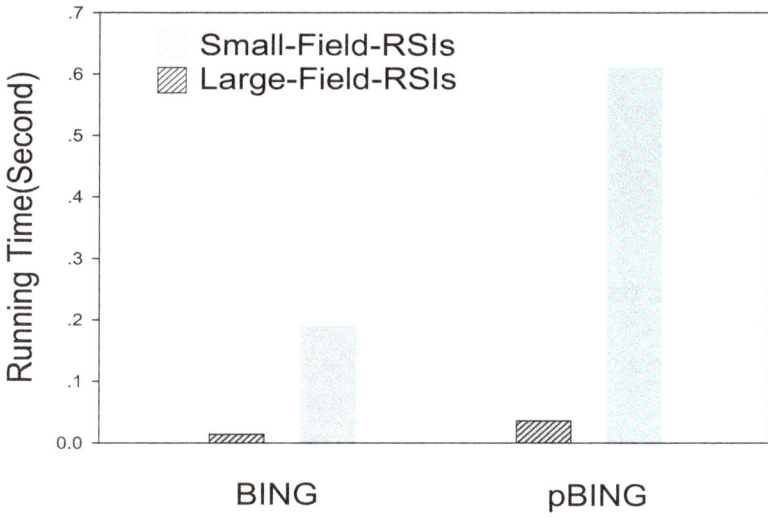

Figure 8. Average running time of the BING algorithm and the pBING algorithm on different datasets.

5. Deep Feature Extraction with Context Scene Feature Fusion and Detection

5.1. Feature Extraction Algorithm CNN

After obtaining the candidate windows that may contain objects, we needed to further determine whether the windows actually contained objects, that is the detection task can be converted into the classification task. To classify the candidate windows, the distinguishing features of each window need to be extracted first. As a base problem of the computer vision field, many classic feature extraction methods have been proposed, such as color histogram, gray level co-occurrence matrix, Haar-like features, HOG features, and SIFT features. These features were designed by scholars according to experience and related prior knowledge and have achieved good results in many fields. However, at the ImageNet LSVRC-2010 competition, Hinton et al. [13] used deep learning to extract features automatically, reducing the top-5 error rate of the image classification task by 15.3%, far beyond the algorithms with traditional feature extraction methods. The reason is that the traditional feature representation can only obtain the shallow features of the image, and it has certain limitations; however, deep learning can automatically learn higher level feature representation from the original image, which has better distinguishing ability and versatility.

Deep learning is developing rapidly, and the most regularly used in the field of object detection is the convolutional neural network (CNN). Compared with traditional neural networks, CNN achieves weight sharing by introducing the convolution layer, which makes the network structure sparser and reduces the complexity of the model. The convolutional algorithm can obtain the feature map of different aspects by difficult convolution kernel parameters. The convolution operation makes the obtained feature map have translation invariance. The convolution layer is usually followed by a pooling layer, which downsamples the obtained feature map layer, preserving useful information while reducing the amount of data to be further processed. The pooling layer usually uses max-pooling to get the maximum response of local features, so that the obtained features have better rotation and light invariance. CNN can learn a higher level feature representation from low-level features through a deep network structure by stacking multiple convolution layers. For different objects, the learned low-level features by CNN differ slightly, which are usually some edge information, and through multi-layer network learning, abstract features of different objects can be obtained finally.

After the multiple convolution layer, a fixed-length feature vector is obtained through the full-connection layer and output to the classifier. Generally, the softmax classifier is employed in the

CNN, and it outputs the probability that the image belongs to each class. CNN combines feature learning with the classification task, which makes the extracted feature more task related.

Model of Neurons

CNN is a type of multi-layer sensor for which each layer is composed of a two-dimensional plane, and each plane is composed of various independent neurons. In the network, some simple and complex cell are marked as cell C and cell S, which is inspired by the vision concept from biology. In the visual cortex, there are two kinds of related cells, simple ones (cell S) and complex ones (cell C). Cell S responds to the stimulation for the modes like margins of images in its maximum receptive field, while cell C has a bigger receptive field, which can locate the modes of stimulation in a spatial way. The merging of C cells forms convolutional layers, denoted as U_C, while the merging of S cells forms the downsampling layers, which can be denoted as U_S. Any intermediate layer in the network is composed of the S-layer and C-layer with series connection. Regularly, U_S is the layer to extract the feature, while U_C is the layer of feature mapping.

In CNN, only the input of cell S is variable, while other inputs are fixed. The first layer can be denoted by $U_{sl}(k_l, n)$, which means a cell s output on the k_l S-plane, and a cell c output on the k_l C-plane can be denoted by $U_{cl}(k_l, n)$. n represent two-dimensional coordinates.

$$U_{sl}(k, n) = r_l(k) \times$$

$$\varphi \left[\frac{1 + \sum_{k_{l-1}}^{K_{l-1}} \sum_{v \in A_l} a_l(v, k_{l-1}, k) U_{cl-1}(k_{l-1}, n+v)}{1 + \frac{r_l(k)}{r_l(k)+1} b_1(k) U_{vl}(n)} - 1 \right] \qquad (12)$$

In the above neuron model formula, $a_l(v, k_{l-1}, k)$ and $b_l(k)$ represent the connection coefficients of positive input and negative input, respectively; $r_l(k)$ is a constant that controls the option of feature extraction; the bigger it is, the worse is at tolerating noise and feature distortion.

Process of convolution: Employ a trainable filter f_X to process the convolution on input images (the c_1 layer is the input, and the inputs of subsequent layers are the outputs of the forward layer), based on an activation function (usually sigmoid) with a offset b_X, to get convolutional layer C_X. M_j is the value of the input feature map:

$$X_j^l = f\left(\sum_{i \in M_j} X_i^{l-1} * k_{ij}^l + b_j^l \right) \qquad (13)$$

Down-sampling process: Each m adjacent pixels sum up to be one pixel (mcan be set), and use j as the weight, add offset b_j, then use the activation function sigmoid to generate feature mapping. The mapping from one plane to another plane can be a convolutional operation, and the layer can be a fuzzy filter functioning as double feature extraction. Spatial resolution decreases with the hidden layer going forward, while the plane number increases for better feature extraction. For the sampling layer, if there are N input features, then there would be N output features, but the size of each feature changes. The details formula are as follows. $down()$ denotes the down-sampling function.

$$X_j^l = f(\beta_j^l \times down(X_i^{l-1}) + b_j^l) \qquad (14)$$

5.2. Context Information

Considering that a specific kind of object can only appear in certain scenes, this a priori knowledge is particularly obvious in remote sensing images; for example, ships can only appear in the port or the sea, and an airplane can only appear on the parking apron or runway. In regular images, the context of the same kind of object varies greatly as a result of different locations, angles, or distances when taking photos. As shown in Figure 9, the car's context scene may be an avenue, the ground, a house, or even water. However, in remote sensing images, the spatial relationship between the object and the background is relatively fixed due to the fixed angle and height of satellites. In addition, as shown in Figure 10, compared to regular images, the objects in high-resolution remote sensing images are usually very small, the details having useful information are few, and objects may not be clear. Therefore, in the remote sensing image, the context scene information of the candidate window may be helpful to determine the objects.

Figure 9. Example of the context scene of a car in regular images.

Figure 10. Airplane objects in high-resolution remote sensing images.

5.3. Deep Feature Extraction and Context Scene Feature Fusion

As objects in remote sensing images have a strong background context, for example, the airplane objects may appear on the runway, parking apron, but not in the forest, the port, etc., and the airplanes may be side by side with another one, how to describe this a priori knowledge and apply them to the detection algorithm comprise a difficult problem. In order to utilize contextual information, conventional algorithms usually define structure and matching constraints, but this manual approach is too subjective and not extensible to different problems.

In this paper, we extract the scene context feature of the candidate window and fuse this feature with the candidate window feature to classify, while making the classifier automatically learn the constraint between the object and the context scene.

In Section 2, it is indicated that the feature extraction based on CNN can avoid the unmanageability and subjectivity of the manual design feature and can obtain a deeper feature representation of the object. Therefore, the feature of the object window and the context scene are both extracted from CNN. For the convenience of description, we named the two networks as objNet and sceNet. In the training process, we trained objNet and sceNet respectively. In the testing process, as shown in Figure 11, we extracted the feature of the candidate window, this being the context scene by the corresponding network, and then fused the two kinds of features to classify. For the feature fusion, taking into account the detection rate, we simply merged the two features and used the faster linear SVM classification.

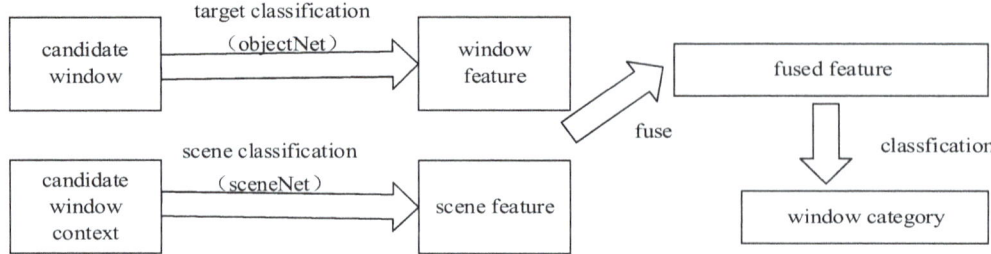

Figure 11. Window feature and context scene feature fusion.

For the scene context of the candidate region, as shown in Figure 12, we extended the candidate region from the center point and obtained a 256 × 256 pixel region as its context scene. When the object was at the edge of the image, we made the scene bounding box a minimum translation, so that it did not exceed the image area.

Figure 12. The definition of the context scene for the candidate region.

For the sceNet network, we used the classic AlexNet network structure, and modified the final output layer number. As shown in Figure 13, the network consisted of eight layers, of which the first five layers were the convolution layer, which can be regarded as multi-stage feature extraction. The latter three layers were fully-connected layers. The parameters of each layer are shown in Table 3. In the detection process, the scene bounding box of the candidate region was input to the network for feed forward calculation, and the output of the fifth layer was used as the scene feature of the candidate region.

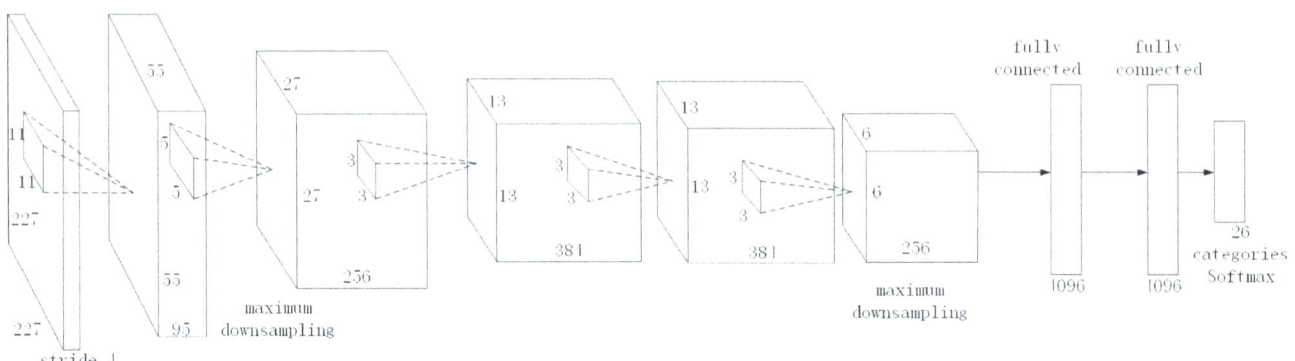

Figure 13. sceNet network structure.

Table 3. The parameters of each layer in the sceNet network.

		1st Layer	2nd Layer	3rd Layer	4th Layer	5th Layer
convolution layer	window size	11×11	5×5	3×3	3×3	3×3
	number of convolution kernels	96	256	384	384	256
	stride	4	1	1	1	1
	pad	0	2	1	1	1
pooling layer	window size	3×3	3×3	–	–	3×3
	stride	2	2	–	–	2

For the feature extraction of the candidate window, the structure of the objNet network is shown in Figure 14. Taking into account that the airplane object is very small and the sizes are more concentrated in 64 × 64 pixels, so the network input layer using 64 × 64, other sizes of candidate windows need to be scaled. Since the object to be detected in this paper is only an airplane, the output of the network is two classes: airplane or background. Compared with the classification of the remote sensing scene using AlexNet, the input size of the objNet network was smaller, and the outputs were fewer, so the network can be considered to need relatively simple feature representation when performing object discrimination. Therefore, when designing the objNet network, we modified the size of the convolution and pool layer windows and reduced the number of convolution cores and neurons in the fully-connected layer. The simplified network had fewer parameters and could reduce the risk of over-fitting properly. However, this does not mean that the network was as simple as possible. In practice, it is found that when the network is too simple, the network classification accuracy is high in the training phase, but it is not good when using the network for detection. The possible reason is that the oversimplified network is not strong enough to abstract the features, leading to poor generalization ability. As shown in Figure 14, the final objNet network consisted of eight layers, and the first five layers were the convolution layers, which can be seen as multi-stage feature extraction. The latter three layers were the fully-connected layers, which can be seen as a classifier. The parameters of each layer are shown in Table 4. In the detection process, the candidate region was input to the network for feed forward calculation, taking the output of the fifth layer of the pool as the feature of the candidate region.

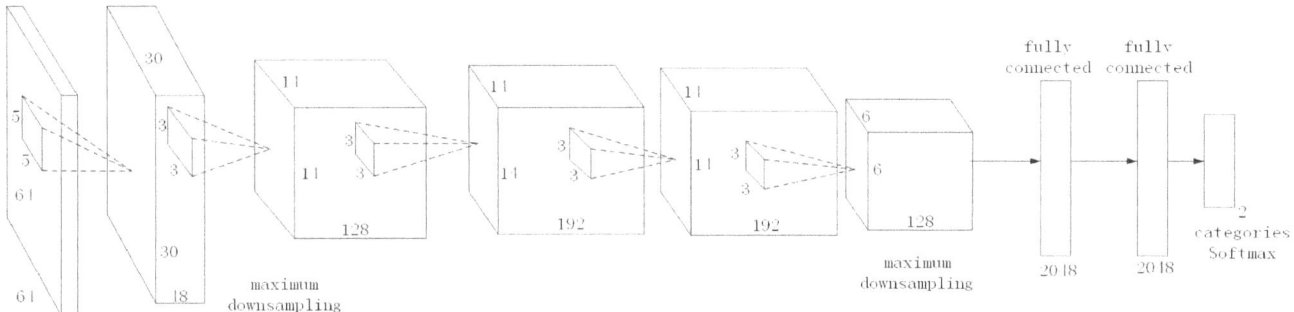

Figure 14. objNet network structure.

Table 4. The parameters of each layer in objNet network.

		1st Layer	2nd Layer	3rd Layer	4th Layer	5th Layer
convolution layer	window size	5×5	3×3	3×3	3×3	3×3
	number of convolution kernels	48	128	192	192	28
	stride	1	1	1	1	1
	pad	0	0	1	1	0
pooling layer	window size	2×2	2×2	–	–	2×2
	stride	2	2	–	–	2

5.4. Training of CNN and Transfer Learning

5.4.1. The Training Process of CNN

The training process of the CNN is shown in Figure 15. It was mainly carried out by iterative updating of the forward propagation and back propagation stages. Specifically, the first stage randomly selected a sample (X_b, Y_b) from the training set and input X_b to the network for feedforward calculation, then obtained the predicted output O_b. At this stage, the data were transformed step-by-step from the input layer through a series of hidden layer levels and finally transmitted to the output layer, which is essentially the process of input multiplication with the weight matrix for each layer, as in Formula (15):

$$O_b = F_n \left(\dots \left(F_2 \left(F_1 \left(X_b W^1 \right) W^2 \right) \dots \right) W^n \right) \tag{15}$$

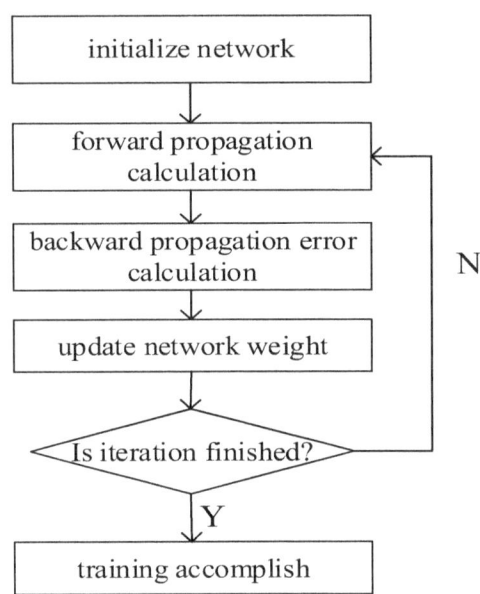

Figure 15. The training process of CNN.

In the second stage, namely backward propagation stages: first, we calculated the error between predicted output O_b and real label Y_b, and then, we continued to pass the error back to the front layer, and each layer updated its weight matrix by minimizing the error of small methods according to the current error situation. For the CNN, the weights usually are updated by the mini-batch gradient descent method, which is an optimization algorithm between the batch gradient descent and the stochastic gradient descent method. When the data volume is large, the mini-batch preserves the advantages of speed in the stochastic gradient descent method, while avoiding the problem of severe congestionin the stochastic gradient descent method. The mini-batch gradient descent method updates the parameters iteratively by randomly selecting small batches of data, as shown in Formula (16), where m represents the number of training samples per iteration in parallel, which is limited by the memory size. For an objNet network, the size of the input sample is small and the network structure is relatively simple, so the value was 1024, and for the sceNet network, it was 64 in the experimental environment used in this section.

$$\omega_k \rightarrow \omega_k' = \omega_k - \frac{\eta}{m} \sum_j \frac{\partial C_{X_j}}{\partial \omega_k}$$

$$b_l \rightarrow b_l' = b_l - \frac{\eta}{m} \sum_j \frac{\partial C_{X_j}}{\partial b_l} \tag{16}$$

where ω_k and b_l are weight parameters and bias parameters in the network, respectively; η is the learning rate; C_{X_j} is the cost loss of the sample X_j.

Before training the network, appropriate training data should be found. For the scene feature extraction network sceNet, we needed to use the remote sensing scene classification data to train. For the remote sensing scene classification, there are two standard datasets: UCMerced-LandUse [29] and WHU-RS [30]. To increase the training set, this paper combined the two datasets, and a detailed description of this dataset can be seen in Section 5.4.1. For the object feature extraction network objNet, we needed to use the sub-images containing the object as the positive sample and the sub-images without the object as the negative sample. In order to make the training process and detection process the same distribution, the network used the pBING algorithm's output on the training set as the training data, in which a candidate window having an ≥ 0.5 intersection-over-union (IoU) overlap with a ground-truth box was labeled as the positive sample and the rest as the negative sample.

5.4.2. Transfer Learning

Although CNN can automatically extract deep features, the network has many parameters to optimize and usually requires large-scale data (big data) in order to form a better network; if not, it easily over-fits. While the reality is the deficiency of labeled training datasets of remote sensing images, in order to fill the gap and make the model have stronger generalization ability, many methods have been proposed, such as data augmentation technology, dropout [31], and so on. However, these methods are still not enough for small remote sensing dataset. Considering the abundant regular image datasets available, we employed transfer learning technology for deep learning, which can break the" deep learning with big data" limitation.

For deep learning, an appropriate applicable transfer learning method is model transfer: firstly, pre-training network parameters through the source field data, then applying these parameters in the object domain, and finally, fine-tuning the network parameters to get better performance. If transfer learning is not employed, it demands initializing the network parameters and then starting to train the entire network using training data. An inappropriate initialization will make the network convergence slow and easily fall into the local minimum. In addition, because of the deepening of the network layer, the problem of gradient disappearance easy occurs: when the hidden layer near the output layer has been trained well, the parameter update near the input layer becomes slow or even stagnates. However, this does not mean that the network is optimal, because the first few layers of the network may not learn anything and may be just a random combination and numerical transformation of the input, but not really a dissociation of features, resulting in the entire network being a linear transformation of higher levels at work. Especially for high-dimensional data such as images, the network does not have good feature dissociation due to the degradation of the lower layer, so that the network is only performing local numerical learning on the input image and the model easily over-fits. Once the input image has changed, such as the direction or color of the airplanes, the network may not identify the object. This problem is mainly due to the fact that in the deep network, the learning rate of different layers is not the same, and the closer to the output layer, the faster the learning rate. This is because the gradient loss of the front layer is based on the product of the gradient loss, and when the number of layers is larger, the gradient loss becomes smaller and smaller.

The problem of gradient disappearance in deep learning networks is an essential problem brought by gradient descent, which is a big obstacle in deep learning. When the training data are large, the parameters can be initialized by a Gaussian distribution or other optimization methods, and the whole network can be fully trained by adjusting the learning rate and some regularization methods, as well as training for a long enough time. However, when the training data are small, the gradient disappearance and over-fitting problem will become more serious. In addition, since the problem in the deep learning network is mainly that the first few layers may not be fully trained, if the first few layers can be initialized by the parameters of other fully-trained networks, which puts the network in a better initial state, this would contribute to the optimization of the network and could accelerate the training process of the network. From another point of view, the first few layers of the deep learning network usually learn the edges of the image, the color, the texture, and other primitive features of the images, which for many visual tasks are typical. Therefore, the network parameters of lower layers can be shared among different image classification tasks. this is equivalent to transferring the feature extraction knowledge carried by these parameters to the object domain, and then, we only need to continue training the network through the training data of the object domain, that is to correct parameter deviation between the object domain and the source domain.

The following details the transfer learning scheme used in the two CNNs used in this paper.

For pre-training of the remote sensing scene classification network sceNet, we can use the scene recognition task in the regular image field as the source field. For the regular scene recognition task, Zhou et al. [32] established a large-scale regular scene dataset, Places, and published the trained network model. By practice and theory, this paper has demonstrated that the deep features learned on the Place dataset are more effective compared with those learned on the ImageNet 2012 dataset. However, for the remote sensing scene classification task, it is necessary to further validate whether the transfer effect of this model is optimal. Figure 16 shows the flowchart of transfer learning on the sceNet network. Firstly, we used the regular scene classification task to pre-train sceNet and then transferred the learned parameters to the remote sensing scene classification. The last layer of parameters was not transferred, just random initialization, and then, we used the remote sensing scene classification data to continue to train the network; then, the network learned the parameters of the last layer through back propagation and corrected the transferred parameters of the first few layers.

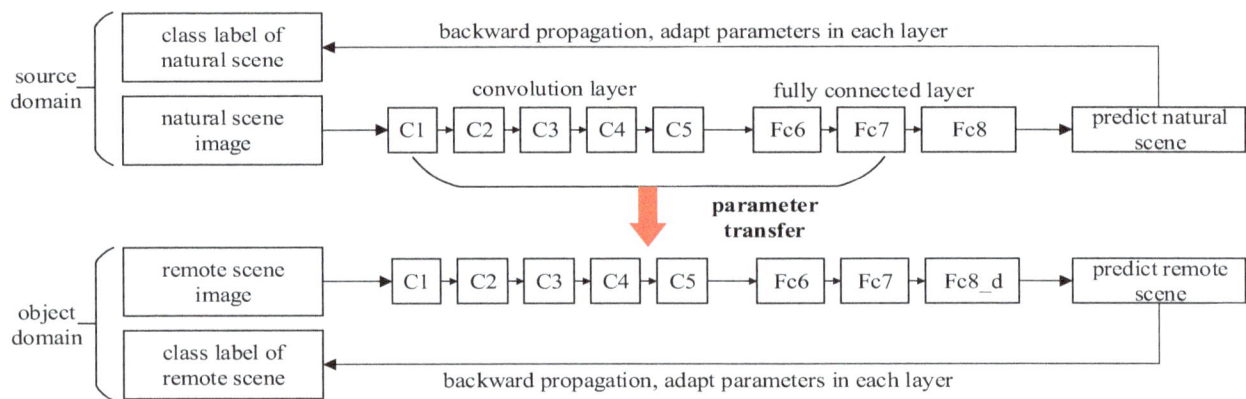

Figure 16. The transfer learning flowchart of the sceNet network.

5.4.3. Training the Classifier and Hard Negative Mining Method

For the classifier, we employed the simple linear SVM. For the training of the SVM classifier, we chose the real object sub-images as the positive sample and the sub-images with ≥ 0.3 IoU overlap with a ground-truth box as negatives. For each training sample, we extracted the object feature by the objNet network and the corresponding context feature by sceNet network, and then, we merged the two features as the input of the SVM classifier.

In general, when training a classifier, the more training samples, the better. However, there is a regular problem in the field of object detection, which is the extreme imbalance of positive and negative samples, that is the positive samples with objects are relatively small, while negative samples with background are very numerous. For remote sensing images, especially large-sized images, the problem is more obvious. Too many negative samples will lead to a very slow training process for classifying algorithms, and will even be detrimental to the performance of the final classifier.

For example, for SVM, many negative samples far away from the separating plane are almost useless for optimization. In addition, too many negative samples will make the algorithm's memory requirement too large. If a negative set with a similar number as the positive set is randomly selected, the algorithm cannot guarantee the best effect on the whole training set. If manually selecting the negative set, the cost is too large, and the subjectivity is too strong. Therefore, it is very important to search for a small representative negative set in the negative sample space. The usual strategy is to initialize a small hard negative set $C_t \in D$ randomly (D denotes the entire negative sample space), train an initial model β_t with all positive samples, and classify the negative sample set C_t. Remove the easy negative sample while searching for hard negative samples from D to add to the C_t, until the memory limit or a threshold L. The iterative updating model β_t and hard negative sample set C_t, until C_t no longer changes or the iteration number reaches a certain limit, stop training. In practice, the hard negative mining method converges very quickly; usually, only a single pass over all images is required.

5.5. Results and Analysis

5.5.1. Dataset

The two high-resolution remote sensing datasets used for object detection are described in Section 4.3.1. This section introduces the remote sensing scene classification datasets used in the training of sceNet networks and the two regular image datasets used for transfer learning. For remote sensing scene classification, there are two public datasets: UCMerced-LandUse [33] and Dataset-WHU-RS [33]. UCMerced-LandUse contains 21 different scene categories, each category containing 100 high-resolution remote sensing images of 256 × 256 pixels. The Dataset-WHU-RS dataset contains 19 scene categories, a total of 950 images of 600 × 600 pixels. In order to expand the training samples, this paper will simply divide one remote sensing image with 600 × 600 pixels into nine 256 × 256 pixel sub-images. In the end, the two remote sensing datasets are merged together, and the data of the same category are merged. In addition, fine-grained similar scenes such as sparse density residential, medium density residential area, and intensive residential area are merged. After merging, there were 26 scene classes, averaging about 320 images per class. In addition, in order to further increase the training data, simple horizontal, vertical flip operations were used.

Next, the regular image scene classification dataset Places used in the transfer learning of sceNet network is introduced briefly. This dataset Places is a large-scale natural scene dataset, containing 205 categories and a total of 2.5 million images.

The data used in the transfer learning of objNet network were mainly extracted from the ImageNet 2012 dataset, in which positive samples were all images in two categories of airplane and military airplane and some airplane images crawled from the Internet, a total of 12,300 images; for negative samples, we picked a category that may appear in the remote sensing images, such as ships, harbors, mountains, etc., and removed the other 980 categories, such as sharks, hens, caps, etc., which might be useless for object identification in remote sensing images. Figure 17 shows examples of the positive and negative samples of the dataset. For the convenience of the following description, this dataset is called NATURE-PLANE.

5.5.2. Environment and Evaluation

The experiment in this section was performed on Caffe. Caffe is widely used in the deep learning domain because of its advantages of being clear, simple, fast, and fully open source. The platform had two NVIDIA GeForce GTX 980 video cards, 16 GB memory, CPU i5-4460. For a detection result with IoU overlap with a real object coincidence degree no less than a threshold (usually set to 0.5), the detection result is considered correct; besides, if there are multiple detections, then only one is considered right, while the rest are false detections. In this paper, we used the precision and recall curve (PR curve) and the average precision (AP) to evaluate the detection performance synthetically. The evaluation method and code used the PASCAL VOC2007 standard. Accuracy and recall are defined as Formulas (17) and (18), respectively:

$$P = \frac{TP}{TP + FP} \tag{17}$$

$$R = \frac{TP}{TP + FN} \tag{18}$$

where TP is true positive, the number of true boxes, that is the number of objects correctly detected; FP is false positive, the number of false positives, that is the number of false detection results; FN is false negative, the number of false negative cases, that is the true number of objects that were missed.

Figure 17. Examples of the NATURE-PLANE dataset used in the transfer learning of the objNet network.

The average detection accuracy of AP can be measured by a single value, which is a representative comprehensive evaluation of the index, called the area under the PR curve, as shown in Formula (19).

$$AP = \int_0^1 P(R)\, dR \tag{19}$$

$$IoU = \frac{DetectionResult \cap GroundTruth}{DetectionResult \cup GroundTruth} \tag{20}$$

5.5.3. Result Analysis

In order to prove the validity of the transfer learning, this section firstly gives the classification accuracy of the object and scene feature extraction network in the training process and then gives the influence of the feature extraction on the detection effect before and after the transfer learning. Furthermore, the effectiveness of scene feature fusion is illustrated by contrasting the detection performance before and after the context scene feature fusion. Finally, we compare the other algorithms to prove the validity of the proposed remote sensing object detection algorithm based on deep learning with scene feature fusion.

Because the sceNet network uses the classic AlexNet network structure, there are many trained parameter models based on the network, which can be used for transfer learning. During the training process, the parameter models of AlexNet, CaffeNet, Places205-AlexNet, and Hybrid-AlexNet were used for transfer learning in this paper. AlexNet and CaffeNet were trained on the ImageNet2012 dataset, and CaffeNet has a very similar architecture to AlexNet, except for two small modifications: training without data augmentation and exchanging the order of pooling and normalization layers. Places205-AlexNet is a parameter model trained on the Places dataset. Hybrid-AlexNet was trained on the dataset combining the Place dataset with the ImageNet2012 dataset for a total of 3.6 million images in 1183 categories. Table 5 shows the classification accuracy of the remote sensing scene using different transfer learning models, where "AlexNet-RSI" indicates the learned network model without transfer learning and "xx-TL" denotes the network transferred from different models.

Table 5. The classification accuracy of the network transferred from different models.

	AlexNet-RSI	AlexNet-TL	CaffeNet-TL	Places205 -AlexNet-TL	Hybrid -AlexNet-TL
accuracy	89.76%	94.04%	93.87%	92.68%	**94.81**%

Table 5 reveals that the accuracy of the network trained with transfer learning was much higher than that of a network trained directly using remote sensing scene data. After transfer learning, the accuracy of the Hybrid-AlexNet network trained on the Place and ImageNet 2012 datasets was the highest, so we used the Hybrid-AlexNet-TL model to extract the feature of the object context scene. In addition, by visualizing the convolution kernel parameters of the first convolution layer, as shown in Figure 18, this shows that the convolution kernel of the network with transfer learning learned more edge features, and without transfer learning, the first layer of the network simply learned some simple fuzzy color information.

Figure 18. Visualization of the first-level convolutions in different network models.

For the training of object classification network objNet, because the network structure was designed in this paper, there was no trained model for transfer learning, so it was necessary to pre-train the transferable model parameters. Therefore, we used the NATURE-PLANE dataset introduced in Section 5.4.1 to pre-train objNet. However, in practice, it appears that if we pre-train the objNet directly using ImageNet2012's complete data and resume training using the NATURE-PLANE dataset, a better classification result could be obtained. Table 6 lists the classification accuracy of different pre-trained objNet networks, where objNet-RSI denotes the network model obtained without using transfer learning.

Table 6. The accuracy of objNet based on different transfer learning methods.

	Accuracy	
	SMALL-FIELD-RSIs	LARGE-FIELD-RSIs
objNet-RSI	93.51%	94.24%
objNet-TL1	96.63%	95.52%
objNet-TL2	97.23%	96.86%

After the training of the objNet network and sceNet network was complete, we used the two networks for feature extraction and classification detection. We first compared the detection performance before and after using transfer learning in objNet when there was no scene feature fusion. Then, to fuse the scene features, in the fusion, we also compared the effect of sceNet before and after transfer. That is, there was in total four groups of experiments, and the four groups of experiments had a progressive relationship, in which the fourth set of experiments was our proposed algorithm. In order to simplify the following description, we named each experiment and the configuration of each set of experiments as listed in Table 7.

Table 7. List pf the configurations of each experiment.

Algorithm	Scene Feature Fusion	sceNet Transfer Learning	objNet Transfer Learning
objNet	×	–	×
objNet-TL	×	–	√
objNet-TL-sceNet	√	×	√
objNet-TL-sceNet-TL (ours)	√	√	√

Figures 19 and 20 show the comparison of the results of the four experiments on two different dataset sizes. From the results of the experiments objNet and objNet-TL, it was revealed that the transfer learning of object the feature extraction network could improve the whole detection performance significantly. Before transfer learning, the accuracy of the curve decreased rapidly with the recall increasing, and the curve decreased slowly after transfer learning, which shows that the extracted feature of network after transfer learning made the classifier discriminate better, which means the transfer learning was effective. It can be concluded from the experiment objNet-TL-sceNet-TL that the detection efficiency on the two remote sensing datasets was better than that for the experiment without context scene feature fusion, indicating the effectiveness of the scene feature fusion. However, compared with objNet-TL-sceNet and objNet-TL, it is shown that when transfer learning was not employed, the improvement was not obvious, the possible reason for which being that the sceNet network has too many parameters for the limited remote sensing scene data, and if we directly trained the network using limited data without transfer learning, it would easily to over-fit, so that the extracted features would not be representative.

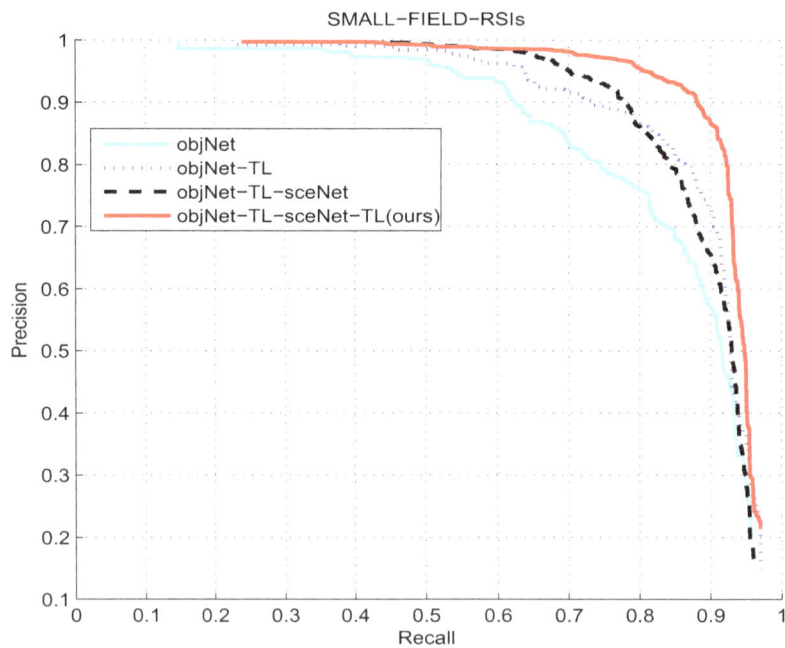

Figure 19. The performance comparison of four experiments on the SMALL-FIELD-RSIs dataset.

Finally, we compared the proposed algorithm objNet-TL-sceNet-TL with the other algorithms. Firstly, in order to prove the effectiveness of the deep features, we used the HOG algorithm [34] to extract the feature of the candidate region and used the SVM algorithm and the hard negative mining method to train the detector; we called it HOG-SVM. In addition, this paper compared the R-CNN algorithm [14]. This algorithm achieved a breakthrough in the PASCAL VOC2007 object detection task. It first uses the selective search algorithm to generate about 2000 candidate regions for each image and then uses the AlexNet network to extract features, and finally classifies each region using linear SVM classification. From the analysis of Section 4.3.2, we can see that the selective search algorithm is not suitable for large-sized remote sensing images. Therefore, this paper replaced the candidate region proposal with the pBING algorithm proposed in this paper. Furthermore, we compared the detection performance of the two methods of R-CNN: directly obtaining classification results by the AlexNet network and extracting features by AlexNet, then classifying by SVM. For convenience, the two algorithms are called pBING-AlexNet and pBING-AlexNet-SVM, respectively. Figures 21 and 22 show the comparison of the detection performance of each algorithm on two different sizes of remote sensing images.

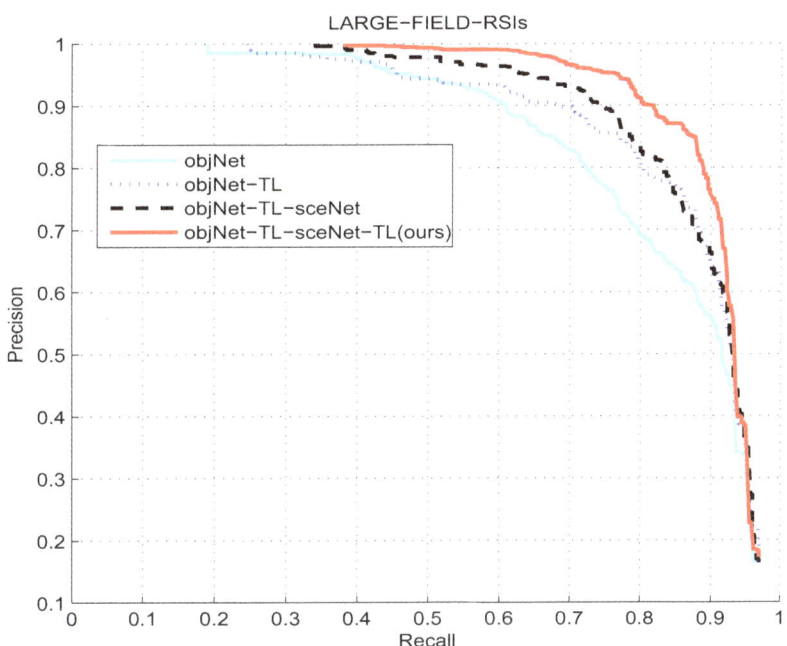

Figure 20. The performance comparison of four experiments on the LARGE-FIELD-RSIs dataset.

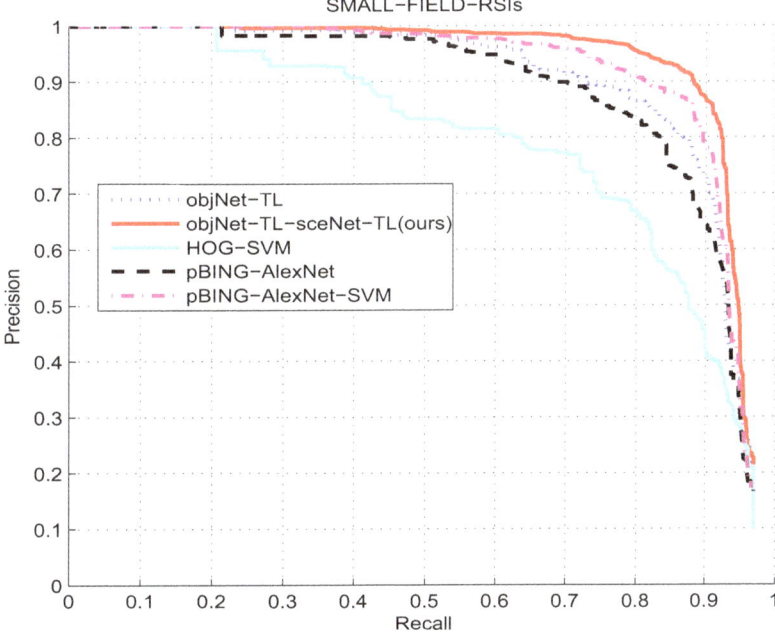

Figure 21. The performance comparison of different algorithms on the SMALL-FIELD-RSIs dataset.

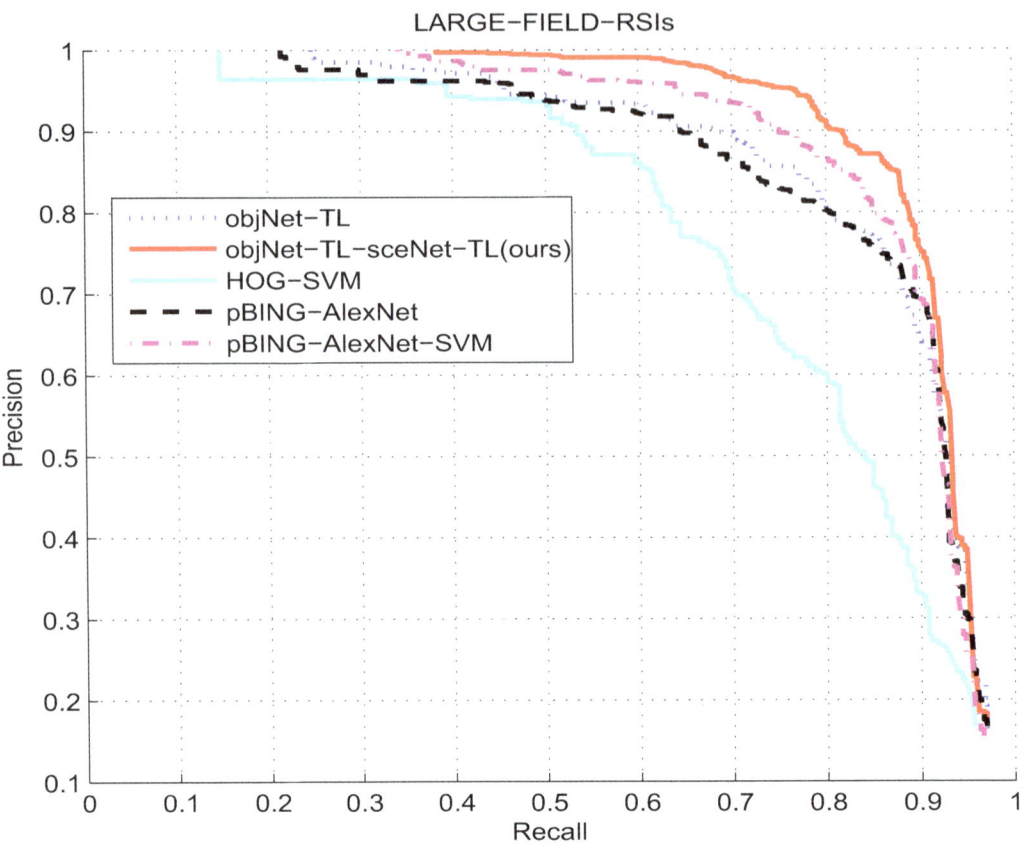

Figure 22. The performance comparison of different algorithms on the LARGE-FIELD-RSIs dataset.

Table 8. The average detection accuracy (average precision (AP)).

Algorithm	SMALL-FIELD-RSIs	LARGE-FIELD-RSIs
objNet	83.38%	82.30%
objNet-TL	87.59%	85.19%
objNet-TL-sceNet	87.67%	86.86%
objNet-TL-sceNet-TL (ours)	**90.77**%	**89.12**%
HOG-SVM	77.52%	76.69%
pBING-AlexNet	85.63%	84.54%
pBING-AlexNet-SVM	88.93%	86.91%

Figures 23 and 24 show the results of the objNet-TL-sceNet-TL algorithm on the SMALL-FIELD-RSIs and LARGE-FIELD-RSIs datasets, respectively. In each image, the red rectangles indicate the real objects marked by the dataset and the blue rectangles the final detection result of our algorithm. It appears that the vast majority of airplane objects can be correctly detected. It is noted that our algorithm can detect one airplane, which was not marked (missed by a human) on the dataset, and it is shown by the blue arrow in Figure 23. The comparison details of average precision (AP) can be checked in Table 8.

However, we found that if the airplane object was ambiguous and small, it may be missed. One missed airplane is pointed out by the red arrow in Figure 24. The reason is that our candidate region proposal algorithm was not strong enough for objects that are too small and ambiguous, and it needs to be further improved in the future.

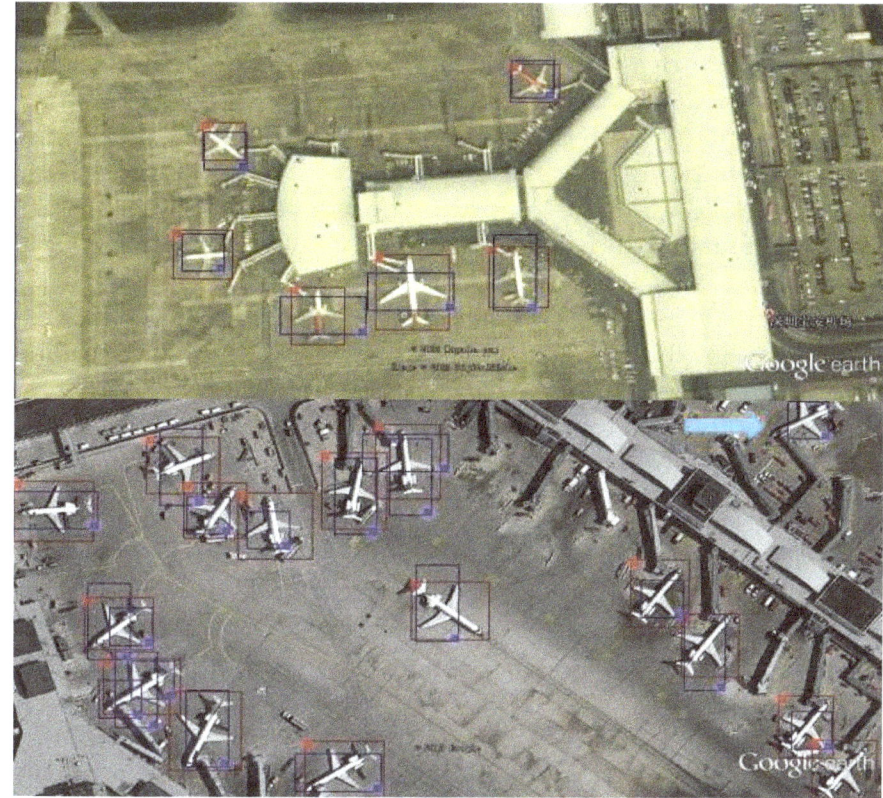

Figure 23. Examples of detection results on the SMALL-FIELD-RSI dataset.

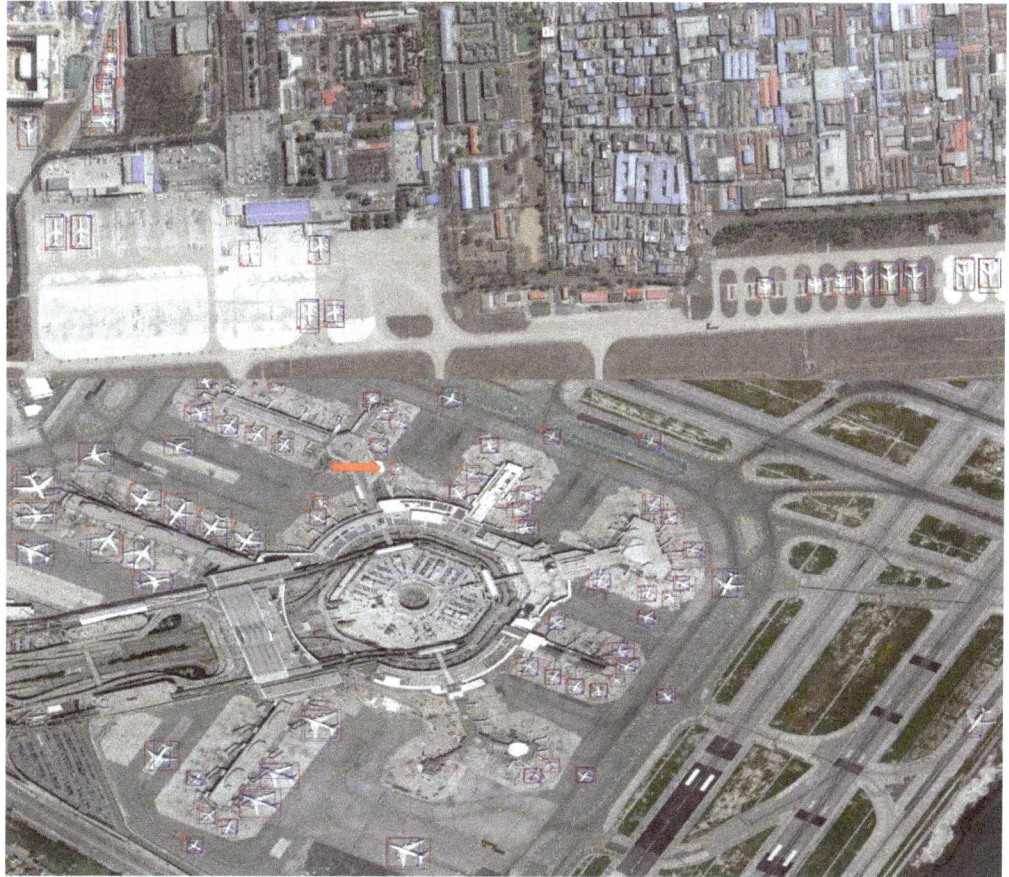

Figure 24. Examples of detection results on the LARGE-FIELD-RSIs dataset.

6. Conclusions

In this paper, we present a deep fusion feature approach to detect objects in high-resolution remote sensing images. Our method is composed of three main steps, which are the candidate region generation, deep feature extraction with fine-tuning, and the SVM classification with deep features. Hence, we re-structured the paper in terms of the three steps. For candidate region generation, we improved the binarized normed gradients algorithm and developed the pBING method. For deep feature extraction, the object feature and scene feature were both extracted for each candidate region, by utilizing the AlexNet model. As the label data in remote sensing are very scarce, we utilized the fine-tuning notion and pre-trained AlexNet on the ImageNet database, then fine-tuned the model with labeled remote sensing data. Finally, the object feature and scene feature were utilized to train an SVM for classification. After introducing the three main steps, we reported the experimental results, which validated the effectiveness of the developed pBING method, the fine-tuning strategy, and the overall detection model.

Author Contributions: E.K.W. is the main writer of the paper. Y.L. responds to the algorithms. Z.N. responds to design the pBing algorithm, J.Y., Z.L. and X.Z. respond to make experiments. S.M.Y. responds to the English Check.

References

1. Weber, J.; Lefevre, S. A multivariate hit-or-miss transform for conjoint spatial and spectral template matching. In Proceedings of the International Conference on Image and Signal Processing, Cherbourg-Octeville, France, 1–3 July 2008; pp. 226–235.
2. Sirmacek, B.; Unsalan, C. Urban-area and building detection using SIFT keypoints and graph theory. *IEEE Trans. Geosci. Remote Sens.* **2009**, *47*, 1156–1167. [CrossRef]
3. Yokoya, N.; Iwasaki, A. Object localization based on sparse representation for remote sensing imagery. In Proceedings of the International Geoscience and Remote Sensing Symposium, Quebec City, QC, Canada, 13–18 July 2014; pp. 2293–2296.
4. Tao, C.; Tan, Y.; Cai, H.; Tian, J. Airport detection from large ikonos images using clustered sift keypoints and region information. *IEEE Geosci. Remote Sens. Lett.* **2011**, *8*, 128–132. [CrossRef]
5. Sun, H.; Sun, X.; Wang, H.; Li, Y.; Li, H. Automatic target detection in high-resolution remote sensing images using spatial sparse coding bag-of-words mode. *IEEE Geosci. Remote Sens. Lett.* **2012**, *9*, 109–113. [CrossRef]
6. Gan, L.; Liu, P.; Wang, L. Rotation sliding window of the hog feature in remote sensing images for ship detection. In Proceedings of the 8th International Symposium on Computational Intelligence and Design, Hangzhou, China, 12–13 December 2015; pp. 401–404.
7. Arastounia, M.; Oude Elberink, S. Application of template matching for improving classification of urban railroad point clouds. *Sensors* **2016**, *16*, 2112. [CrossRef] [PubMed]
8. Aytekin, O.; Zongur, U.; Halici, U. Texture-based airport runway detection. *IEEE Geosci. Remote Sens. Lett.* **2013**, *10*, 471–475. [CrossRef]
9. Felzenszwalb, P.F.; Girshick, R.B.; McAllester, D.; Ramanan, D. Object detection with discriminatively trained part-based models. *IEEE Trans. Pattern Anal. Mach. Intell.* **2010**, *32*, 1627–1645. [CrossRef]
10. Han, J.; Zhang, D.; Cheng, G.; Guo, L.; Ren, J. Object detection in optical remote sensing images based on weakly supervised learning and high-level feature learning. *IEEE Trans. Geosci. Remote Sens.* **2015**, *53*, 3325–3337. [CrossRef]
11. Mottaghi, R.; Chen, X.; Liu, X.; Cho, N.-G.; Lee, S.-W.; Fidler, S.; Urtasun, R.; Yullie, A. The role of context for object detection and semantic segmentation in the wild. In Proceedings of the IEEE Conference on Computer Vision and Pattern Recognition, Columbus, OH, USA, 23–28 June 2014; pp. 891–898.
12. Ren, X.; Ramanan, D. Histograms of sparse codes for object detection. In Proceedings of the IEEE Conference on Computer Vision and Pattern Recognition, Portland, OR, USA, 23–28 June 2013; pp. 3246–3253.
13. Krizhevsky, A.; Sutskever, I.; Hinton, G.E. Imagenet Classification with Deep CNNs. In Proceedings of the Advances in Neural Information Processing Systems, Lake Tahoe, NV, USA, 3–6 December 2012; pp. 1097–1105.
14. Uijlings, J.; Sande, K.E.; Gevers, T.; Smeulders, A.W.M. Selective search for object recognition. *Int. J. Comput. Vis.* **2013**, *104*, 154–171. [CrossRef]

15. Cheng, M.; Zhang, Z.; Lin, W.-Y.; Torr, P. BING: Binarized Normed Gradients for Objectness Estimation at 300fps. In Proceedings of the Computer Vision and Pattern Recognition, Columbus, OH, USA, 24–27 June 2014; pp. 3286–3293.

16. Zitnick, C.L.; Dollar, P. Edge boxes: Locating object proposals from edges. In Proceedings of the European Conference on Computer Vision, Zurich, Switzerland, 6–12 September 2014; pp. 391–405.

17. Humayun, A.; Li, F.; Rehg, J.M. RIGOR: Reusing inference in graph cuts for generating object regions. In Proceedings of the Computer Vision and Pattern Recognition, Columbus, OH, USA, 24–27 June 2014; pp. 336–343.

18. Girshick, R.; Donahue, J.; Darrell, T.; Malik, J. Rich feature hierarchies for accurate object detection and semantic segmentation. In Proceedings of the Computer Vision and Pattern Recognition, Columbus, OH, USA, 24–27 June 2014; pp. 580–587.

19. Simonyan, K.; Zisserman, A. Very deep convolutional networks for large-scale image recognition. *arXiv* **2014**, arXiv:1409.1556.

20. Oquab, M.; Bottou, L.; Laptev, I.; Sivic, J. Learning and transferring mid-level image representations using convolutional neural networks. In Proceedings of the IEEE Computer Vision and Pattern Recognition, Columbus, OH, USA, 24–27 June 2014; pp. 1717–1724.

21. Girshick, R.; Donahue, J.; Darrell, T.; Malik, J. Region-based convolutional networks for accurate object detection and segmentation. *IEEE Trans. Pattern Anal. Mach. Intell.* **2016**, *38*, 142–158. [CrossRef] [PubMed]

22. Kong, T.; Yao, A.; Chen, Y.; Sun, F. HyperNet: Towards Accurate Region Proposal Generation and Joint Object Detection. Accurate region proposal generation and joint object detection. In Proceedings of the Computer Vision and Pattern Recognition, Las Vegas, NV, USA, 26 June–1 July 2016; pp. 845–853.

23. Ouyang, W.; Wang, X.; Zeng, X.; Qiu, S.; Luo, P.; Tian, Y.; Li, H.; Yang, S.; Wang, Z.; Loy, C.-C.; et al. DeepID-Net: Deformable deep convolutional neural networks for object detection. In Proceedings of the Computer Vision and Pattern Recognition, Boston, MA, USA, 7–12 June 2015; pp. 2403–2412.

24. Redmon, J.; Divvala, S.; Girshick, R.; Farhadi, A. You only look once: Unified, real-time object detection. In Proceedings of the Computer Vision and Pattern Recognition, Boston, MA, USA, 7–12 June 2015; pp. 779–788.

25. Hosang, J.; Benenson, R.; Dollar, P.; Schiele, B. What makes for effective detection proposals? *IEEE Trans. Pattern Anal. Mach. Intell.* **2016**, *38*, 814–830. [CrossRef] [PubMed]

26. Hare, S. Efficient online structured output learning for keypoint-based object tracking. In Proceedings of the IEEE Conference on Computer Vision and Pattern Recognition, Providence, RI, USA, 16–21 June 2012; pp. 1894–1901.

27. Zheng, S.; Sturgess, P.; Torr, P.H.S. Approximate structured output learning for constrained local models with application to real-time facial feature detection and tracking on low-power devices. In Proceedings of the IEEE International Conference and Workshops on Automatic Face and Gesture Recognition, Shanghai, China, 22–26 April 2013; pp. 1–8.

28. Kegl, B. The return of AdaBoost.MH: Multi-class hamming trees. *arXiv* **2013**, arXiv:1312.6086.

29. Wang, X.; Han, T.X.; Yan, S. An HOG-LBP human detector with partial occlusion handling. In Proceedings of the International Conference on Computer Vision, Lisboa, Portugal, 5–8 February 2009; pp. 32–39.

30. Srivastava, N.; Hinton, G.; Krizhevsky, A.; Sutskever, I.; Salakhutdinov, R. Dropout: A simple way to prevent neural networks from overfitting. *J. Mach. Learn. Res.* **2014**, *15*, 1929–1958.

31. Zhou, B.; Garcia, A.L.; Xiao, J.; Torralba, A.; Olivia, A. Learning deep features for scene recognition using places database. In Proceedings of the Advances in Neural Information Processing Systems, Montreal, QC, Canada, 7–12 December 2015; pp. 487–495.

32. Yang, Y.; Newsam, S. Bag-of-visual-words and spatial extensions for land-use classification. In Proceedings of the Advances in Geographic Information Systems, San Jose, CA, USA, 2–5 November 2010; pp. 270–279.

33. Xia, G.S.; Yang, W.; Delon, J.; Gousseau, Y.; Sun, H.; Maitre, H. Structural high-resolution satellite image indexing. In Proceedings of the ISPRS TC VII Symposium-100 Years ISPRS, Vienna, Austria, 5–7 July 2010; pp. 298–303.

34. Dalal, N.; Triggs, B. Histograms of oriented gradients for human detection. In Proceedings of the Computer Vision and Pattern Recognition, San Diego, CA, USA, 20–25 June 2005; pp. 886–893.

8

Spatial Data Reconstruction via ADMM and Spatial Spline Regression

Bang Liu [1,*], Borislav Mavrin [2], Linglong Kong [2] and Di Niu [1]

[1] Electrical and Computer Engineering, University of Alberta, 9211-116 Street NW, Edmonton, AB T6G 1H9, Canada; dniu@ualberta.ca

[2] Mathematical and Statistical Sciences, University of Alberta, 632 Central Academic Building, Edmonton, AB T6G 2G1, Canada; mavrin@ualberta.ca (B.M.); lkong@ualberta.ca (L.K.)

*   Correspondence: bang3@ualberta.ca

†   This paper is an extended version of our paper published in the 2017 IEEE International Conference on Data Mining (ICDM), New Orleans, LA, USA, 18–21 November 2017.

Abstract: Reconstructing fine-grained spatial densities from coarse-grained measurements, namely the aggregate observations recorded for each subregion in the spatial field of interest, is a critical problem in many real world applications. In this paper, we propose a novel Constrained Spatial Smoothing (CSS) approach for the problem of spatial data reconstruction. We observe that local continuity exists in many types of spatial data. Based on this observation, our approach performs sparse recovery via a finite element method, while in the meantime enforcing the aggregated observation constraints through an innovative use of the Alternating Direction Method of Multipliers (ADMM) algorithm framework. Furthermore, our approach is able to incorporate external information as a regression add-on to further enhance recovery performance. To evaluate our approach, we study the problem of reconstructing the spatial distribution of cellphone traffic volumes based on aggregate volumes recorded at sparsely scattered base stations. We perform extensive experiments based on a large dataset of Call Detail Records and a geographical and demographical attribute dataset from the city of Milan, and compare our approach with other methods such as Spatial Spline Regression. The evaluation results show that our approach significantly outperforms various baseline approaches. This proves that jointly modeling the underlying spatial continuity and the local features that characterize the heterogeneity of different locations can help improve the performance of spatial recovery.

Keywords: spatial sparse recovery; constrained spatial smoothing; spatial spline regression; alternating direction method of multipliers

---

1. Introduction

The problem of reconstructing fine-grained spatial data from its coarse-grained aggregate observations of each subregions lies in the core of many real world applications. For example, the reconstruction of fine-grained spatial distribution of cell phone activities is of particular interest to telecommunication and information technology companies, where the recovered data can be used for device installation, capacity planning, the study of urban ecology [1–3], population density estimation [4–6], and human mobility prediction [7–11]. However, the companies may only have access to the aggregate mobile traffic volumes on each base station, as either privacy issues or additional technical overhead is involved to get fine-grained spatial data of users. Similarly, it is also highly valuable if we can infer the spatial distribution of population (e.g., the population vote for a certain party) densities based on the total population recorded at polling stations that sparsely scattered at different subregions. Internet media providers or retailers, such as Google, Tencent, Amazon,

Facebook, etc., may want to recover a fine-grained geographical distribution of their users based on the aggregated user counts observed at different points of presence (PoPs) or data centers. Note that, in all the above-mentioned cases, it is impossible or not allowed to track the position of each individual due to either privacy concerns or technical overhead. Therefore, reconstructing the spatial data from coarse aggregation will be highly useful in such cases.

In this paper, we study such spatial sparse recovery problem, that is, to infer the fine-grained distribution of certain spatial data in a region given the aggregate observations recorded for each of its subregions. However, it is an extremely challenging problem and has seldom been studied. A straightforward idea is assuming the density is uniformly distributed within each subregion. Based on the based on the obtained aggregate observation, we can calculate a patched piece-wise constant estimation for each subregion. However, the densities estimated by this method will jump between neighboring subregions and disregard the local continuity or similarity of the studied spatial distribution across subregion boundaries. In addition, the piece-wise constant spatial field given by this approach provides little value for applications such as hot spot discovery. Many spatial data presents local continuity, e.g., Internet activity or cell phone activity. This is because the data often highly depend on underlying factors which are usually smoothly changing, like area functionality, urban geographical features, population density and so on. To exploit the smoothness, we may utilize spatial smoothing techniques such as Thin Plate Splines [12], Soap film smoothing [13], Spline smoothing [14], Bivariate Spline Regression [15], or Spatial Spline Regression [16] developed in statistics to smoothen the patched estimation. However, nearly all existing spatial smoothing techniques [12–16] are designed to recover a spatial field of densities according to sampled observations, e.g., reconstruct a spatial field of temperatures based on the temperature records at some sample points. In contrast, our problem needs to recovery a spatial field based on coarse-grained aggregate observations. Therefore, existing spatial smoothing techniques are not directly applicable to our new problem. Without modification, these smoothing techniques will violate the necessary constraint that the estimated spatial data in each subregion must sum up to its corresponding aggregate observation in the first place, leading to systematic errors.

To overcome the difficulties mentioned above, in this paper, we propose a new technique named Constrained Spatial Smoothing (CSS) for the problem of spatial data reconstruction. Specifically, given a region, we aim to reconstruct a spatial field of densities over that region based on observed aggregate values in patched subregions. Our approach penalizes the "roughness" of the reconstructed spatial field subject to the constraint that the aggregation of discretized values of the spatial field in each patched subregion equals the aggregate observation made in that subregion. It is distinct from previous spatial smoothing techniques due to the additional constraint in our problem. We propose an Alternating Direction Method of Multipliers (ADMM) [17,18] algorithm to decouple the problem into the alternated minimizations of a quadratic program (QP) [19] subproblem and a spatial smoothing subproblem, where we use the QP to iteratively enforce the observation constraints, while solving the spatial smoothing subproblem with a recently proposed finite element technique called Spatial Spline Regression (SSR) [16]. In addition, our approach not only leverages the intrinsic smoothness from local continuity to reconstruct a spatial field, but is also able to incorporate additional external information, such as the number of schools, number of bus stops, population, etc., in the underlying geographical region as a regression add-on component to further enhance recovery performance. Last but not least, our algorithm can be applied to a variety of sparse recovery problem where intrinsic smoothness exists.

Another important contribution of the paper is that we conduct extensive evaluation to compare our proposed algorithms with a variety of baseline methods. In our evaluation, we are trying to reconstruct the mobile phone activity distributions in Milan, Italy from base station observations. The Telecom Italia Big Data Challenge dataset is a multi-source dataset that contains a variety of informations, including aggregation of telecommunication activities, news, social networks, weather, and electricity data from the city of Milan. With the important information about human activities contained in the dataset, especially the cellphone activity records, researchers utilized the data to study different problems, such as modeling human mobility patterns [20–22], population density estimation [4,5], models the spread of diseases [23,24], modeling city structure [3] and city ecology [2], etc. Specifically, our evaluation is based on the Milan Call Detail Records (CDR) dataset, a part of the Telecom Italia Big Data Challenge dataset [25] which contains the phone call and Short Message Service (SMS) activity records of two months in each grid square of 235 m × 235 m in the city of Milan, Italy.

Given the Milan Call Detail Records (CDR) dataset, we consider a region that consists of 2726 grid squares in an irregularly bounded region in the city of Milan. To stress-test the algorithm performance, we assume we only know the aggregate phone activities observed on 100 or 200 base stations and aim to recover the entire spatial field of phone activities. We also use another geographical attribute dataset available from the Municipality of Milan's Open Data website [1] as the additional external attribute data to improve performance. Extensive evaluation shows that our proposed approach achieves significant improvement, compared to various state-of-the-art baseline methods, including the spatial spline regression (SSR) [16] approach. Our technique can recover the fine-grained cell phone activity distribution of 2726 data points only from 200 data points of base stations, with a mean absolute percentage error of 0.309, representing a 26.3% improvement from the SSR baseline scheme.

The remainder of this paper is organized as follows. In Section 2, we formulate the problem of spatial field reconstruction from coarse aggregate observations. In Section 3, we describe existing solutions, including a state-of-the-art Spatial Spline Regression (SSR) technique for spatial smoothing. In Section 4, we propose our Constrained Spatial Smoothing method which respects both the local continuity in the spatial field and the aggregation constraints at the same time. In Section 5, we conduct extensive evaluation in comparison with various other methods through a solid and extensive case study of cell phone activity density estimation in the city of Milan. We discuss related literature in Section 6 and conclude the paper in Section 7.

2. Problem Formulation

In this section, we formally introduce the problem of spatial field reconstruction from coarse aggregations observed at sparse scattered points in that field. Our problem can be formulated as a new type of sparse recovery problems. To ease the presentation, we may use cell phone activity recovery as an example.

Let $\Omega \subset \mathbb{R}^2$ denote an irregularly bounded domain, which is the entire region of interest in our problem. Usually, it excludes the uninhabited areas such as hills, ocean coasts, rivers, and so on. Suppose $f(\mathbf{p})$ is a real-valued function that represents certain spatial densities field (e.g., cell phone activities), where $\mathbf{p} = (x, y) \in \Omega$ denotes different geographical positions in Ω. Let $B = \{B_1, \ldots, B_m\}$ denote m observation points (e.g., base stations) that scattered in Ω. Each point B_i is located in a position $\mathbf{p}_{B_i} \in \Omega$ and in charge of a subregion Ω_{B_i}. In our problem, we are given the *aggregated volume* z_i in Ω_{B_i} that B_i is in charge of. Our goal is to reconstruct the spatial field $f(\mathbf{p})$ based on the observed aggregated volumes z_i.

To give an instance, consider the problem of recover cell phone activity distribution. In this case, each user will connect to a base station (cell tower) that is closest to his/her cell phone. Therefore, we can observe the aggregated volume for each base station

$$z_i = \int_{\Omega_{B_i}} f(\mathbf{p})\, \mathrm{d}\mathbf{p}, \quad i = 1, \ldots, m,$$

where Ω_{B_i} denotes the subregion that B_i is in charge of, and is given by

$$\Omega_{B_i} = \{\mathbf{p} \in \Omega : \|\mathbf{p} - \mathbf{p}_{B_i}\| < \|\mathbf{p} - \mathbf{p}_{B_{i'}}\|, \forall B_{i'} \in B, \ i' \neq i\}.$$

Given the aggregated activity volumes $z_1, \ldots, z_m$ recorded on m base stations, our goal is to reconstruct the entire cell phone activity densities distribution f, which is a spatial field in the domain Ω. We may call $z_1, \ldots, z_m$ base station volumes in this case. However, reconstructing a continuous spatial field is almost computationally infeasible as a personal computer can not handle the continuous nature of Ω_{B_i}.

In reality, we only need to recover f to a certain granularity required by the operator (e.g., 235 m $\times$ 235 m squares in the dataset provided by Telecom Italia Mobile). To fix notations, suppose Ω is discretized into n small grid squares $\mathbf{p}_1, \ldots, \mathbf{p}_n$, where $\mathbf{p}_j = (x_j, y_j) \in \Omega, j = 1, \ldots, n$ are the center positions of each square j in Ω. We can assume the area of each square is $\Delta = 1$ without loss of generality. In addition, the number of aggregate observations is much smaller than the total number of squares to be reconstructed, therefore we have $m \ll n$.

After domain discretization, we can get the aggregate volume on each base station B_i by

$$z_i = \sum_{\mathbf{p}_j \in \Omega_{B_i}} f(\mathbf{p}_j) \cdot \Delta, \quad i = 1, \ldots, m, \tag{1}$$

where the subregion that B_i represents is given by

$$\Omega_{B_i} = \{\mathbf{p}_j : 1 \le j \le n, \|\mathbf{p}_j - \mathbf{p}_{B_i}\| < \|\mathbf{p}_j - \mathbf{p}_{B_{i'}}\|, \forall i' \ne i\}. \tag{2}$$

Therefore, our goal is to reconstruct the underlying spatial field f, and especially the activity densities

$$\mathbf{f} := (f(\mathbf{p}_1), \ldots, f(\mathbf{p}_n))^\mathsf{T}$$

in all n grid squares if the desired granularity is on a per-square level, with only access to the aggregated observations z_i in Label (1).

The problem defined above is broadly applicable to characterize a variety of applications other than the recovery of cell phone activity density distribution, e.g., inferring a fine-grained geographical user distribution for a certain app or website based on aggregated user counts collected at sparsely distributed Presence of Points (PoPs) or data centers, and recovering the voter distribution for a certain party based on aggregate voting statistics at different polling stations. The nonessential difference is that the definition of subregion Ω_{B_i}, from which volume z_i is aggregated, is different for each specific application.

Constrained Spatial Smoothing Problem

Denote $\mathbf{z} = (z_1, \ldots, z_m)^\mathsf{T}$. Since all Ω_{B_i} are predetermined, e.g., from Label (2) for the problem of cell phone activity distribution recovery, and z_i are known, reconstructing spatial field $\mathbf{f}$ from (1) is essentially solving a linear system of equations for $\mathbf{f}$, i.e.,

$$\mathbf{z} = \mathbf{A}\mathbf{f},$$

where the matrix $\mathbf{A} \in \mathbb{R}^{m \times n}$ is given by

$$A_{ij} = \begin{cases} 1, & \text{if } \mathbf{p}_j \in \Omega_{B_i}, \\ 0, & \text{otherwise.} \end{cases} \tag{3}$$

Since $m \ll n$, i.e., the number of equations is far smaller than the number of the unknowns, reconstructing $f(\mathbf{p}_1), \ldots, f(\mathbf{p}_n)$ from $z_1, \ldots, z_m$ is essentially a sparse recovery problem.

Directly solving the linear system of Equation (1) is infeasible, as it is an underdetermined system which has an infinite number of solutions. However, the spatial property of f can be utilized as constraints to make the sparse recovery problem feasible and has a unique solution. We observe that spatial data usually exhibit local continuity or correlation within domain Ω. For example, in the problem of cell phone activity density recovery, the activity density of a certain location highly depends on the population and activity at that place, e.g., the downtown has more population and cell phone activity than suburban residential areas. In addition, the underlying area functionality and the spatial distributions of human activity density are often slowly changing over the domain Ω rather than suddenly jumping between different subregions.

Therefore, we can formulate our constrained spatial sparse recovery problem as the following:

$$
\begin{aligned}
\underset{f}{\text{minimize}} \quad & \int_{\Omega} (\nabla^2 f)^2 \, d\mathbf{p}, \\
\text{subject to} \quad & \mathbf{z} = \mathbf{A} f, \\
& \mathbf{f} \geq 0,
\end{aligned}
\tag{4}
$$

by taking into account the non-negative property and the local spatial continuity (smoothness) of f. $\nabla^2 f = \frac{\partial^2 f}{\partial x^2} + \frac{\partial^2 f}{\partial y^2}$ is the Laplacian of f, and is utilized to encourage local similarity and penalize the roughness of the spatial field f. It is worth noting that once f is reconstructed, we have not only recovered the densities $\mathbf{f}$ at the square centers $\mathbf{p}_1, \ldots, \mathbf{p}_n$, but can also recover the density $f(\mathbf{p})$ of any point $\mathbf{p} \in \Omega$, e.g., between the centers of two neighboring grid squares, although such a fine-grained recovery may not be needed in every application.

To further improve the recovery performance, we can utilize additional external demographic or social features at each location. In the problem of cell phone activity density reconstruction, cell phone activities are often correlated with the underlying population density and social functionalities (e.g., the percentage of green area, the number of schools, the number of businesses/restaurants, the number of sport facilities, and the number of bus stops, etc.) of the considered regions.

Specifically, suppose $\mathbf{w}_j = (w_{j1}, \ldots, w_{jq})^\mathsf{T}$ represents the feature vector consisting of q external feature values of square j. When $\mathbf{w}_j$ is available as additional input, we can estimate the spatial density data in square j by

$$
f(\mathbf{p}_j) = f'(\mathbf{p}_j) + \mathbf{w}_j^\mathsf{T} \boldsymbol{\beta},
\tag{5}
$$

where $f'(\mathbf{p})$ is an underlying spatial field functional that preserves local spatial continuity, while $\mathbf{w}_j^\mathsf{T} \boldsymbol{\beta}$ is a linear regression part based on the attributes of square $\mathbf{p}_j$ that allows position-specific variation or jumps.

In the presence of attributes, we can formulate the constrained spatial sparse recovery problem as

$$
\begin{aligned}
\underset{f', \boldsymbol{\beta}}{\text{minimize}} \quad & \int_{\Omega} (\nabla^2 f')^2 \, d\mathbf{p}, \\
\text{subject to} \quad & f(\mathbf{p}_j) = f'(\mathbf{p}_j) + \mathbf{w}_j^\mathsf{T} \boldsymbol{\beta}, \quad j = 1, \ldots, n, \\
& \mathbf{z} = \mathbf{A} f, \\
& \mathbf{f} \geq 0.
\end{aligned}
\tag{6}
$$

Once we get the spatial field f' and $\boldsymbol{\beta}$, we can reconstruct $f(\mathbf{p}_j)$ for all the squares using (5). For example, we can calculate the cell phone activity at a specific place by the summation of an underlying smooth spatial field $f'(\mathbf{p}_j)$ and a linear regression of location attributes, where the add-on regression helps to model the jump between two subregions if the two regions are quite different and have distinct functionalities or attributes.

3. Patched Estimation and Spatial Spline Regression

In this section, we present some tentative solutions and then show their limitations in solving our constrained spatial sparse recovery problem.

3.1. Patched Piece-Wise Constant Estimation

In our problem, we only have access to the aggregated volumes z_i at locations $\mathbf{p}_{B_i}$. To infer the fine-grained spatial distribution of z_i over subregion Ω_{B_i} that covers the point B_i, a first intuitive heuristic is estimating $f(\mathbf{p}_j)$ as the volume z_i divided by its area by assuming the density is distributed uniformly:

$$\bar{f}(\mathbf{p}_j) = \frac{z_i}{|\Omega_{B_i}|}, \text{ for each } \mathbf{p}_j \in \Omega_{B_i}, \tag{7}$$

where $|\Omega_{B_i}|$ is the area of Ω_{B_i}. This method gives us a patched piece-wise constant estimation. Note that we use *patch* to refer to Ω_{B_i} in this paper, which is the subregion covered B_i.

However, the patched estimation gives an oversimplified solution. The reconstructed spatial field $\bar{f}(\mathbf{p}_j)$ may have jumps on the borders of neighboring patches, which is far from smooth. In reality, the spatial field $f(\mathbf{p}_j)$ should change smoothly over the domain, as the underlying characteristics also change smoothly across different regions. Hence, $f(\mathbf{p}_j)$ should not be constant within each patch Ω_{B_i}.

3.2. Spatial Spline Regression

Given the above observation, we can naturally come up with a second idea, which is learning a smooth estimation of $\bar{f}(\mathbf{p}_j)$ by spatial smoothing techniques. In the following, we introduce the powerful smoothing technique named Spatial Spline Regression (SSR) proposed in Sangalli et al. [16]. We will show how it can be applied to our particular spatial data reconstruction problem, as well as point out its limitations in solving the problem.

Given l data points in Ω, which contains the following information: (1) their positions $\{\mathbf{p}_j\}_{j=1}^l$; (2) the values of these l points: $\{h_j\}_{j=1}^l$; and (3) their feature vectors $\{\mathbf{w}_j\}_{j=1}^l$, SSR is able to fit a smooth spatial field f by minimizing the following equation [14,16], i.e.,

$$\underset{\boldsymbol{\beta}, f}{\text{minimize}} \sum_{j=1}^l (h_j - \mathbf{w}_j^\mathsf{T}\boldsymbol{\beta} - f(\mathbf{p}_j))^2 + \lambda \int_\Omega (\nabla^2 f)^2 \, d\mathbf{p}, \tag{8}$$

where f is assumed to be twice-differentiable over Ω, and $\nabla^2 f = \frac{\partial^2 f}{\partial x^2} + \frac{\partial^2 f}{\partial y^2}$ denotes the *Laplacian* of f to penalize the roughness of f. The hyper parameter λ is used to trade the smoothness of f off for a better approximation to data value h_j.

However, the challenge to solving problem (8) is that it involves searching for a functional f over a possibly non-convex domain Ω that may have strong concavities, complicated boundaries, and even interior holes. Although kernel-based methods [26] are also a commonly used smoothing technique, their major drawback is that, by using uniformly damping weights in distance-based kernels, they tend to link data points across unrelated or weakly related subregions in an irregularly shaped non-convex domain.

We now briefly describe how spatial spline regression [16] can solve problem (8) via finite element analysis for any irregularly shaped domain Ω. SSR splits a domain Ω by transforming it into a triangular mesh with triangulation methods (e.g., Delaunay triangulation [27]). After triangulation, it defines a polynomial function on each triangle, such that the summation of these polynomial functions defined on different pieces closely approximates the desired spatial field f.

Specifically, let $\zeta_1, \ldots, \zeta_K$ denote the vertices of all the small triangles, which are called control points and can be adaptively selected by available data points. Define a piecewise linear or quadratic basis function $\psi_k(x, y)$ called *Lagrangian finite element* with $(x, y) \in \Omega$, associated with each control point ζ_k such that ψ_k evaluates to 1 at ζ_k and is equal to 0 at all other control points. Therefore, according to the *Lagrangian property of the basis*, we can approximate $f(x, y)$ for any $(x, y) \in \Omega$ only using the values of f on the K control points, i.e., $\mathbf{f}_K := (f(\zeta_1), \ldots, f(\zeta_K))^\mathsf{T}$. That is, if we let $\psi(x, y) := (\psi_1(x, y), \ldots, \psi_K(x, y))^\mathsf{T}$ denote the K predefined basis functions, each corresponding to a control point, then we have

$$f(x, y) = \sum_{k=1}^K f(\zeta_k)\psi_k(x, y) = \mathbf{f}_K^\mathsf{T}\psi(x, y). \tag{9}$$

Since $\psi_1(x,y),\ldots,\psi_K(x,y)$ are predefined and known a priori, the variational estimation of f in problem (8) boils down to the estimation of only K scalar values, i.e., $\mathbf{f}_K = (f(\zeta_1),\ldots,f(\zeta_K))^\top$.

In fact, it is shown in Sangalli et al. [16] that with the piece-wise approximation given by (9), solving (8) is simply solving a set of linear equations for $\hat{f}(\zeta_1),\ldots,\hat{f}(\zeta_K)$. The estimator $\hat{f}(x,y)$ for f can then be derived from (9) as

$$\hat{f}(x,y) = \hat{\mathbf{f}}_K^\top \psi(x,y).$$

It is worth noting that commodity triangulation software for finite element analysis is readily available in many free and commercial finite element packages. For example, Delaunay triangulations of a set of data location points (e.g., [27]) V are such that no point in V is inside the circumcircle of any triangle; they maximize the minimum angle of all the triangle angles, avoiding stretched triangles.

Now, we can see that if $l = n$ and we plug $h_j = \bar{f}(\mathbf{p}_j), j = 1,\ldots,n$ into problem (8), we will get a new density surface $\hat{f}$ as a solution to the SSR problem (8) that is a smoothened approximation of the patched estimates $\bar{f}(\mathbf{p}_j)$.

However, SSR given by (8) can not accommodate any constraints, which is the major limitation in solving our problem. Specially, in our case, SSR does not enforce the aggregated volume constraint (1) (or $\mathbf{z} = \mathbf{A}\mathbf{f}$ in (4)). Therefore, SSR gives no guarantee that the estimated densities in each patch Ω_{B_i} will sum up to the observed volume z_i on the point B_i. In this way, SSR would likely cause large estimation errors as it violates the constraint.

4. An ADMM Algorithm for Constrained Spatial Smoothing

Our spatial sparse recovery problem (4) is different from (8) from two aspects: the additional constraints and the loss function. As a consequence, we can not directly apply the previous SSR method to solve it. A new approach is needed to handle our new loss function with constraints.

In this section, we propose to utilize the Alternating Direction Method of Multipliers (ADMM) [28], to decompose our constrained optimization problem into two sub-problems that can be solved effectively by SSR and Quadratic Programming (QP), respectively. Algorithm 1 presents the proposed ADMM algorithm to learn our model parameters.

Algorithm 1: Constrained Spatial Smoothing by ADMM

Input: The m observed volume of base stations $\mathbf{z} = (z_1,\ldots,z_m)^\top$, smoothing parameter λ,
      penalty parameter β, initialize $\boldsymbol{\alpha} = \boldsymbol{\alpha}^0$, $\mathbf{f} = \mathbf{f}^0$.
Output: Spatial field and parameters $\hat{f}$, $\hat{\beta}$. Estimation values on n locations
      $\hat{f} = (\hat{f}(\mathbf{p}_1),\ldots,\hat{f}(\mathbf{p}_n))$.

1: **for** $iter = 1,\ldots,$ maxIter **do**
2:     Update f by solving (18) using Quadratic Programming.
3:     Update g by solving (19) using Spatial Spline Regression.
4:     Update α according to (17).
5: **end for**

First, we introduce the following indicator function $\mathbb{1}_{\mathbf{f}}$,

$$\mathbb{1}_{\mathbf{f}} = \begin{cases} 0, & \text{if } \mathbf{f} \geq 0 \text{ and } \mathbf{z} = \mathbf{A}\mathbf{f}, \\ \infty, & \text{otherwise.} \end{cases} \tag{10}$$

With the indicator function, the original problem (4) is equivalent to

$$\underset{f}{\text{minimize}} \quad \lambda \int_\Omega (\nabla^2 f)^2 \, \mathrm{d}\mathbf{p} + \mathbb{1}_{\mathbf{f}}, \tag{11}$$

where λ is a hyper parameter that controls the smoothness of f.

Second, we introduce an auxiliary variable $\mathbf{g}$ that is defined as

$$\mathbf{g} := (g(\mathbf{p}_1), \ldots, g(\mathbf{p}_n))^{\mathsf{T}}. \tag{12}$$

This variable is utilized to split the convex optimization problem into two sub-convex problems. With $\mathbf{g}$, we can formulate the problem as the standard ADMM format,

$$\begin{aligned} \underset{f}{\text{minimize}} \quad & \lambda \int_{\Omega} (\nabla^2 f)^2 \, d\mathbf{p} + \mathbb{1}_{\mathbf{g}}, \\ \text{subject to} \quad & \mathbf{f} = \mathbf{g}. \end{aligned} \tag{13}$$

The *augmented Lagrangian* for (13) is

$$\begin{aligned} \text{minimize} \quad \mathcal{L}_{\rho}(\mathbf{f}, \mathbf{g}, \boldsymbol{\alpha}) = & \lambda \int_{\Omega} (\nabla^2 f)^2 \, d\mathbf{p} + \mathbb{1}_{\mathbf{g}} \\ & + \boldsymbol{\alpha}^{\mathsf{T}} (\mathbf{g} - \mathbf{f}) + \frac{\rho}{2} \|\mathbf{g} - \mathbf{f}\|_2^2, \end{aligned} \tag{14}$$

where $\boldsymbol{\alpha} = (\alpha_1, \ldots, \alpha_n)^{\mathsf{T}}$ is the dual variable, and $\rho > 0$ is the penalty parameter in ADMM. Then, the ADMM consists of the following iterations:

$$\mathbf{g}^{k+1} := \underset{\mathbf{g}}{\text{argmin}} \, \mathcal{L}_{\rho}(\mathbf{f}^k, \mathbf{g}, \boldsymbol{\alpha}^k), \tag{15}$$

$$\mathbf{f}^{k+1} := \underset{\mathbf{f}}{\text{argmin}} \, \mathcal{L}_{\rho}(\mathbf{f}, \mathbf{g}^{k+1}, \boldsymbol{\alpha}^k), \tag{16}$$

$$\boldsymbol{\alpha}^{k+1} := \boldsymbol{\alpha}^k + \rho(\mathbf{f} - \mathbf{g}). \tag{17}$$

For the $\mathbf{g}$-update step in each iteration, Label (16) is equivalent to

$$\begin{aligned} \underset{\mathbf{g}}{\text{minimize}} \quad & \frac{\rho}{2} \|\mathbf{g}\|_2^2 + (\boldsymbol{\alpha}^{\mathsf{T}} - \rho \mathbf{f}^{\mathsf{T}}) \mathbf{g}, \\ \text{subject to} \quad & \mathbf{g} \geq 0, \\ & \mathbf{z} = \mathbf{A}\mathbf{g}. \end{aligned} \tag{18}$$

We can solve this convex problem efficiently by Quadratic Programming (QP). For the $\mathbf{f}$-update step in each iteration, Equation (15) is equivalent to

$$\underset{f}{\text{minimize}} \quad \left\| \left(\boldsymbol{\alpha}^{\mathsf{T}} + \rho \mathbf{g}^{\mathsf{T}} \right) / 2 - \mathbf{f} \right\|_2^2 + \lambda \int_{\Omega} (\nabla^2 f)^2 \, d\mathbf{p}, \tag{19}$$

which is exactly the form of (8) with $h_j = (\alpha_j + \rho g(\mathbf{p}_j)) / 2$ and $\mathbf{w}_j = 0$, thus can be solved efficiently by SSR. It should be noted that λ is the penalty parameter which controls the smoothness of f. If it is small, we put little emphasis on the smoothness, and the estimated surface f will be over fitted. If it is too big, the surface will be too smooth, which can cause underfitting.

For the case with attributes, the algorithm does not require major changes. We just need to replace $\mathbf{f}$ by $\mathbf{f} + \mathbf{W}\boldsymbol{\beta}$ in (19), where $\mathbf{W} := (\mathbf{w}_1, \ldots, \mathbf{w}_n)^{\mathsf{T}}$ represents the attributes and $\boldsymbol{\beta}$ is the corresponding contributions.

Our proposed ADMM training algorithm is able to efficiently reconstruct the spatial field and fit the covariates for our constrained spatial sparse recovery problem. In $\mathbf{g}$-update step, it enforces the constraints by solving a constrained QP with no need to worry about smoothing; in a $\mathbf{f}$-update step, it approximates the obtained $\mathbf{g}$ with a smooth f using the SSR-based smoothing technique. In this way, we decouple the handling of smoothing and constraints which was not possible in pure SSR previously.

5. Performance Evaluation

In this section, we perform an extensive case study of the approach we described above in order to demonstrate its applicability. We picked the cell phone data as an example of how the model can solve empirical problem and compare the model's performance to other approaches.

5.1. Dataset Description

The model in (13) is general and not attached to any particular empirical problem, and it does not contain many implicit assumptions. However, in order to measure its performance, we evaluate the model using real-world data. Due to generality of the proposed learning algorithm, the range of possible data sets is potentially big. For our empirical case study, we chose cell phone data, where there exists a problem of recovering a spatial field from coarse aggregations observed at sparse cell phone towers. We do not overestimate the problem, but rather see this particular data set suitable for an extensive case study.

The Milan Call Description Records (CDR) dataset is a part of the Telecom Italia Big Data Challenge dataset provided by Telecom Italia Mobile. It contains the telecommunications activity records from 1 November 2013 to 31 December 2013 in the city of Milan [25]. The dataset divides Milan into a 100×100 square grid, where each square is size of about 235 m $\times$ 235 m. In the dataset, each record consists of six entries: square ID, incoming call activity, outgoing call activity, incoming SMS activity, outgoing SMS activity, and time-stamp of 10-minute time slot. The values of the four types of activities are normalized to the same scale.

Another dataset we utilized is the Milan geographical attribute dataset available from the Municipality of Milan's Open Data website [1]. This dataset consist of features of central 2726 squares among the whole 10,000 squares. The features of each square include: population, green area percentage, number of sport centers, number of universities, number of businesses, and number of bus stops. Figure 1 shows the area covered by these grid squares. The 2726 squares covers the central part of the Milan city and contains the majority of telecommunication activities in the dataset. We refer to [2] for more detailed description about this dataset. In our experiments, we compare the performance of different approaches on these squares.

The general problem of recovering a spatial field from coarse aggregations observed at sparse points in the field in this particular case study is reformulated into the problem of recovering the distribution of cell phone activities over the whole 2726 square regions given that only aggregated activity observations in base stations are known. We need to further process the Milan CDR dataset to study this problem.

First, we sum up the four types of activities during 1 November 2013 to 28 November 2013 and 1 December 2013 to 28 December 2013, respectively, to come up with the activity volume of each squares during November or December. These two datasets are served as the ground-truth datasets of Milan cell phone activity distributions. Figure 1a,b show the heat maps of activity volumes in each square during November and December.

Second, after we aggregated the two months' activities for each square, we need to set the locations of base stations (BSs). According to [29], there are roughly 200 base stations in Milan. However, the exact locations are not available. Thus, we assume the 200 BSs are randomly distributed according to the following probability distribution

$$\Pr(\text{Set square } i \text{ as BS}) = f(\mathbf{p}_i) / \sum_{j=1}^{N} f(\mathbf{p}_j), \tag{20}$$

where $f(\mathbf{p}_i)$ is the cell phone activity volume in square i, $i = \{1, \ldots, N\}$, $N = 2726$ is the number of squares we are focusing on. Notice that, when we have 200 base station's aggregated observations, they only cover 7.34% of the whole 2726 squares region. This is extremely sparse and makes our problem highly challenging. In addition, we also assume $n_{BS} = 100$ and choose 100 squares as BSs according to the same probability distribution to stress-test our algorithm's capability under even sparser observations. Figure 2a,b show the base station distributions for $n_{BS} = 200$ and $n_{BS} = 100$, respectively.

After sampling the location of base stations, for each square, we assign the activity of it to its closest base station. When multiple base stations are equidistant from a square, the activity of the square will be evenly distributed among these base stations. We then assume we only know the aggregated activities in base station squares, which is usually the true case in reality. Figure 2c shows the regions split by 100 base stations, where each colour patch is a region charged by one base station. To save space, we don't present the figure for 200 base stations.

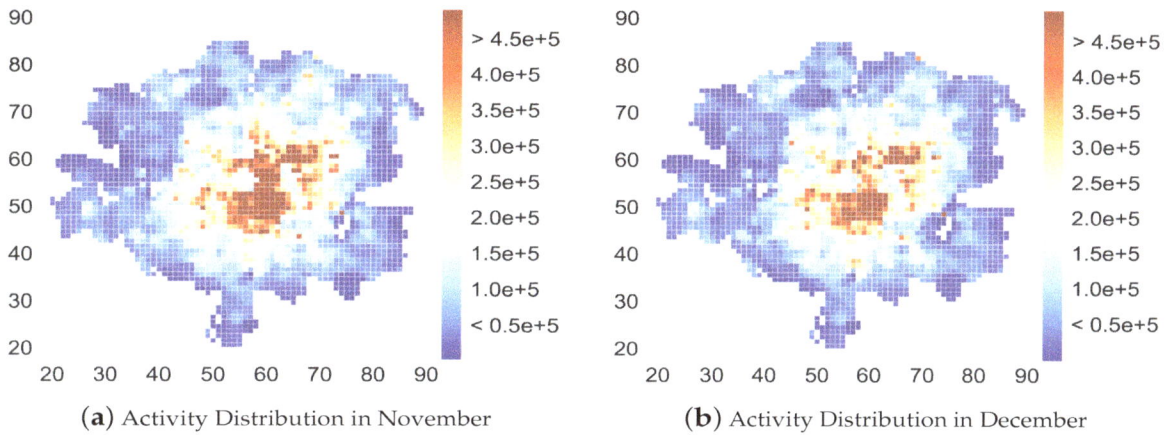

(a) Activity Distribution in November   (b) Activity Distribution in December

Figure 1. The cell phone activity distributions of Milan. It shows the metropolitan area of Milan, Italy, and the area covered by the 2726 grid squares. (**a**,**b**) show the heat map of cell phone activities (Call + SMS) during November and December respectively.

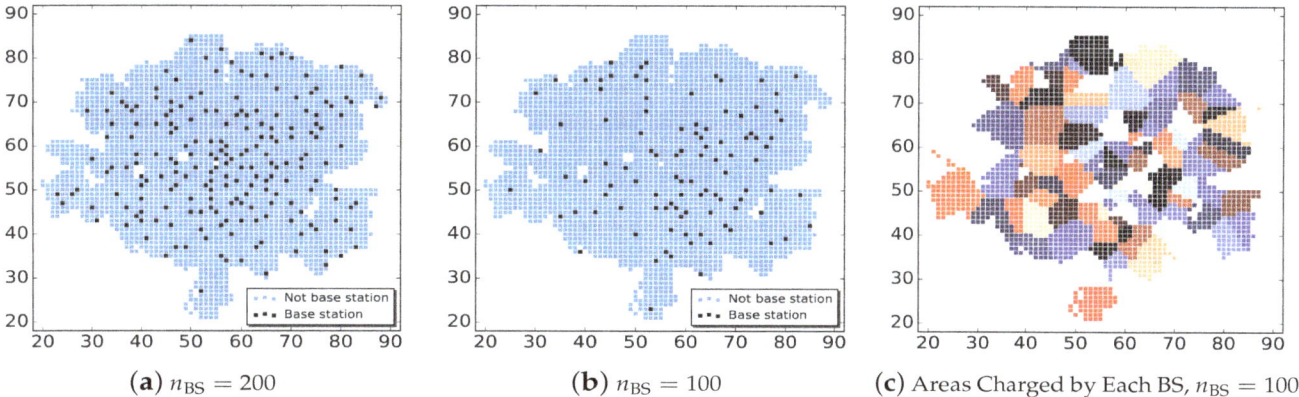

(a) $n_{BS} = 200$ (b) $n_{BS} = 100$ (c) Areas Charged by Each BS, $n_{BS} = 100$

Figure 2. The location distributions of sampled base stations and the areas in charged by them. (**a**,**b**) shows the sampled base station distributions for $n_{BS} = 200$ and $n_{BS} = 100$; (**c**) shows the areas in charged by different base stations for $n_{BS} = 100$.

5.2. Experimental Setup

Algorithms Evaluated

We test our proposed approach and compare it with three baseline methods. In particular, we evaluate and compare the following models using the aggregated November and December datasets, with number of base stations $n_{BS} = 200$ or $n_{BS} = 100$ for stress testing.

- **Patched Estimation**: estimate the cell phone activity distribution by patched piece-wise constant estimation, that is, assume cell phone activity density is distributed uniformly within each sub-region Ω_{B_i}, i.e., the area covered by base station B_i, and estimate each square's activity volume by (7).

- **Patched Estimation + SSR 1**: first estimate *only base station* activity volumes by (7). Use these sparse points to fit a smooth surface by running Spatial Spline Regression to obtain the estimated cell phone activity in all squares.

- **Patched Estimation + SSR 2**: as opposed to the previous model, get the initial estimation of the activity volume of *all squares* by Patched Estimation. Then, use all these points to fit a smooth surface by running Spatial Spline Regression to obtain the final estimated cell phone activity in all squares.

- **Constrained Spatial Smoothing**: first get the initial estimation of the activity volume of all squares by Patched Estimation, then run Constrained Spatial Smoothing algorithm to get the final activity volumes estimation of all squares.

- **Constrained Spatial Smoothing + Features**: in this case, we incorporate the geographical features into Constrained Spatial Smoothing algorithm.

We set the penalty parameter $\lambda = 1$ when $n_{BS} = 200$ and $\lambda = 10$ when $n_{BS} = 100$, for all methods that utilizes SSR. The geographical features of Milan are only incorporated in the last algorithm described above. In addition, for the implementation of Spatial Spline Regression, we use the *fdaPDE* R Package [30].

To compare different approaches, we evaluate the performance by the Mean Relative Error (MRE) of the produced activity estimates for the true activity values. The relative error of an estimation $\hat{f}(\mathbf{p}_j)$ compared to the true value $f(\mathbf{p}_j)$ is defined as $|\hat{f}(\mathbf{p}_j) - f(\mathbf{p}_j)|/f(\mathbf{p}_j)$.

5.3. Performance Evaluation

5.3.1. Comparison of Different Algorithms

We show the cumulative distribution function (CDF) of Relative Errors given by different approaches in Figures 3 and 4. In addition, we compare the estimation's Mean Relative Error of different approaches in Figure 5. It is quite clear that our proposed algorithms outperform other three baseline approaches significantly in all cases ($n_{BS} = 200$ and $n_{BS} = 100$, data aggregated in November and in December).

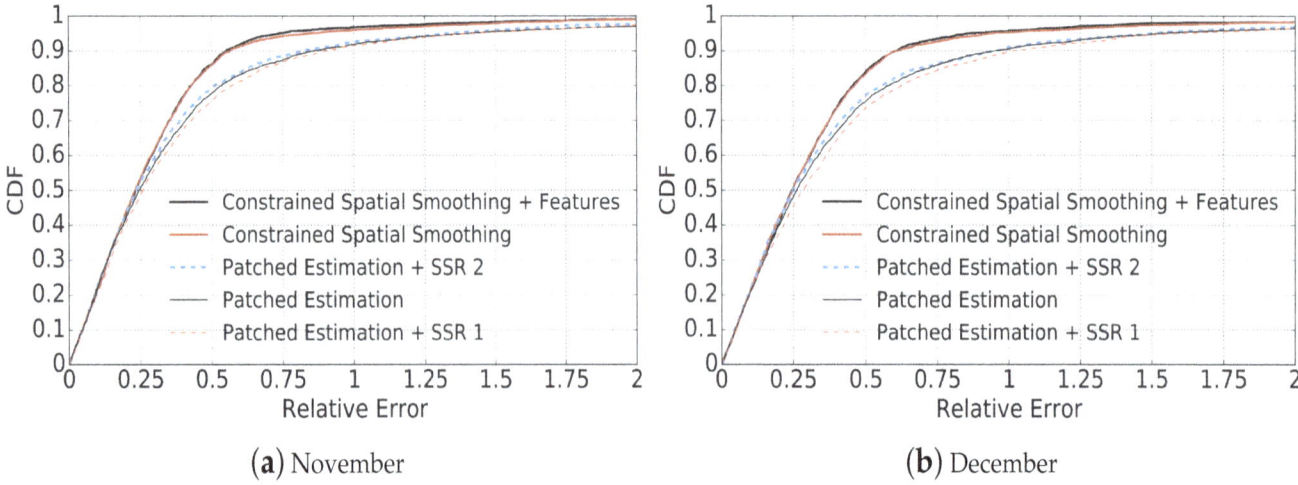

(**a**) November   (**b**) December

Figure 3. Comparison of the CDFs of estimation relative errors given by different methods when $n_{BS} = 200$. The legends follow the same order as the curves at relative error $= 0.5$. (**a**) compares the CDFs based on the data aggregated in November; (**b**) compares the CDFs based on the data aggregated in December.

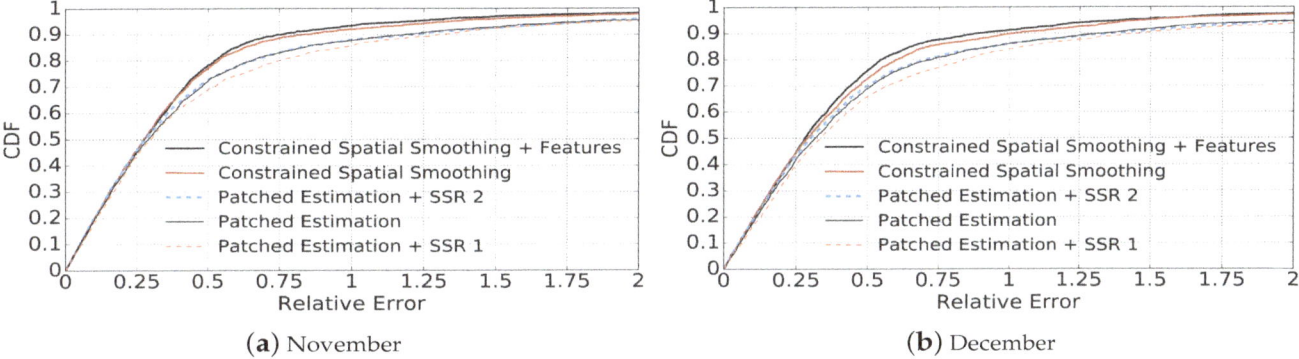

Figure 4. Comparison of the CDFs of estimation relative errors given by different methods when $n_{BS} = 100$ for stress-testing. The legends follow the same order as the curves at relative error $= 0.5$. (**a**) compares the CDFs based on the data aggregated in November; (**b**) compares the CDFs based on the data aggregated in December.

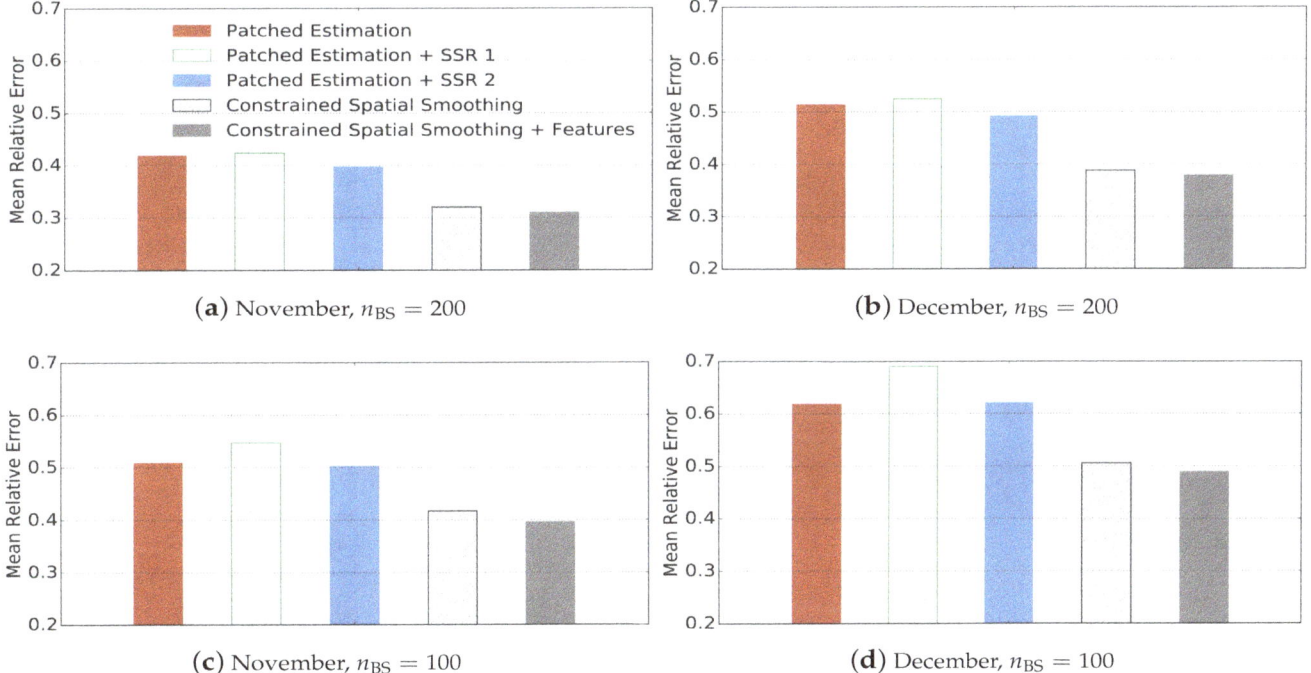

Figure 5. Comparison of the estimation's Mean Relative Error of different methods when $n_{BS} = 200$ or $n_{BS} = 100$ for stress-testing. In each figure, the bars from left to right stands for Patched Estimation, Patched Estimation + SSR 1, Patched Estimation + SSR 2, Constrained Spatial Smoothing, and Constrained Spatial Smoothing + Features respectively. (**a**) we use the data aggregated in November, and set number of base stations to be 200, similarly for (**b**–**d**).

By comparing Patched Estimation + SSR 1 with Patched Estimation approach, we can see that using spatial smoothing based on only base station squares' observations leads to worse performance than patched estimation. This can be explained by the smoothing property of SSR and how we set the values of base station squares. As we described, we set the activity value of base stations by averaging the total activity amount of each base station on all squares it covers. Thus, given the activity $\frac{z_i}{|\Omega_{B_i}|}$, ($|\Omega_{B_i}|$ denotes the number of squares within region Ω_{B_i}) of a base station B_i, the true activities of itself and its surrounding squares within region B_i are distributed with a mean of $\frac{z_i}{|\Omega_{B_i}|}$. Given two base stations B_1 and B_2 that are close to each other, with aggregated activities of z_1 and z_2 respectively, the Spatial Smoothing approach will fit a smooth surface between the two base stations. Suppose

$z_1 > z_2$, in this case, overall, the activities of B_1's neighbour squares will be underestimated, and that of B_2 will be overestimated. Therefore, Patched Estimation + SSR 1's performance is not as good as Patched Estimation.

By comparing Patched Estimation + SSR 2 with Patched Estimation and Patched Estimation + SSR 1, we can observe that applying spatial smoothing on the results of patched estimation improves the performance. This proves the rationality and effectiveness of introducing smoothness into the estimated cell phone activity distribution surface.

Our proposed approach achieves much better performance compared with the three baseline methods. By using Constrained Spatial Smoothing instead of applying Spatial Spline Regression directly, we are able to fit a smooth activity distribution while forcing it to match the observations of base station squares (the aggregated activity volumes) at the same time. By comparing Constrained Spatial Smoothing that incorporates additional features of each square with the version without features, we can see that the performance is further improved. The reason is that the heterogeneity of different locations will influence the telecommunication activity distribution, therefore making the distribution not smooth everywhere. Incorporating additional features into our model can help to explain the residuals between estimated smooth distribution and the true activity distribution, therefore further increasing estimation accuracy. By comparing Figure 3 and Figure 4, we also can see that incorporating additional features into Constrained Spatial Smoothing becomes more important when the base stations are more sparse.

The performance of different methods on the December dataset is worse than on the November dataset. This is because there are multiple holidays in December. The cell phone activities will become more irregular than usual during holidays, as discussed in Cici et al. [2] and Ratti et al. [29].

Figure 6a–c show the distribution surfaces of true cell phone activity volumes, estimated volumes by Patched Estimation, and estimated volumes by Constrained Spatial Smoothing with features when $n_{BS} = 200$ using the November dataset. We can see that the Patched Estimation approach fits a stepped surface, while our approach gives a much smoother surface.

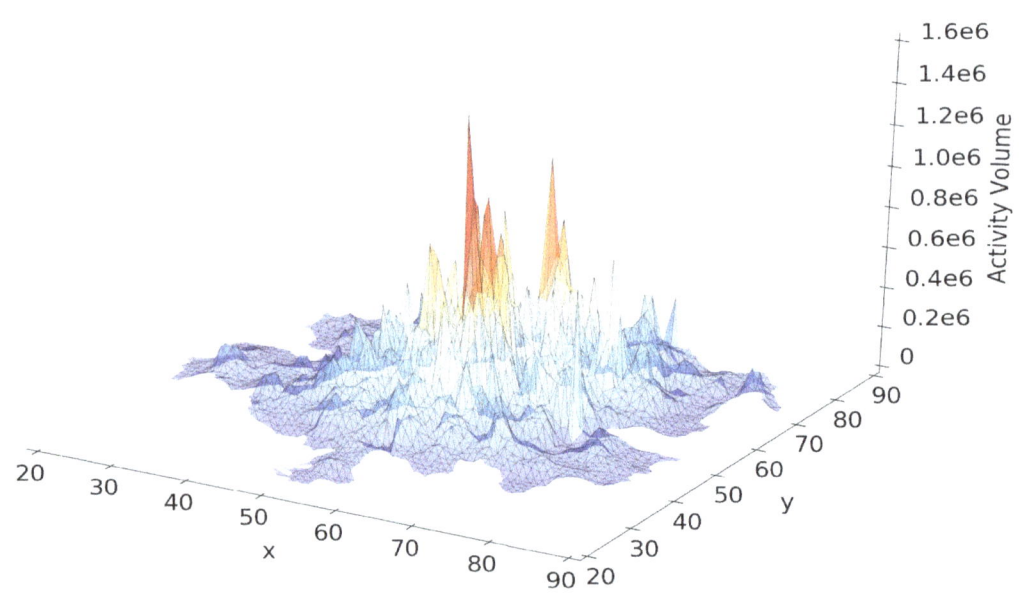

(**a**) Real cell phone activity distribution.

Figure 6. *Cont.*

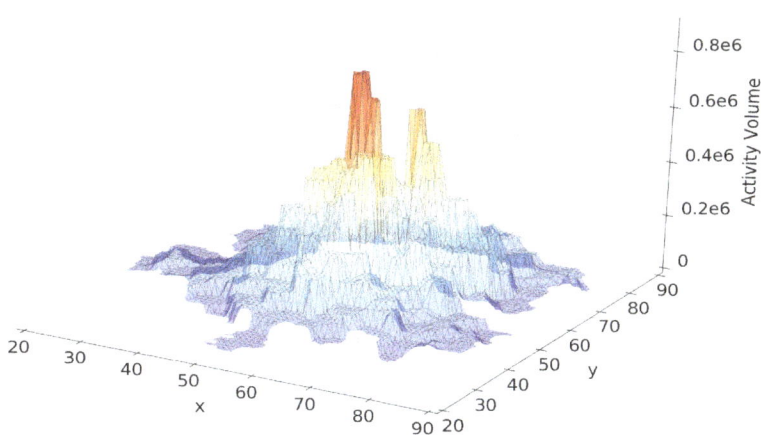

(**b**) Estimated cell phone activity distribution by Patched Estimation.

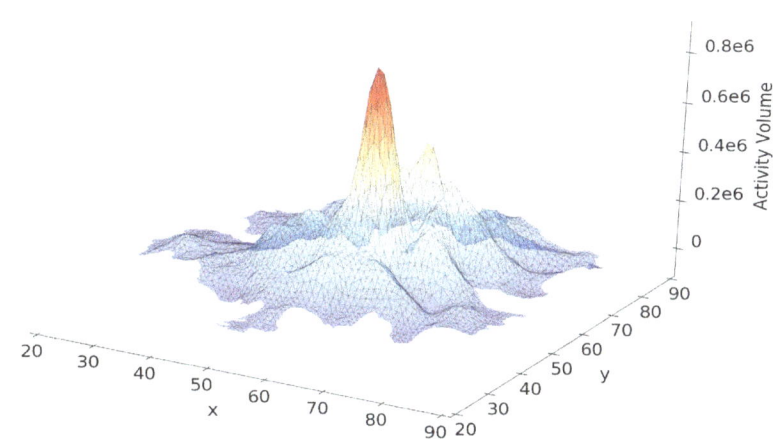

(**c**) Estimated cell phone activity distribution by Constrained Spatial Smoothing + Features.

Figure 6. Comparison of the activity distributions. (**a**) real cell phone activity distribution; (**b**) estimated distribution by Patched Estimation method; (**c**) estimated distribution by our method.

For time efficiency, experiments based on the Milan Call Description Records (CDR) dataset show that the average time for our approach to converge is less than five minutes on a MacBook Pro with a 2 GHz Intel Core i7 processor, and 8 GB memory. This proves that our system is highly efficient and practical.

5.3.2. Impact of Smooth Penalty Parameter λ

Figure 7 shows how the the estimation's Mean Relative Error varies when λ increases from 10^{-4} to 10^3. We make two interesting observations. First, λ around $1\sim10$ usually gives the best performance. Too big or too small λ will decrease the estimation accuracy. This is reasonable, as when λ is too small, we put little emphasis on the smoothness of estimated surface, thus the performance will suffer. If λ is too big, it enforces a smooth surface, which also doesn't match the reality. Second, when we have less base stations, λ that gives the best performance will increase (from 1 to 10). In addition, we can see that the performance of the model with λ between $1\sim100$ does not significantly change when $n_{BS} = 100$. That indicates the following: when the base station distribution is more sparse, the estimation performance is less sensitive to λ when it is around the best value (1 for $n_{BS} = 200$ and 10 for $n_{BS} = 100$).

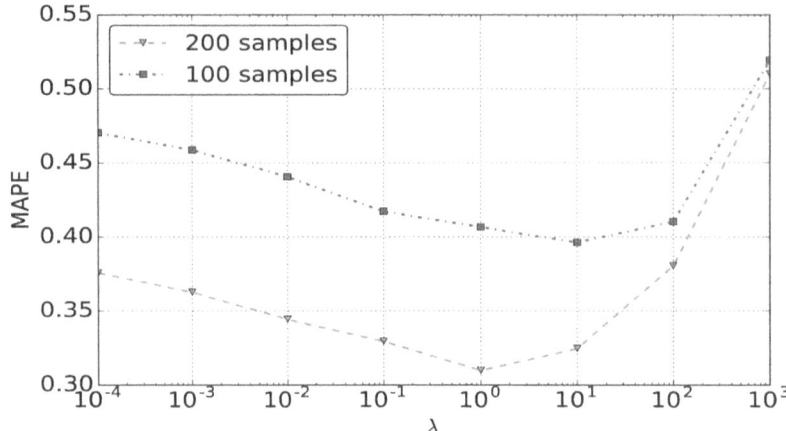

Figure 7. Influence of λ to estimation's Mean Relative Error when $n_{BS} = 200$ and $n_{BS} = 100$ for stress-testing. The figure is based on the November dataset. Results on the December dataset are similar.

6. Related Work

The Telecom Italia Big Data Challenge dataset is widely used to study different problems [2–5,20–24]. However, little research work has been done to estimate the spatial distribution of cellphone activity itself, despite the great value of this problem.

There are various tasks where the key problem is estimating a spatial field over a region based on observations of sampled points, such as house price estimation and population density estimation. Chopra et al. [26] model the underlying surface of land desirability using kernel-based interpolation. However, it is hard to choose the form of kernel functions and tune a large number of hyper-parameters. Spatial Spline Regression technique is applied to the problem of population density estimation in Sangalli et al. [16]. However, in our problem, we only get the accumulated activity density in base stations, rather than real densities in each base station location. In addition, BS locations distribution is highly sparse in our case.

Although a range of kernel-based methods [26,31,32] can be applied to fit a spatial field, the common drawback of these approaches is that, by using uniformly damping weights in distance-based kernels, they tend to link weakly related data points across areas in a non-convex domain. Spatial spline regression [16] on the other hand uses finite-element analysis approach to jointly solve for f and β from the model described by Equation (8) over any irregularly shaped domain Ω.

As it was discussed earlier, the fine-grained data for the distribution of the volume of calls and SMS are not usually available. A common type of data is the data collected by cell phone base stations. Sometimes, cell phone providers interpolate the data collected by the base stations as is discussed in Manfredini et al. [33]. Some researchers interpolate the data to obtain fine grained distributions as in Ratti et al. [29]. However, in Ratti et al. [29], authors do not evaluate the performance of the interpolated distribution. To the best of our knowledge, there is no extensive work done in trying to obtain optimal reconstructions of fine grained cell phone data distribution. We are the first to apply the latest spatial functional analysis techniques to cellphone activity distribution modeling, assuming the activity densities consist of a regression part based on social or demographical statistic features and a spatial field that captures the underlying smoothness property of cellphone activities. In particular, we leverage the idea of spatial spline regression to handle any irregularly shaped geographic regions. We have developed a novel Constrained Spatial Smoothing approach and corresponding training algorithm to recover spatial distribution of cellphone activities from highly sparse observations.

7. Conclusions

In this paper, we study the problem of inferring the fine-grained spatial distribution of certain density data in a region based on the aggregate observations recorded for each of its subregions, which is extremely challenging and seldom visited before, and analyze the challenges of it. We propose the Constrained Spatial Smoothing (CSS) approach that exploits both the intrinsic smooth property of underlying factors and the additional features from external social or domestic statics. We further propose a training algorithm which combines the Spatial Spline Regression (SSR) technique and ADMM technique to learn our model parameters efficiently.

To evaluate our algorithm and compare it with various other approaches, we run extensive evaluations based on the Milan Call Detail Records dataset provided by Telecom Italia Mobile. The simulation results on the dataset show that our algorithm significantly outperforms other baseline approaches by a great percentage. (Note that cross validation and statistical testing are techniques that are usually applied in experiments. However, both techniques require sampling effectively from the sparse spatial data while keeping the intrinsic spatial structure, which is difficult in our problem.) This shows that jointly modeling the underlying spatial continuity and the local features that characterize the heterogeneity of different locations can effectively improve the performance of spatial recovery.

Although we use the data on cell phone activities to illustrate our methodology, our algorithm is not limited to solving the problem of inferring the distribution of cell phone activities, but is also applicable to a variety of problems where estimating an implicit or explicit smooth surface is required, such as inferring the spatial distribution of population densities based on the aggregate population observed at sparsely scattered polling stations, reconstructing a fine-grained geographical distribution of users for an Internet media provider or retailer only from aggregated user counts observed at certain datacenters or points of presence (PoPs), and so on.

Author Contributions: Conceptualization, B.L., B.M., L.K. and D.N.; Data curation, B.L. and B.M.; Formal analysis, B.L., B.M., L.K. and D.N.; Investigation, B.L., B.M., L.K. and D.N.; Methodology, B.L., B.M., L.K. and D.N.; Project administration, B.L., B.M., L.K. and D.N.; Software, B.L. and B.M.; Supervision, B.L., B.M., L.K. and D.N.; Validation, B.L., B.M., L.K. and D.N.; Visualization, B.L., B.M. and L.K.; Writing—original draft, B.L., B.M., L.K. and D.N.; Writing—review and editing, B.L., B.M. and D.N.

Abbreviations

The following abbreviations are used in this manuscript:

CSS	Constrained Spatial Smoothing
SSR	Spatial Spline Regression
SMS	Short Message Service
ADMM	Alternating Direction Method of Multipliers
QP	Quadratic Programming
CDR	Call Detail Records
PoPs	Presence of Points
CDF	Cumulative Distribution Function

References

1.   Barlacchi, G.; De Nadai, M.; Larcher, R.; Casella, A.; Chitic, C.; Torrisi, G.; Antonelli, F.; Vespignani, A.; Pentland, A.; Lepri, B. A multi-source dataset of urban life in the city of Milan and the Province of Trentino. *Sci. Data* **2015**, *2*, 150055. [CrossRef] [PubMed]

2.   Cici, B.; Gjoka, M.; Markopoulou, A.; Butts, C.T. On the decomposition of cell phone activity patterns and their connection with urban ecology. In Proceedings of the 16th ACM International Symposium on Mobile Ad Hoc Networking and Computing, ACM MobiHoc'15, Hangzhou, China, 22–25 June 2015.

3. Louail, T.; Lenormand, M.; Ros, O.G.C.; Picornell, M.; Herranz, R.; Frias-Martinez, E.; Ramasco, J.J.; Barthelemy, M. From mobile phone data to the spatial structure of cities. *Sci. Rep.* **2014**, *4*, 5276. [CrossRef]

4. Douglass, R.W.; Meyer, D.A.; Ram, M.; Rideout, D.; Song, D. High resolution population estimates from telecommunications data. *EPJ Data Sci.* **2015**, *4*, 4. [CrossRef]

5. Deville, P.; Linard, C.; Martin, S.; Gilbert, M.; Stevens, F.R.; Gaughan, A.E.; Blondel, V.D.; Tatem, A.J. Dynamic population mapping using mobile phone data. *Proc. Natl. Acad. Sci. USA* **2014**, *111*, 15888–15893. [CrossRef] [PubMed]

6. Liu, X.; Kyriakidis, P.C.; Goodchild, M.F. Population-density estimation using regression and area-to-point residual kriging. *Int. J. Geogr. Inf. Sci.* **2008**, *22*, 431–447. [CrossRef]

7. Cho, E.; Myers, S.A.; Leskovec, J. Friendship and mobility: User movement in location-based social networks. In Proceedings of the 17th ACM SIGKDD International Conference on Knowledge Discovery and Data Mining, San Diego, CA, USA, 21–24 August 2011; ACM: New York, NY, USA, 2011; pp. 1082–1090.

8. Li, X.; Pan, G.; Wu, Z.; Qi, G.; Li, S.; Zhang, D.; Zhang, W.; Wang, Z. Prediction of urban human mobility using large-scale taxi traces and its applications. *Front. Comput. Sci.* **2012**, *6*, 111–121.

9. Lu, X.; Wetter, E.; Bharti, N.; Tatem, A.J.; Bengtsson, L. Approaching the limit of predictability in human mobility. *Sci. Rep.* **2013**, *3*, 2923. [CrossRef] [PubMed]

10. Wang, D.; Pedreschi, D.; Song, C.; Giannotti, F.; Barabasi, A.L. Human mobility, social ties, and link prediction. In Proceedings of the 17th ACM SIGKDD International Conference on Knowledge Discovery and Data Mining, San Diego, CA, USA, 21–24 August 2011; ACM: New York, NY, USA, 2011; pp. 1100–1108.

11. De Domenico, M.; Lima, A.; Musolesi, M. Interdependence and predictability of human mobility and social interactions. *Pervasive Mob. Comput.* **2013**, *9*, 798–807. [CrossRef]

12. Wood, S.N. Thin plate regression splines. *J. R. Stat. Soc. Ser. B Stat. Methodol.* **2003**, *65*, 95–114. [CrossRef]

13. Wood, S.N.; Bravington, M.V.; Hedley, S.L. Soap film smoothing. *J. R. Stat. Soc. Ser. B Pervasiv Methodol.* **2008**, *70*, 931–955. [CrossRef]

14. Ramsay, T. Spline smoothing over difficult regions. *J. R. Stat. Soc. Ser. B Pervasiv Methodol.* **2002**, *64*, 307–319. [CrossRef]

15. Guillas, S.; Lai, M.J. Bivariate splines for spatial functional regression models. *J. Nonparametr. Stat.* **2010**, *22*, 477–497. [CrossRef]

16. Sangalli, L.M.; Ramsay, J.O.; Ramsay, T.O. Spatial spline regression models. *J. R. Stat. Soc. Ser. B Stat. Methodol.* **2013**, *75*, 681–703. [CrossRef]

17. Boyd, S.; Parikh, N.; Chu, E.; Peleato, B.; Eckstein, J. Distributed optimization and statistical learning via the alternating direction method of multipliers. *Found. Trends Mach. Learn.* **2011**, *3*, 1–122. [CrossRef]

18. Parikh, N.; Boyd, S. Proximal algorithms. *Found. Trends Optim.* **2014**, *1*, 127–239. [CrossRef]

19. Bertsekas, D.P. *Nonlinear Programming*; Athena Scientific: Belmont, MA, USA, 1999.

20. Gonzalez, M.C.; Hidalgo, C.A.; Barabasi, A.L. Understanding individual human mobility patterns. *Nature* **2008**, *453*, 779–782. [CrossRef]

21. Csáji, B.C.; Browet, A.; Traag, V.A.; Delvenne, J.C.; Huens, E.; Van Dooren, P.; Smoreda, Z.; Blondel, V.D. Exploring the mobility of mobile phone users. *Phys. A Stat. Mech. Its Appl.* **2013**, *392*, 1459–1473. [CrossRef]

22. Song, C.; Qu, Z.; Blumm, N.; Barabási, A.L. Limits of predictability in human mobility. *Science* **2010**, *327*, 1018–1021. [CrossRef]

23. Blondel, V.D.; Decuyper, A.; Krings, G. A survey of results on mobile phone datasets analysis. *EPJ Data Sci.* **2015**, *4*, 10.

24. Grauwin, S.; Sobolevsky, S.; Moritz, S.; Gódor, I.; Ratti, C. Towards a comparative science of cities: Using mobile traffic records in new york, london, and hong kong. In *Computational Approaches For Urban Environments*; Springer: Berlin/Heidelberg, Germany, 2015; pp. 363–387.

25. Telecom. Telecom Italia Big Data Challenge. 2014. Available online: https://dandelion.eu/datamine/open-big-data/ (accessed on 27 July 2016).

26. Chopra, S.; Thampy, T.; Leahy, J.; Caplin, A.; LeCun, Y. Discovering the hidden structure of house prices with a non-parametric latent manifold model. In Proceedings of the 13th ACM SIGKDD International Conference on Knowledge Discovery and Data Mining, San Jose, CA, USA, 12–15 August 2007; ACM: New York, NY, USA, 2007; pp. 173–182.

27. Hjelle, Ø.; Dæhlen, M. *Triangulations and Applications*; Springer Science & Business Media: Berlin, Germany, 2006.

28. Douglas, J.; Rachford, H.H. On the numerical solution of heat conduction problems in two and three space variables. *Trans. Am. Math. Soc.* **1956**, *82*, 421–439. [CrossRef]
29. Ratti, C.; Frenchman, D.; Pulselli, R.M.; Williams, S. Mobile landscapes: Using location data from cell phones for urban analysis. *Environ. Plan. B Plan. Des.* **2006**, *33*, 727–748. [CrossRef]
30. Lila, E.; Sangalli, L.M.; Ramsay, J.; Formaggia, L. *fdaPDE: Functional Data Analysis and Partial Differential Equations; Statistical Analysis of Functional and Spatial Data, Based on Regression with Partial Differential Regularizations*; R Package Version 0.1-4; The Comprehensive R Archive Network. 2016. Available online: https://cran.r-project.org/web/packages/fdaPDE/index.html (accessed on 26 April 2019).
31. Clapp, J.M. A semiparametric method for estimating local house price indices. *Real Estate Econ.* **2004**, *32*, 127–160. [CrossRef]
32. Caplin, A.; Chopra, S.; Leahy, J.V.; LeCun, Y.; Thampy, T. Machine Learning and the Spatial Structure of House Prices and Housing Returns. 2008. Available online: https://ssrn.com/abstract=1316046 (accessed on 26 April 2019).
33. Manfredini, F.; Pucci, P.; Tagliolato, P. Toward a systemic use of manifold cell phone network data for urban analysis and planning. *J. Urban Technol.* **2014**, *21*, 39–59. [CrossRef]

9

Impact of Texture Information on Crop Classification with Machine Learning and UAV Images

Geun-Ho Kwak and No-Wook Park *

Department of Geoinformatic Engineering, Inha University, Incheon 22212, Korea; root0109@inha.edu
* Correspondence: nwpark@inha.ac.kr

Abstract: Unmanned aerial vehicle (UAV) images that can provide thematic information at much higher spatial and temporal resolutions than satellite images have great potential in crop classification. Due to the ultra-high spatial resolution of UAV images, spatial contextual information such as texture is often used for crop classification. From a data availability viewpoint, it is not always possible to acquire time-series UAV images due to limited accessibility to the study area. Thus, it is necessary to improve classification performance for situations when a single or minimum number of UAV images are available for crop classification. In this study, we investigate the potential of gray-level co-occurrence matrix (GLCM)-based texture information for crop classification with time-series UAV images and machine learning classifiers including random forest and support vector machine. In particular, the impact of combining texture and spectral information on the classification performance is evaluated for cases that use only one UAV image or multi-temporal images as input. A case study of crop classification in Anbandegi of Korea was conducted for the above comparisons. The best classification accuracy was achieved when multi-temporal UAV images which can fully account for the growth cycles of crops were combined with GLCM-based texture features. However, the impact of the utilization of texture information was not significant. In contrast, when one August UAV image was used for crop classification, the utilization of texture information significantly affected the classification performance. Classification using texture features extracted from GLCM with larger kernel size significantly improved classification accuracy, an improvement of 7.72%p in overall accuracy for the support vector machine classifier, compared with classification based solely on spectral information. These results indicate the usefulness of texture information for classification of ultra-high-spatial-resolution UAV images, particularly when acquisition of time-series UAV images is difficult and only one UAV image is used for crop classification.

Keywords: unmanned aerial vehicle; texture; gray-level co-occurrence matrix; machine learning; crop

1. Introduction

Agricultural environments are known to be sensitive to abnormal weather conditions and climatic disasters such as drought and flood [1,2], thus rendering essential systematic monitoring of crop conditions and crop yield forecasting [3,4]. Remote sensing technology received attention in the agriculture community due to its ability to provide periodic and regional information for crop monitoring and thematic mapping [5,6].

Crop type maps derived from classification of remote sensing images are important resources for crop yield estimation and forecasting. Since any error in the crop type maps affects outputs of crop yield and forecasting models, it is critical to generate reliable crop type maps [6]. The most important elements of input remote sensing images for crop classification are their spatial and temporal resolutions. Since each individual crop has its own growth cycle, time-series images are necessary to fully account for variations of physical characteristics that accompany crop growth [7,8]. According to the scale of the target area of interest, satellite images with proper spatial resolution should be used as input for crop classification. If coarse-resolution satellite images are used, mixed pixel problems are likely and classification performance decreases [9,10]. This is a common issue in Korea, where various types of crops are cultivated in small areas. The use of high-resolution satellite images and aerial photos can contribute to resolving the mixed pixel issues [11,12]. Despite the increased discrimination capability of high-resolution images, it is difficult to collect time-series datasets over the full growth cycles of crops. Acquisition of optical satellite images depends heavily on atmospheric conditions; thus, the images are often contaminated and masked by clouds. In addition, it is difficult to acquire time-series aerial photos at desired times due to cost issues.

In recent years, there was a growing interest in imaging of unmanned aerial vehicles (UAV) [11–15]. The advantage of UAV images over satellite images is their ability to provide local thematic information with much higher spatial and temporal resolutions [15]. UAV images with ultra-high spatial resolution [16,17] can improve the discrimination capability of various surface objects, leading to an increase in the number of detectable targets. Compared with satellite images, low-cost flexible control of unmanned aerial systems (UAS) enables easier acquisition of images at the desired times between sowing and harvesting of crops [12,15,16].

Despite the great potential of UAV imaging, the technique has several practical issues. Firstly, the ultra-high spatial resolution of UAV images usually causes noise effects due to increased detectable targets when conventional pixel-based approaches are applied for classification [11,18,19]. The common approach to mitigate noise effects is to either use spatial contextual information or apply an object-oriented classification approach. For the spatial contextual information approach, texture information is firstly extracted from a gray-level co-occurrence matrix (GLCM) [20] and then combined with spectral information for classification [21–23]. The utilization of such texture information can reduce the impacts of isolated pixels within the pixel-based approach. The object-oriented approach first extracts meaningful objects via multi-resolution segmentation [24] and classification is then carried out on object units [25–27]. These two approaches are known to achieve better classification accuracy than the pixel-based approach based purely on spectral information [19,22]. The second issue is heavy computational load related to data preprocessing and processing [11]. Most UAV images are acquired using a narrow field-of-view, which requires mosaicking of many sub-images to obtain a complete image set. If the sub-images are taken at different solar conditions and flight altitudes, radiometric calibration should be employed during mosaicking. The ultra-high spatial resolution of UAV images makes preprocessing complex and requires much processing time for classification [11].

Another important issue is that it is not always possible to construct a time-series UAV image set for crop classification. Although the acquisition of UAV images is less affected by atmospheric conditions than satellite images, it may be difficult to take UAV images in some season [12], particularly the rainy season which coincides with the growing season of crops in Korea. From an operational viewpoint, the acquisition of time-series UAV images for crop classification essentially has a prerequisite that operators make several visits to the area of interest. From a practical point of view, it is necessary to acquire optimal images at certain times, achieving classification accuracy comparable to the use of a complete time-series image set. Crop classification using UAV images is primarily conducted using a single UAV image [21,28], but accuracy comparisons with the case using a time-series image set are yet to be considered fully.

In addition to data acquisition issues, selection of proper classification methodology is important in order to generate reliable crop classification results. Since the 2000s, machine learning algorithms such as random forest (RF) and support vector machine (SVM) were widely applied to crop classification with remote sensing data [29–34].

Along with the aforementioned issues related to crop classification with UAV images and selection of appropriate classification methodology, this paper focuses on the evaluation of the effectiveness of texture information for crop classification with UAV images. In particular, the classification performance using a single-date UAV image is compared with that of a time-series image set.

In this study, two machine learning algorithms, RF and SVM, are applied as classification models, and the GLCM-based texture features [20] are used as additional features to reduce noise effects. From a practical viewpoint, we also investigate how much the utilization of texture information can improve classification accuracy when only a single-time UAV image is available. A case study of crop classification with UAV images in Anbandegi, a highland Kimchi cabbage cultivation area in Korea, was conducted to illustrate and discuss the two issues including the limited use of the UAV image and the impact of GLCM-based texture features on classification performance.

2. Materials and Methods

2.1. Study Area

Anbandegi, in the Gangwon Province of Korea, a major highland Kimchi cabbage cultivation area, was selected as the case study area (Figure 1). Summer Kimchi cabbage is usually cultivated in highlands in Korea because high temperature and humidity causes physiological disorders, insect pests, and diseases [35]. The altitude of the study area is about 1000 meters above mean sea level and is relatively higher than the surrounding terrain, which is suitable for highland Kimchi cabbage cultivation [35]. In the study area, cabbage and potatoes are also grown along with highland Kimchi cabbage. The total area of all crop parcels in the study area is 42.5 ha and the average size of each crop parcel is about 0.6 ha. The land-cover type of non-crop areas is mainly forest.

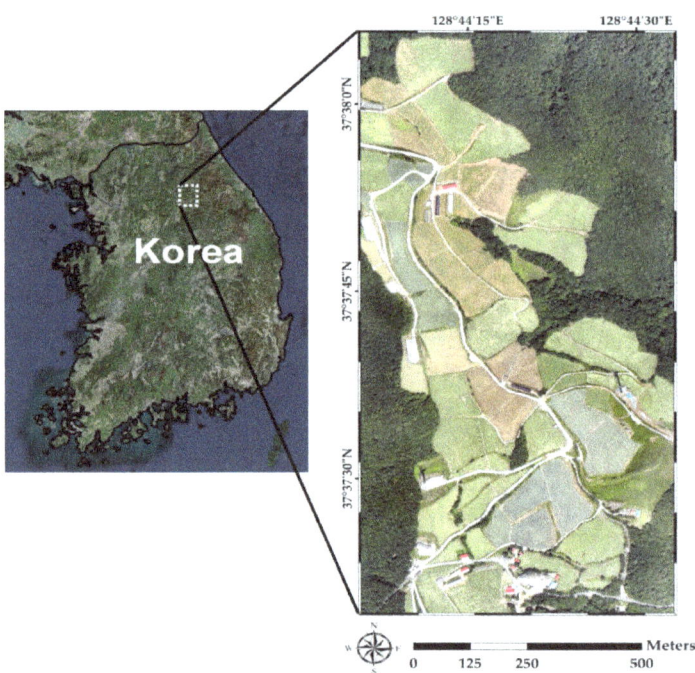

Figure 1. Location of the study area and the unmanned aerial vehicle (UAV) image mosaic acquired in the study area.

2.2. Datasets

2.2.1. UAV Images

We used six UAV image mosaics taken from June to September 2017, by considering the growth cycle of highland Kimchi cabbage (Table 1). The preprocessed UAV image mosaics provided by the National Institute of Agricultural Sciences (NAAS) were acquired from a fixed-wing unmanned aerial system (UAS; eBee, Sensefly, Swiss) equipped with a Cannon S110 camera that includes green (550 nm), red (625 nm), and near-infrared (NIR; 850 nm) spectral bands (hereafter referred to as VNIR). The UAV image mosaics with a ground sampling distance of 12 cm were upscaled to 25 cm resolution to facilitate data processing without loss of information. Upscaling may result in loss of textural image information. However, a significant change in the generation of texture features and classification results was not observed in our preliminary experiment at subareas in the study area, which was attributed to the size of crop parcels in the study area. Hence, the image mosaics with a 25 cm resolution were used as inputs for classification. To examine the applicability of a single-date image with texture information, the UAV image mosaic acquired on 25 August was selected due to the peak in vitality of highland Kimchi cabbage. This selection is explained in detail in Section 3.

Table 1. List of unmanned aerial vehicle (UAV) image mosaics acquired in the study area in 2017.

No.	Acquisition Date
1	29 June
2	12 July
3	27 July
4	25 August
5	13 September
6	21 September

2.2.2. Ground-Truth Data and Land-Cover Map

Ground-truth crop types were obtained by field surveys, which were also provided by NAAS. These data were used to both extract training data and to evaluate the classification performance. Table 2 presents crop classes for supervised classification and area information of each crop type. To mimic a case with limited available training data, 20,000 pixels (about 0.3% of ground-truth data) were randomly selected and used for training data for supervised learning. The remaining 6,710,210 pixels (99.7% of ground-truth data) were used as reference data. Note that a relatively small training dataset and a large reference dataset are used for classification and evaluation, respectively. Since the main targets of classification were crops in the study area, non-crop areas, including forests, were masked out prior to classification using land-cover maps from the Ministry of Environment [36].

Table 2. Crop classes and their respective area information in the study area.

Classes	Total Area (ha)	Average Area per Parcel (ha)
Highland Kimchi Cabbage	22.38	0.59
Cabbage	8.35	0.59
Potato	8.65	0.86
Fallow	3.08	0.31

2.3. Classification Methods and Feature Extraction

2.3.1. Random Forest

The RF classifier developed by Breiman [37] performs classification by extending decision trees to multiple trees rather than a single tree. Its classification performance is superior to a single decision tree due to its ability to maximize diversity through tree ensembles. It also demonstrates greater stability due to the synthesis of classification results from a large number of trees and the determination of final class labels through majority voting. In addition, RF requires a few parameters (i.e., the number of variables for node partitioning and the number of trees to be grown) to be set, unlike other machine learning algorithms.

The RF classifier applies bootstrap aggregating (bagging) to tree learners. Bagging repeatedly selects a random sample to replace the training data and fits trees to these samples. The remaining training data, the out-of-bag (OOB) data, are used to validate trees [37]. The OOB error that is the error rate of the OOB classifiers is often used as a measure of the generalization error on the training data [37]. To avoid overfitting the training data, each node of the trees determines the partitioning condition, and each tree chooses the random predictor variable and divide node using a genie index, as a measure of heterogeneity. An additional function of the RF classifier is to compute quantitative measures for variable importance using mean decrease impurity (MDI) and mean decrease accuracy (MDA) [28]. When constructing a large number of trees, MDI and MDA can be calculated by averaging the weighted impurity of each tree and the degree of accuracy improvement, respectively, by randomly changing the variable. In this study, the variable importance was used to quantify how useful texture information is for crop classification.

2.3.2. Support Vector Machine

SVM is a machine learning algorithm for finding the optimal decision boundary of training data located at the boundary of classes [38]. The SVM classifier is known to be effective for classification with a limited amount of training data [39]. The main concept of SVM is to solve the optimization problem which maximizes the margin between decision boundaries [40]. To solve non-linear optimization problems, kernel functions such as radial basis function (RBF) are commonly used [39]. When the RBF kernel is used, the parameters of cost and gamma should be optimally determined. Large values of cost and gamma result in overfitting to the training data, yielding poor generalization ability of the classifier [41]. In this study, these two hyper-parameters were determined using a grid search based on 10-fold cross-validation of training data [42].

2.3.3. Texture Information

To reduce the noise effects of isolated pixels in classification results, texture information is considered as an auxiliary feature for classification. Image texture analysis methods can be divided into four categories: statistical, geometric, model-based, and signal processing [43]. GLCM, developed by Haralick et al. [20], is a widely applied statistical method for remote sensing data processing such as vegetation structure modeling [44] and land-cover classification [45]. The original image is first converted to gray-scale. Then, the spatial features of the gray-scale image are extracted using the relationship of the brightness values between the center pixel and its neighborhood within the predefined kernel. The relationship of the brightness values is represented by a matrix which consists of the occurrence frequency of sequential pairs of pixel values existing simultaneously along the defined direction. By using this relationship, the GLCM can generate different texture information according to gray-scale level, kernel size, and direction. Fourteen texture features defined by Haralick et al. [20] are correlated, indicating that using all possible texture features provides redundant spatial contextual information which is not useful for classification. In this study, six texture features [46] were considered: (1) mean (ME), (2) standard deviation (STD), (3) homogeneity (HOM), (4) dissimilarity (DIS), (5) entropy (ENT), and (6) angular second moment (ASM), presented in Equations (1) to (6):

$$\text{ME} = \sum_{i=0}^{N-1} \sum_{j=0}^{N-1} i \times P(i,j), \tag{1}$$

$$\text{STD} = \sqrt{\sum_{i=0}^{N-1} \sum_{j=0}^{N-1} P(i,j) \times (i - \text{ME})^2}, \tag{2}$$

$$\text{HOM} = \sum_{i=0}^{N-1} \sum_{j=0}^{N-1} \frac{P(i,j)}{1 + (i-j)^2}, \tag{3}$$

$$\text{DIS} = \sum_{i=0}^{N-1} \sum_{j=0}^{N-1} P(i,j) \times |i - j|, \tag{4}$$

$$\text{ENT} = \sum_{i=0}^{N-1} \sum_{j=0}^{N-1} -P(i,j) \times \ln(p(i,j)), \tag{5}$$

$$\text{ASM} = \sum_{i=0}^{N-1} \sum_{j=0}^{N-1} P(i,j)^2, \tag{6}$$

where N denotes gray-scale level, while $P(i,j)$ is the normalized gray-scale value at positions i and j within the kernel, and its sum is 1.

The above texture features were generated from omnidirectional 64-shade gray-scale images. To test the impacts of the kernel size, we used three different kernel sizes: 3×3 (GK3), 15×15 (GK15), and 31×31 (GK31).

2.4. Classification Procedures

The entire procedure for crop classification with UAV images is presented in Figure 2. For each classifier, optimal parameters were first sought during a training phase. To investigate the impacts of both the number of input images and texture features, we tested eight combination cases for each classifier: UAV images (two cases: with the August image and with six multi-temporal images), and texture features (four cases: with texture features from three different kernel sizes (GK3, GK15, and GK31), and without texture features). These combinations were considered for comparison purposes since the main objective of this study was to evaluate the effectiveness of using texture information when a single-date UAV image is used for crop classification. The classification accuracy was assessed using quantitative measures based on a confusion matrix such as overall accuracy (OA), producer's accuracy (PA), and user's accuracy (UA).

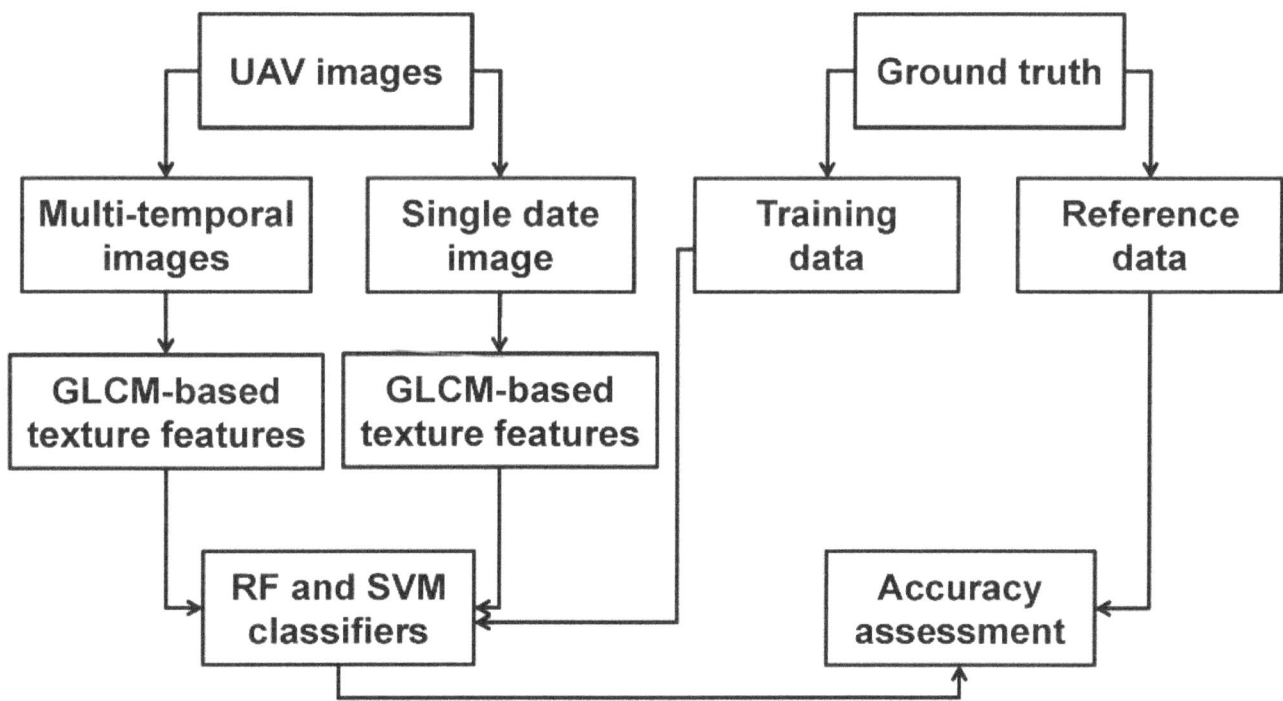

Figure 2. Schematic diagrams of all crop classification procedures applied in this study. GLCM: gray-level co-occurrence matrix; RF: random forest; SVM: support vector machine.

2.5. Implementation

ENVI software version 4.8 was used for generation of GLCM-based features and visualization of classification results. All procedures for classification and evaluation were done within the R software environment [47]. SVM and RF models were built using the R packages e1071 [42] and randomForest [48], respectively.

3. Results and Discussion

3.1. Parameterization of RF and SVM Classifiers

For the RF classifier, two parameters, the number of variables required for node partitioning and the number of trees to be grown, have to be selected. Firstly, the number of variables for node partitioning was set to $\sqrt{n}$ of the total number of variables. For example, for the case using the August image with texture information, there were nine variables (three spectral bands and six texture features); thus, the number of variables for node partitioning was set to 3. To determine the number of trees to be grown, variations of OOB errors with respect to the number of trees were investigated. From the variations of OOB errors, one can judge whether a sufficient number of trees were used for the RF modeling. In general, the OOB errors tend to decrease as the number of trees increases, and then converge to a certain value at the specific number of trees. When six multi-temporal UAV images were used as inputs, no distinctive differences in OOB errors were observed, and the error values were also very low for different texture feature cases. Figure 3 shows the variations of OOB errors when using the August image without and with texture features. The four combination cases showed different convergence values, but the variation patterns were very similar. As the number of trees increased to about 50, the OOB errors of all four combination cases decreased sharply. Then, the OOB errors reached the convergence values when the number of trees was about 150. By considering the convergence of OOB errors and processing time, the number of trees to be grown was set to 150.

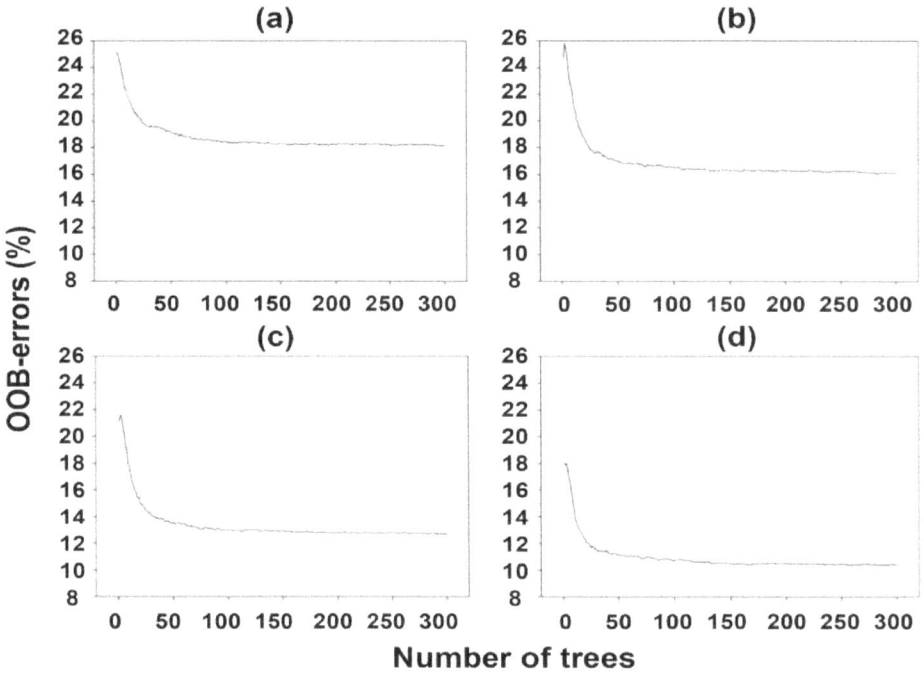

Figure 3. Variations of out-of-bag (OOB) errors of RF models with respect to the number of trees for the case using the August UAV image without and with texture features: (**a**) visible and near infrared (VNIR) spectral bands only; (**b**) VNIR + 3 × 3 kernel (GK3); (**c**) VNIR + 15 × 15 kernel (GK15); and (**d**) VNIR + 31 × 31 kernel (GK31).

Two parameters (cost and gamma) of the RBF kernel for the SVM classifier were tuned using a grid search. The optimal combination of the two parameters was determined through 10-fold cross-validation of training data. The optimal cost and gamma values were similar for combination cases of different kernel sizes of GLCM and input UAV images. Figure 4 presents the grid search results for the cases using the August image and six multi-temporal UAV images with texture feature GK31, showing the different training accuracy values. The training accuracy obtained by the grid search ranged between 52 and 82.4% for the case using the August image with texture features, while the maximum training accuracy for the case using six UAV images increased to 94%. It should be noted that this accuracy was obtained during the training phase; hence, higher training accuracy may fail to achieve higher prediction performance. It was found that the performance difference with respect to variations of the model parameters for the SVM classifier was also great, compared to the RF classifier, which indicates the importance of optimal parameter search for the SVM classifier.

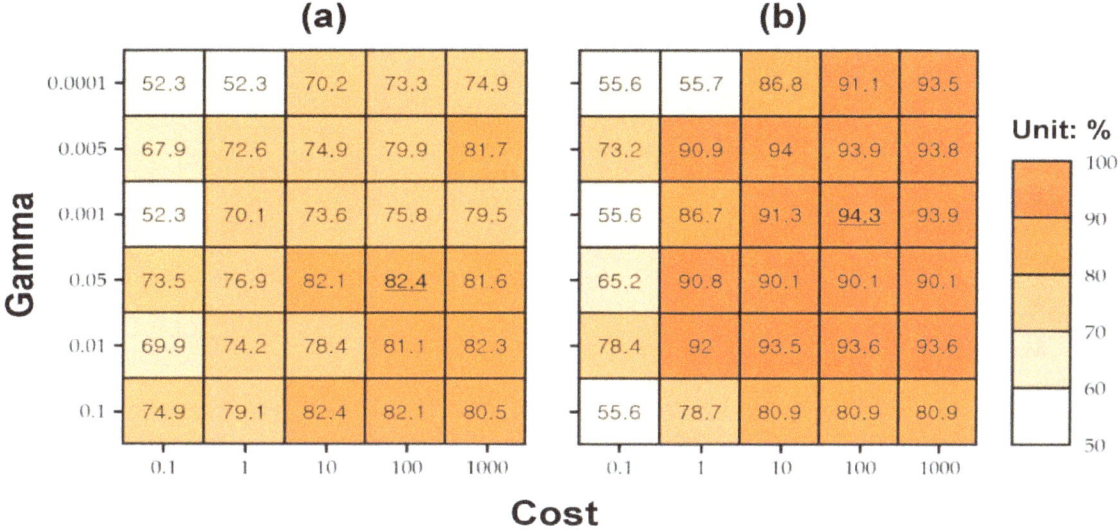

Figure 4. Cross-validation accuracy of SVM classifiers through a grid search. The case with the best accuracy is underlined: (**a**) using a single August image with VNIR and GK31 texture features; and (**b**) using six UAV images with VNIR and GK31 texture features.

3.2. Visual Assessment of Classification Results

Once optimal parameters were determined, the RF and SVM classifiers were applied to the different case combinations. Prior to quantitative accuracy assessment, the visual assessment of classification results was first conducted. When the RF and SVM classification results were compared for different combinations of input images and kernel sizes, the RF classifier showed misclassifications at some parcels in the southeastern parts of the study area, but significant differences in classification patterns were not observed. Figure 5 shows some classification results using the SVM classifier. When three spectral bands of the August image were used for classification, misclassification and noise effects by isolated pixels were the greatest in visual inspection of classification results. Confusion between highland Kimchi cabbage and cabbage was most common, as shown in Figure 5b, mainly due to their similar spectral characteristics in August (this is further discussed in Section 3.5). When texture features were combined with spectral information for the case using the August image only, the number of misclassified and isolated pixels decreased, but some misclassified pixels were still shown (Figure 5c). Using multi-temporal images greatly reduced misclassified pixels within each parcel, except for some around the parcel boundaries (Figure 5d). As expected, the use of texture features as additional information with multi-temporal spectral information showed the best agreement with the ground-truth data from visual inspection (Figure 5e), indicating the necessity of time-series images and texture features for crop classification.

The impacts of texture features generated by different kernel sizes on the classification results were also visually compared. The classified patterns were significantly affected by kernel size. When a very small kernel size, such as GK3, was used to extract texture features, the classification result was very similar to the case with spectral information only. As the kernel size increased, the noise effect was greatly alleviated. When multi-temporal images were used for classification, however, the combination of texture features with multi-temporal spectral information was less affected by the change in kernel size. The increase in kernel size resulted in the reduction of isolated noise patterns, but the difference was subtle compared to the case using the August UAV image.

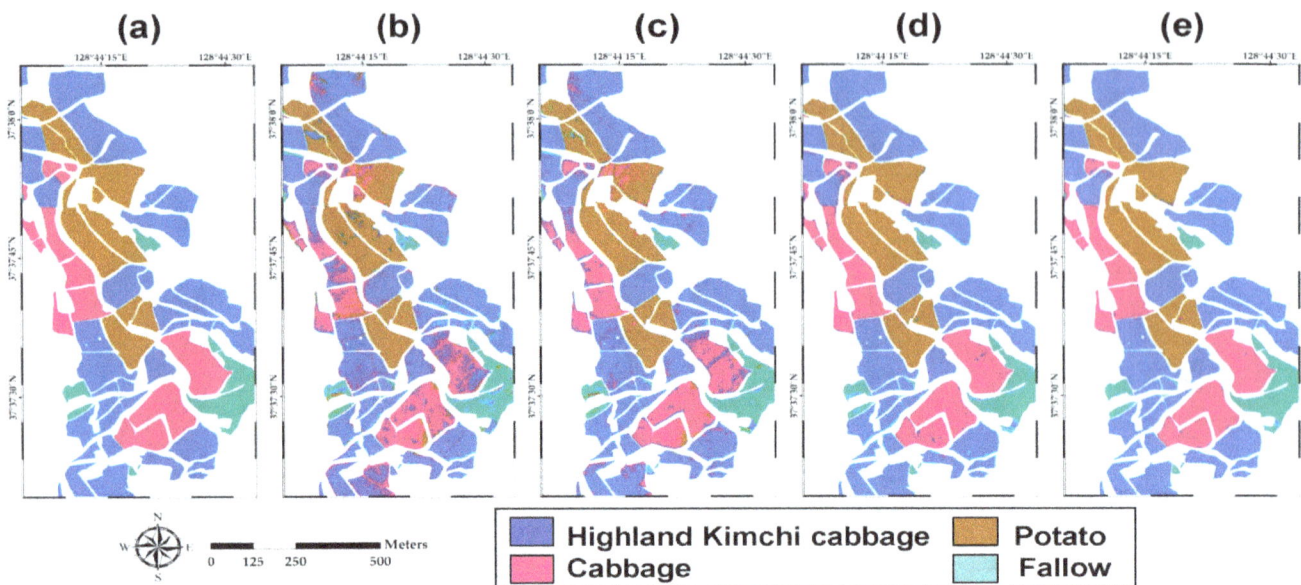

Figure 5. Comparison of SVM-based classification results with ground-truth data: (**a**) ground-truth data; (**b**) August image with VNIR; (**c**) August image with VNIR and GK31 texture features; (**d**) six multi-temporal images with VNIR; and (**e**) six multi-temporal with VNIR and GK31 texture features.

3.3. Quantitative Accuracy Assessment

The aforementioned visual and qualitative comparison results were further evaluated quantitatively by computing and comparing accuracy statistics. Confusion matrices were first prepared for all combination cases of each classifier, and related accuracy statistics were calculated by comparing classification results with reference data that were not used for training.

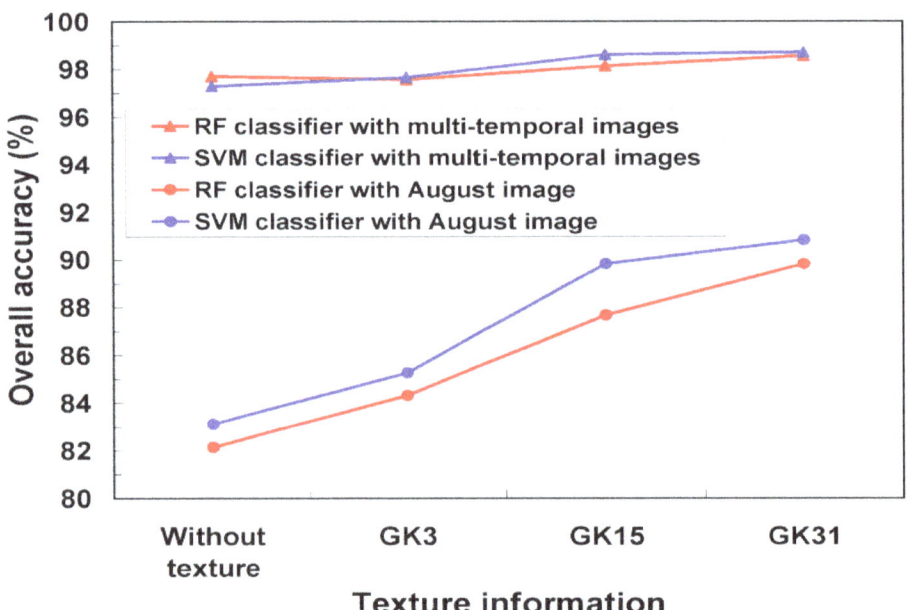

Figure 6. Overall accuracy of classification results without texture features and with texture features generated from different kernel sizes for the cases using the August image and six multi-temporal images.

Figure 6 shows variations in OA of classification results without texture features (VNIR) and with texture features generated from different kernel sizes (GK3, GK15, and GK31) using the August image and multi-temporal images. Although a very small portion of ground-truth data were used as training data, the OA values for the two classifiers were notably high (i.e. over 97%) when multi-temporal images were used for classification. Regardless of the number of input images and the classifier type, the combination of texture features and spectral information led to an increase in OA. The OA also increased with kernel size; however, only a slight improvement of OA was achieved for classification with multi-temporal images as kernel size increased. This result can be explained by the fact that most useful information for the discrimination of crops was already provided by time-series spectral information; hence, the contribution of texture features was minimal. In contrast, the improvement in OA by accounting for texture features was much more significant in the classification result using the August image only than using the multi-temporal images. Furthermore, the kernel size of GLCM greatly affected the OA using the August image. As kernel size increased, OA increased for both SVM and RF classifiers, and the use of GK31 texture features showed the best classification accuracy.

When comparing classification performance of both classifiers, the SVM classifier exhibited better OA than the RF classifier for the classification with the August image, indicating the superiority of the SVM classifier for the classification of crops in this study area. The difference in OA between SVM and RF classifiers was significant at the 5% significance level from the McNemar test [49], regardless of kernel sizes. It is noteworthy that the small difference in OA between two classifiers was significant at the 5% significance level even for all classification results based on multi-temporal images. Despite the similar OA values between two classifiers in the classification of multi-temporal images, this statistically significant difference was mainly due to evaluation with a very large amount of reference data (6,710,210 pixels). Even though parameter tuning is more demanding in the SVM classifier than the RF classifier, the optimal two parameters of the SVM classifier which were determined during a training stage with a relatively small training dataset could avoid overfitting the training data, leading to generalization ability for the large amount of reference data in this study.

Some confusion matrices for typical combination cases of the SVM classifier (one image versus multi-temporal images and with or without texture features) are listed in Table 3. Considering only the August image, combining texture features (GK31) with spectral information led to an increase of 7.72%p in OA, compared with the classification result with spectral information only (from 83.13% to 90.85%). The increase of class-wise accuracies was also achieved, as well as the improvement in OA. As discussed in the visual analysis of classification results, the confusion among four classes in Table 3 (particularly between highland Kimchi cabbage and cabbage) was significantly reduced, yielding increases in both PA and UA for all classes. When the August image with VNIR only was considered for classification, the similar vegetation vitality of highland Kimchi cabbage, cabbage, and weeds within the fallow class resulted in severe confusion. By accounting for texture features with spectral information, the confusion could be reduced. However, PA and UA of cabbage were relatively lower than that of other crops, indicating a persistent misclassification of cabbage to highland Kimchi cabbage. When multi-temporal images were used for classification, the accuracy values of all classes increased, particularly with cabbage. Texture features with multi-temporal spectral information proved most useful in the cabbage class because it alleviated the misclassification of cabbage to highland Kimchi cabbage.

Based on all evaluation results in Figure 6 and Table 3, it can be concluded that texture information extracted by the proper kernel size can improve classification performance, and the impact of using texture features is most significant when using a single image for crop classification. The latter finding implies the usefulness of texture information when only one UAV image is available for crop classification, due to difficulty acquiring time-series UAV images in the area of interest.

Table 3. Confusion matrices and accuracy statistics of some combination cases for the support vector machine (SVM) classifier. VNIR: visible and near infrared; UA: user's accuracy; PA: producer's accuracy; OA: overall accuracy; GK31: kernel size of 31 × 31.

August Image: VNIR Spectral Information

Classification \ Reference	Highland Kimchi Cabbage	Cabbage	Potato	Fallow	UA (%)
Highland Kimchi cabbage	3,074,131	342,355	49,838	84,726	86.57
Cabbage	230,661	869,250	65,288	6627	74.18
Potato	107,897	124,020	1,259,483	31,883	82.68
Fallow	67,045	12,343	9497	375,166	80.85
PA (%)	88.34	64.49	91.00	75.27	
OA (%)	83.13				

August Image: VNIR Spectral Information and GK31 Texture Features

Classification \ Reference	Highland Kimchi Cabbage	Cabbage	Potato	Fallow	UA (%)
Highland Kimchi cabbage	3,317,807	237,669	22,404	44,455	91.59
Cabbage	128,847	1,050,991	57,648	2636	84.75
Potato	16,522	54,544	1,294,756	18,465	93.53
Fallow	16,558	4764	9298	432,846	93.39
PA (%)	95.35	77.97	93.54	86.85	
OA (%)	90.85				

Multi-Temporal Images: VNIR Spectral Information

Classification \ Reference	Highland Kimchi Cabbage	Cabbage	Potato	Fallow	UA (%)
Highland Kimchi cabbage	3,421,871	46,150	11,031	41,618	97.19
Cabbage	15,143	1,294,009	10,189	4185	97.77
Potato	2092	4562	1,360,566	200	99.50
Fallow	40,628	3247	2320	452,399	90.73
PA (%)	98.34	96.00	98.30	90.77	97.30
OA (%)	97.30				

Multi-Temporal Images: VNIR Spectral Information and GK31 Texture Features

Classification \ Reference	Highland Kimchi Cabbage	Cabbage	Potato	Fallow	UA (%)
Highland Kimchi cabbage	3,461,811	35,558	5349	16,199	98.38
Cabbage	7159	1,309,847	5323	2337	98.88
Potato	1346	792	1,372,686	186	99.83
Fallow	9418	1771	748	479,680	97.57
PA (%)	99.48	97.17	99.17	96.24	98.72
OA (%)	98.72				

3.4. Comparison of Spectral and Texture Information

To examine which variable was most influential for classification performance, quantitative measures for variable importance were computed using the MDA in the RF classifier. MDA values of input variables with respect to different kernel sizes of GLCM are shown in Figure 7. Since 54 input variables were used for the classification of six multi-temporal images, only the top nine variables with the highest MDA values are presented for illustration purposes. Regardless of input images and the kernel size of GLCM, NIR and green bands were the most influential variables of the RF classifier. In particular, the NIR bands from July to September were included as important variables for the classification of multi-temporal images. Note that spectral information was more useful than texture information, and only one texture feature, such as ME, was helpful for multi-temporal images. ME, which is an estimate of the intensity of all pixels in spatial relationships that contribute to the GLCM, was the most important variable among the six texture features, irrespective of input images.

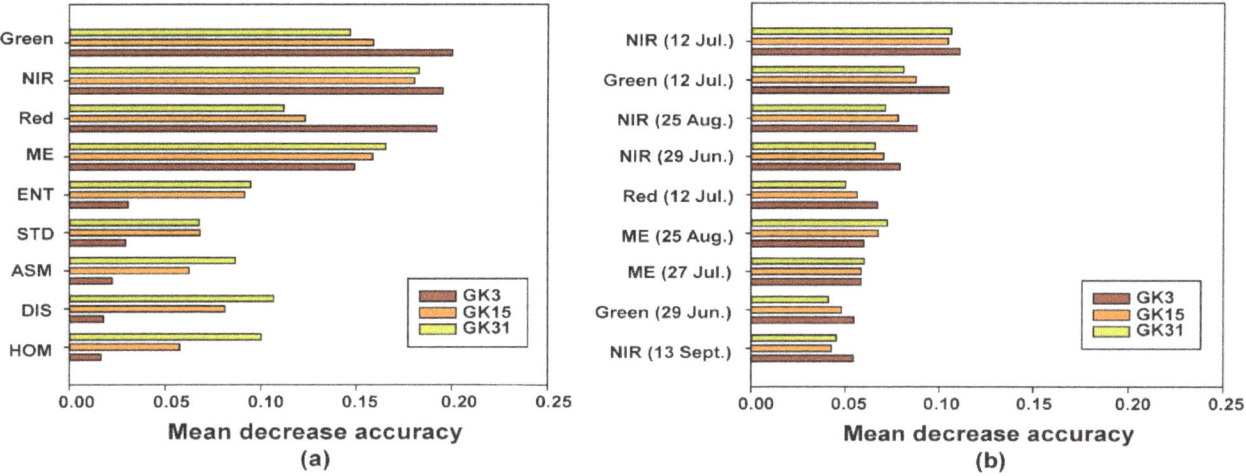

Figure 7. Mean decrease accuracy (MDA) values of input spectral and texture variables with respect to kernel size: (**a**) August image; and (**b**) multi-temporal images. ME: mean; ENT: entropy; ASM: angular second moment; STD: standard deviation; HOM: homogeneity; DIS: dissimilarity.

The MDA values of input variables were quite different according to the input images. When six multi-temporal images were used for classification, the MDA value for each variable was relatively small due to contributions of many input variables, but information content provided by many input variables led to very high classification accuracy, as shown in Table 3. Although multi-temporal spectral bands were considered the most informative, the influence of ME increased with kernel size (see the MDA value of ME for GK31 in Figure 7). With classification using only the August image, ME was the second most important variable for GK15 and GK31, indicating that the ME feature is very useful for the classification of crops in the study area. The contribution of other texture features increased with kernel size. For GK3, MDA values of texture features were much smaller than those of spectral bands. With increasing kernel size, gains in MDA values were most significant for texture features, including DIS and ENT. Texture information extracted from the GLCM with the proper kernel size can fill gaps in multi-spectral information, leading to an improvement in classification accuracy, as shown in Figure 6 and Table 3.

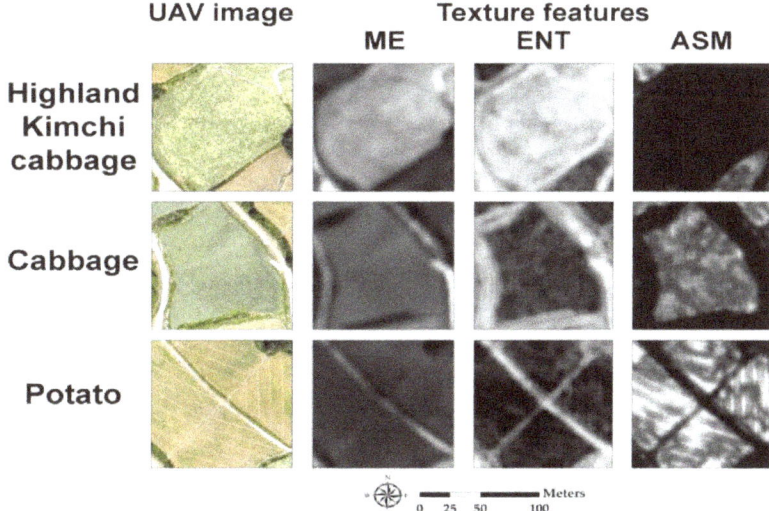

Figure 8. Some texture features (GK31) in subareas of the August image.

For further qualitative inspection of texture features, some texture features in four subareas of GK31 are provided in Figure 8. Brighter colors represent larger values in each texture feature. ME, which is regarded as the GLCM mean, provides low-pass filtered spatial information that is useful to mitigate noise effects in the ultra-high-resolution UAV image. As an index for measuring the randomness of contrast distributions, ENT increased with greater change of brightness values between the center pixel and its neighboring pixels. ENT values for different classes appear in Figure 8. ASM, which measures uniformity of contrast, also changed with the four classes. This visual inspection of texture features further confirmed the usefulness of texture information.

When considering the spatial resolution of the UAV image used for crop classification (i.e., 25 cm), GK3 and GK31 texture information represents 0.75 m and 7.75 m on the ground, respectively. The GK31 texture features are likely to represent the serial line patterns of crop cultivation well, consequently leading to superior OA. However, this is the particular result in the study area. If the spatial resolution of input images and the crop types change, the optimal kernel size of GLCM should be determined by considering spatial resolution, as well as cultivation patterns and crop characteristics such as size and shape.

3.5. Time-Series Analysis of Normalized Difference Vegetation Index for Selection of Optimal UAV Image

Spectral characteristics of crops depend on crop type and health conditions, but different crops may exhibit similar spectral response [35,50]. Accordingly, time-series images acquired during growth cycles of crops are often used to examine how well these images account for temporal variations of spectral response. For example, if temporal patterns in spectral responses of crops in the study area are significantly different, classification based on multi-temporal images can achieve satisfactory classification accuracy. Conversely, discrimination of crops with similar temporal variations of spectral responses may be difficult, even when multi-temporal images are used.

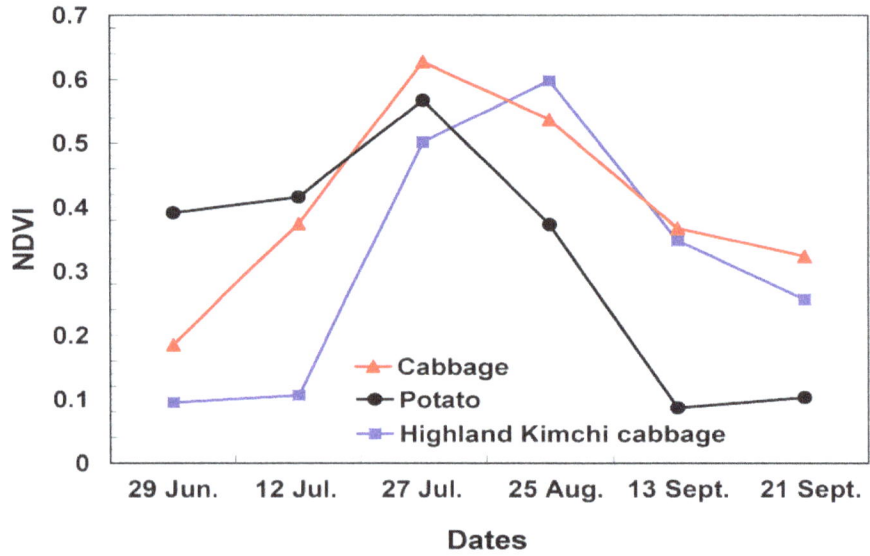

Figure 9. Temporal profiles of average normalized difference vegetation index (NDVI) values for three crops.

Figure 9 shows temporal variations in the average of normalized difference vegetation index (NDVI) values at pixels belonging to each crop. NDVI is a standardized index that quantifies greenness by using the difference in reflectance between NIR and red bands [51]. The average NDVI value of highland Kimchi cabbage was significantly lower than other crops on 12 July, and peaked in late August. In late July, cabbage had the highest NDVI value, followed by potatoes. The difference in average NDVI values between highland Kimchi cabbage and cabbage was not great in the August image (Figure 9), which led to difficulty in discerning the two crops. Although the difference was greater on 27 July, as shown in Figure 9, the lowest NDVI value of highland Kimchi may have resulted in the confusion with fallow and other small vegetation in the classification result using the 27 July image. If only one UAV image should be acquired, the image needs to be acquired when the vegetation vitality of the crop of interest reaches its maximum. Since highland Kimchi cabbage reached its maximum NDVI value in the 25 August image, we selected that image as the optimal single image. Actually, the classification accuracy using either the 12 July or the 27 July image was either similar to or lower than that using the August image. Despite the risk of misclassification using only the August image, similar spectral responses of different crops highlight the necessity of using additional information such as texture features, as applied in this study. Since the time to reach the maximum peak in NDVI may differ every year depending on weather conditions, however, the selection of the most appropriate acquisition date should be made by considering conditions and types of crops. Therefore, more extensive experiments should be carried out in other areas with different crop types. In addition, if phenological characteristics can be estimated from the entire time-series image set [52,53], a single-image acquisition date can be determined more optimally.

3.6. Classification Methods

In this study, two machine learning algorithms including RF and SVM were applied to crop classification. Recently, deep learning algorithms including convolutional neural network (CNN) were widely applied to remote sensing data classification [54–56]. Despite the promising performance of CNN, Kim et al. [57] reported that the training sample size has greater effects on the accuracy of CNN than that of SVM in crop classification, indicating a need for numerous training samples for improved CNN classification performance. Furthermore, Yu et al. [58] also reported that SVM with adjacent region features showed better accuracy than CNN for moderate-resolution land-cover classification. Therefore, deep learning is not always superior for all cases, and conventional machine learning algorithms can achieve classification performance comparable to, or even better than deep learning algorithms if proper spatial contextual features are combined with spectral information. To further evaluate the usefulness of texture features for crop classification, comparison with a patch-based CNN classifier will be conducted.

4. Conclusions

This study investigated the potential of GLCM-based texture information for crop classification with time-series UAV images and machine learning algorithms. The main focus was on the evaluation of the benefit of utilization of texture features along with spectral information when using a single UAV image. A case study of crop classification in the highland Kimchi cabbage cultivation area demonstrated the most accurate classification of multi-temporal UAV images with GLCM-based texture features. However, the utilization of texture features with spectral information from multi-temporal images did not lead to a significant improvement in classification accuracy. In contrast, when only a single UAV image was used, the utilization of texture features could significantly improve the classification accuracy. Therefore, when only one UAV image should be used for crop classification due to a difficulty in constructing a time-series UAV dataset, the information deficiency in spectral information can be complemented by structural information from texture features. Furthermore, the impact of texture information on classification accuracy was dependent on the kernel size of GLCM. Texture information extracted from the GLCM with larger kernel size improved classification performance in the study area. Therefore, proper kernel size selection is critical for the extraction of GLCM-based texture features. This indicates that both spatial resolution of input UAV images and shape characteristics of individual crops of interest should be considered in selection of optimal kernel size. However, these findings may be specific to this study area with particular crop types and not applicable to other areas. Therefore, more experiments on other areas with different combinations of crops should be carried out to strengthen the potential benefit of texture information from UAV images for crop classification. Experiments regarding determination of the minimum number of UAV images in crop classification with texture features, and comparison with deep learning algorithms will also be carried out in the future to extend key findings and recommendations presented herein.

Author Contributions: Conceptualization, G.-H.K. and N.-W.P.; methodology development and experiments, G.-H.K.; manuscript writing, G.-H.K. and N.-W.P.; supervision, N.-W.P.

Acknowledgments: The authors thank Kyung-Do Lee, Sang-Il Na, and Chan-Won Park at the National Institute of Agricultural Sciences for providing UAV images and ground-truth data used in this study. Constructive comments from the three anonymous reviewers are gratefully acknowledged.

References

1. Rosenzweig, C.; Elliott, J.; Deryng, D.; Ruane, A.C.; Müller, C.; Arneth, A.; Boote, K.J.; Folberth, C.; Glotter, M.; Khabarov, N.; et al. Assessing agricultural risks of climate change in the 21st century in a global gridded crop model intercomparison. *Proc. Natl. Acad. Sci. USA* **2014**, *111*, 3268–3273. [CrossRef] [PubMed]

2. Zhong, L.; Gong, P.; Biging, G.S. Efficient corn and soybean mapping with temporal extendability: A multiyear experiment using Landsat imagery. *Remote Sens. Environ.* **2014**, *140*, 1–13. [CrossRef]

3. Na, S.I.; Park, C.W.; So, K.H.; Ahn, H.Y.; Lee, K.D. Development of biomass evaluation model of winter crop using RGB imagery based on unmanned aerial vehicle. *Korean J. Remote Sens.* **2018**, *34*, 709–720, (In Korean with English Abstract).

4. Tilman, D.; Balzer, C.; Hill, J.; Befort, B.L. Global food demand and the sustainable intensification of agriculture. *Proc. Natl. Acad. Sci. USA* **2011**, *108*, 20260–20264. [CrossRef] [PubMed]

5. Lee, J.; Seo, B.; Kang, S. Development of a biophysical rice yield model using all-weather climate data. *Korean J. Remote Sens.* **2018**, *33*, 721–732, (In Korean with English Abstract).

6. Kim, Y.; Park, N.-W.; Lee, K.-D. Self-learning based land-cover classification using sequential class patterns from past land-cover maps. *Remote Sens.* **2018**, *9*, 921. [CrossRef]

7. Clark, M.L. Comparison of simulated hyperspectral HyspIRI and multispectral Landsat 8 and Sentinel-2 imagery for multi-seasonal, regional land-cover mapping. *Remote Sens. Environ.* **2017**, *200*, 311–325. [CrossRef]

8. Sonobe, R.; Yamaya, Y.; Tani, H.; Wang, X.; Kobayashi, N.; Mochizuki, K.-I. Mapping crop cover using multi-temporal Landsat 8 OLI imagery. *Int. J. Remote Sens.* **2017**, *38*, 4348–4361. [CrossRef]

9. Friedl, M.A.; McIver, D.K.; Hodges, J.C.F.; Zhang, X.Y.; Muchoney, D.; Strahler, A.H.; Woodcock, C.E.; Gopal, S.; Schneider, A.; Cooper, A.; et al. Global land cover mapping from MODIS: Algorithms and early results. *Remote Sens. Environ.* **2002**, *83*, 287–302. [CrossRef]

10. Yang, C.; Wu, G.; Ding, K.; Shi, T.; Li, Q.; Wang, J. Improving land use/land cover classification by integrating pixel unmixing and decision tree methods. *Remote Sens.* **2017**, *9*, 122. [CrossRef]

11. Hall, O.; Dahlin, S.; Marstorp, H.; Archila Bustos, M.; Öborn, I.; Jirström, M. Classification of maize in complex smallholder farming systems using UAV imagery. *Drones* **2018**, *2*, 22. [CrossRef]

12. Böhler, J.; Schaepman, M.; Kneubühler, M. Crop classification in a heterogeneous arable landscape using uncalibrated UAV data. *Remote Sens.* **2018**, *10*, 1282. [CrossRef]

13. Pajares, G. Overview and current status of remote sensing applications based on unmanned aerial vehicles (UAVs). *Photogramm. Eng. Remote Sens.* **2015**, *81*, 281–330. [CrossRef]

14. Shahbazi, M.; Théau, J.; Ménard, P. Recent applications of unmanned aerial imagery in natural resource management. *GISci. Remote Sens.* **2014**, *51*, 339–365. [CrossRef]

15. Latif, M.A. An agricultural perspective on flying sensors. *IEEE Geosci. Remote Sens. Mag.* **2018**, *6*, 10–22. [CrossRef]

16. Poblete-Echeverría, C.; Olmedo, G.F.; Ingram, B.; Bardeen, M. Detection and segmentation of vine canopy in ultra-high spatial resolution RGB imagery obtained from unmanned aerial vehicle (UAV): A case study in a commercial vineyard. *Remote Sens.* **2017**, *9*, 268. [CrossRef]

17. Melville, B.; Fisher, A.; Lucieer, A. Ultra-high spatial resolution fractional vegetation cover from unmanned aerial multispectral imagery. *Int. J. Appl. Earth Obs. Geoinf.* **2019**, *78*, 14–24. [CrossRef]

18. Im, J.; Jensen, J.R.; Tullis, J.A. Object-based change detection using correlation image analysis and image segmentation. *Int. J. Remote Sens.* **2008**, *29*, 399–423. [CrossRef]

19. Li, M.; Zang, S.; Zhang, B.; Li, S.; Wu, C. A review of remote sensing image classification techniques: The role of spatio-contextual information. *Eur. J. Remote Sens.* **2014**, *47*, 389–411. [CrossRef]

20. Haralick, R.M.; Shanmugam, K.; Dinstein, I. Textural features for image classification. *IEEE Trans. Syst. Man Cybern.* **1973**, SMC-3, 610–621. [CrossRef]

21. Feng, Q.; Liu, J.; Gong, J. UAV remote sensing for urban vegetation mapping using random forest and texture analysis. *Remote Sens.* **2015**, *7*, 1074–1094. [CrossRef]

22. Yang, M.-D.; Huang, K.-S.; Kuo, Y.-H.; Tsai, H.P.; Lin, L.-M. Spatial and spectral hybrid image classification for rice lodging assessment through UAV imagery. *Remote Sen.* **2017**, *9*, 583. [CrossRef]

23. Zhang, X.; Cui, J.; Wang, W.; Lin, C. A study for texture feature extraction of high-resolution satellite images based on a direction measure and gray level co-occurrence matrix fusion algorithm. *Sensors* **2017**, *17*, 1474. [CrossRef] [PubMed]

24. Benz, U.C.; Hofmann, P.; Willhauck, G.; Lingenfelder, I.; Heynen, M. Multi-resolution, object-oriented fuzzy analysis of remote sensing data for GIS-ready information. *ISPRS J. Photogramm. Remote Sens.* **2004**, *58*, 239–258. [CrossRef]

25. Laliberte, A.S.; Rango, A.; Havstad, K.M.; Paris, J.F.; Beck, R.F.; McNeely, R.; Gonzalez, A.L. Object-oriented image analysis for mapping shrub encroachment from 1937 to 2003 in southern New Mexico. *Remote Sens. Environ.* **2004**, *93*, 198–210. [CrossRef]

26. Liu, H.; Zhang, J.; Pan, Y.; Shuai, G.; Zhu, X.; Zhu, S. An efficient approach based on UAV orthographic imagery to map paddy with support of field-level canopy height from point cloud data. *IEEE J. Sel. Top. Appl. Earth Obs. Remote Sens.* **2018**, *11*, 2034–2046. [CrossRef]

27. de Castro, A.I.; Torres-Sánchez, J.; Peña, J.M.; Jiménez-Brenes, F.M.; Csillik, O.; López-Granados, F. An automatic random forest-OBIA algorithm for early weed mapping between and within crop rows using UAV imagery. *Remote Sens.* **2018**, *10*, 285. [CrossRef]

28. Ahmed, O.S.; Shemrock, A.; Chabot, D.; Dillon, C.; Williams, G.; Wasson, R.; Franklin, S.E. Hierarchical land cover and vegetation classification using multispectral data acquired from an unmanned aerial vehicle. *Int. J. Remote Sens.* **2017**, *38*, 2037–2052. [CrossRef]

29. Tatsumi, K.; Yamashiki, Y.; Morante, A.K.M.; Fernández, L.R.; Nalvarte, R.A. Pixel-based crop classification in Peru from Landsat 7 ETM+ images using a random forest model. *J. Agric. Meteorol.* **2016**, *72*, 1–11. [CrossRef]

30. Moeckel, T.; Dayananda, S.; Nidamanuri, R.R.; Nautiyal, S.; Hanumaiah, N.; Buerkert, A.; Wachendorf, M. Estimation of vegetable crop parameter by multi-temporal UAV-borne images. *Remote Sens.* **2018**, *10*, 805. [CrossRef]

31. Song, Q.; Xiang, M.; Hovis, C.; Zhou, Q.; Lu, M.; Tang, H.; Wu, W. Object-based feature selection for crop classification using multi-temporal high-resolution imagery. *Int. J. Remote Sens.* **2018**, 1–16. [CrossRef]

32. Ma, L.; Fu, T.; Blaschke, T.; Li, M.; Tiede, D.; Zhou, Z.; Ma, X.; Chen, D. Evaluation of feature selection methods for object-based land cover mapping of unmanned aerial vehicle imagery using random forest and support vector machine classifiers. *ISPRS Int. J. Geo-Inf.* **2017**, *6*, 51. [CrossRef]

33. Yuan, Y.; Lin, J.; Wang, Q. Hyperspectral image classification via multitask joint sparse representation and stepwise MRF optimization. *IEEE Trans. Cybern.* **2016**, *46*, 2966–2977. [CrossRef]

34. Xie, F.; Li, F.; Lei, C.; Ke, L. Representative band selection for hyperspectral image classification. *ISPRS Int. J. Geo-Inf.* **2018**, *7*, 338. [CrossRef]

35. Lee, K.-D.; Park, C.-W.; So, K.-H.; Kim, K.-D.; Na, S.-I. Characteristics of UAV aerial images for monitoring of highland Kimchi cabbage. *Korean J. Soil Sci. Fertil.* **2017**, *50*, 162–178, (In Korean with English Abstract). [CrossRef]

36. EGIS (Environmental Geographic Information Service). Available online: https://egis.me.go.kr (accessed on 15 October 2018).

37. Breiman, L. Random forests. *Mach. Learn.* **2001**, *45*, 5–32. [CrossRef]

38. Gehler, P.V.; Schölkopf, B. An introduction to kernel learning algorithms. In *Kernel Methods for Remote Sensing Data Analysis*; Camps-Valls, G., Bruzzone, L., Eds.; Wiley: Chichester, UK, 2009; pp. 25–48.

39. Foody, G.M.; Mathur, A. A relative evaluation of multiclass image classification by support vector machines. *IEEE Trans. Geosci. Remote Sens.* **2004**, *42*, 1335–1343. [CrossRef]

40. Brereton, R.G.; Lloyd, G.R. Support vector machines for classification and regression. *Analyst* **2010**, *135*, 230–267. [CrossRef]

41. Foody, G.M.; Mather, A. Toward intelligent training of supervised image classifications: Directing training data acquisition for SVM classification. *Remote Sens. Environ.* **2004**, *93*, 107–117. [CrossRef]

42. Meyer, D.; Dimitriadou, E.; Hornik, K.; Weingessel, A.; Leisch, F. e1071: Misc Functions of the Department of Statistics, Probability Theory Group (Formerly: E1071), TU Wien. Available online: https://CRAN.R-project.org/package=e1071 (accessed on 28 December 2018).

43. Tuceryan, M.; Jain, A.K. Texture analysis. In *Handbook of Pattern Recognition & Computer Vision*, 2nd ed.; Chen, C.H., Pau, L.F., Wang, P.S.P., Eds.; World Scientific Publishing: Singapore, 1999; pp. 207–248.

44. Castillo-Santiago, M.A.; Ricker, M.; de Jong, B.H.J. Estimation of tropical forest structure from SPOT-5 satellite images. *Int. J. Remote Sens.* **2010**, *31*, 2767–2782. [CrossRef]

45. Johansen, K.; Coops, N.C.; Gergel, S.E.; Stange, Y. Application of high spatial resolution satellite imagery for riparian and forest ecosystem classification. *Remote Sens. Environ.* **2007**, *110*, 29–44. [CrossRef]

46. Szantoi, Z.; Escobedo, F.; Abd-Elrahman, A.; Smith, S.; Pearlstine, L. Analyzing fine-scale wetland composition using high resolution imagery and texture features. *Int. J. Appl. Earth Obs.* **2013**, *23*, 204–212. [CrossRef]

47. R Core Team. R: A Language and Environment for Statistical Computing. Available online: https://www.R-project.org (accessed on 28 December 2018).

48. Liaw, A.; Wiener, M. Classification and regression by randomForest. *R News* **2002**, *2*, 18–22.

49. Foody, G.M. Thematic map comparison: Evaluating the statistical significance of differences in classification accuracy. *Photogramm. Eng. Remote Sens.* **2004**, *70*, 627–633. [CrossRef]

50. Na, S.I.; Park, C.W.; So, K.H.; Ahn, H.Y.; Lee, K.D. Application method of unmanned aerial vehicle for crop monitoring in Korea. *Korean J. Remote Sens.* **2018**, *34*, 829–846, (In Korean with English Abstract).

51. Lillesand, T.M.; Kiefer, R.W.; Chipman, J.W. *Remote Sensing and Image Interpretation*, 6th ed.; Wiley: Hoboken, NJ, USA, 2008.

52. Atzberger, C.; Klisch, A.; Mattiuzzi, M.; Vuolo, F. Phenological metrics derived over the European continent from NDVI3g data and MODIS time series. *Remote Sens.* **2014**, *6*, 257–284. [CrossRef]

53. Lee, K.D.; Park, C.W.; So, K.H.; Kim, K.D.; Na, S.I. Estimating of transplanting period of highland Kimchi cabbage using UAV imagery. *J. Korean Soc. Agric. Eng.* **2017**, *59*, 39–50, (In Korean with English Abstract).

54. Ji, S.; Zhang, C.; Xu, A.; Shi, Y.; Duan, Y. 3D convolutional neural networks for crop classification with multi-temporal remote sensing images. *Remote Sens.* **2018**, *10*, 75. [CrossRef]

55. Kampffmeyer, M.; Salberg, A.-B.; Jenssen, R. Urban land cover classification with missing data modalities using deep convolutional neural networks. *IEEE J. Sel. Top. Appl. Earth Obs. Remote Sens.* **2018**, *11*, 1758–1768. [CrossRef]

56. Tan, K.; Wang, X.; Zhu, J.; Hu, J.; Li, J. A novel active learning approach for the classification of hyperspectral imagery using quasi-Newton multinomial logistic regression. *Int. J. Remote Sens.* **2018**, *39*, 3029–3054. [CrossRef]

57. Kim, Y.; Kwak, G.-H.; Lee, K.-D.; Na, S.-I.; Park, C.-W.; Park, N.-W. Performance evaluation of machine learning and deep learning algorithms in crop classification: Impact of hyper-parameters and training sample size. *Korean J. Remote Sens.* **2018**, *34*, 811–827, (In Korean with English Abstract).

58. Yu, L.; Su, J.; Li, C.; Wang, L.; Luo, Z.; Yan, B. Improvement of moderate resolution land use and land cover classification by introducing adjacent region features. *Remote Sens.* **2018**, *10*, 414. [CrossRef]

10

A Critical Review of Spatial Predictive Modeling Process in Environmental Sciences with Reproducible Examples in R

Jin Li

National Earth and Marine Observations Branch, Environmental Geoscience Division, Geoscience Australia, Canberra 2601, Australian Capital Territory, Australia; Jin.Li@ga.gov.au

Abstract: Spatial predictive methods are increasingly being used to generate predictions across various disciplines in environmental sciences. Accuracy of the predictions is critical as they form the basis for environmental management and conservation. Therefore, improving the accuracy by selecting an appropriate method and then developing the most accurate predictive model(s) is essential. However, it is challenging to select an appropriate method and find the most accurate predictive model for a given dataset due to many aspects and multiple factors involved in the modeling process. Many previous studies considered only a portion of these aspects and factors, often leading to sub-optimal or even misleading predictive models. This study evaluates a spatial predictive modeling process, and identifies nine major components for spatial predictive modeling. Each of these nine components is then reviewed, and guidelines for selecting and applying relevant components and developing accurate predictive models are provided. Finally, reproducible examples using *spm*, an R package, are provided to demonstrate how to select and develop predictive models using machine learning, geostatistics, and their hybrid methods according to predictive accuracy for spatial predictive modeling; reproducible examples are also provided to generate and visualize spatial predictions in environmental sciences.

Keywords: spatial predictive models; predictive accuracy; model assessment; variable selection; feature selection; model validation; spatial predictions; reproducible research

1. Introduction

Spatial predictions of environmental variables are increasingly required in environmental sciences and management. Accurate spatially continuous data are required for environmental modeling, and for evidence-based environmental management and conservation. Such data are, however, usually not readily available and they are difficult and expensive to acquire, especially in areas that are difficult to access (e.g., mountainous or marine regions). In many cases, the spatial data of environmental variables are collected from point locations. Thus, spatial predictive methods are essential for generating spatially continuous predictions of environmental variables from the point samples. Moreover, predictive methods are increasingly being used to generate spatial predictions across various disciplines in environmental sciences [1–6] in parallel to recent advances in (1) computing technology and modeling techniques [7–9], and (2) data acquisition and data processing using remote-sensing techniques and geographic information systems. These advancements resulted in increasingly more environmental variables available for spatial predictive modeling. Consequently, more sophisticated spatial predictive modeling approaches are needed to deal with a large number of predictive variables.

Accuracy of spatial predictive model(s) is critical as it determines the quality of their predictions that form the scientific evidence to inform decision- and policy-making. Therefore, improving the accuracy by choosing an appropriate method and then identifying and developing the most accurate predictive model(s) is essential. It is often difficult to select an appropriate method for any given dataset because spatial predictive methods may be data- or even variable-specific and many factors need to be considered [10,11]. Although the development of hybrid methods of machine learning and geostatistics, and their application considerably improved predictive accuracy, these methods may also be data- or variable-specific [12–14]. For spatial predictive modeling, "no free lunch theorems" [15] are also applicable.

Furthermore, even with the right predictive method, it is a challenging task to identify and develop the most accurate predictive model(s). This is because the spatial predictive modeling process involves many factors or components [10,16,17]. In fact, only a portion of such factors were considered in many previous studies, which often led to sub-optimal or even misleading predictive models [11,18]. This not only presents an opportunity for scientists to develop and improve their predictive models, but also highlights the challenge of selecting relevant predictive variables from a large number of available predictive variables to form the most accurate predictive model. Heavy computations are often involved in identifying and selecting accuracy-improved predictive models for given datasets when the number of predictive variables is large, although high-performance computing facilities may be able to significantly alleviate this challenge.

This study aims to assist spatial modelers and scientists by critically reviewing the spatial predictive modeling process, developing guidelines for selecting the most appropriate spatial predictive methods and identifying and developing the most accurate predictive model to generate spatial predictions. In this study, I focus on spatial predictive models or spatial predictive modeling for generating spatially continuous predictions rather than on other models (e.g., inferential model) as discussed previously [19]. Consequently, the term accuracy in this study refers to the accuracy of predictive model(s) based on validation, and the term uncertainty refers to prediction uncertainty generated by predictive model(s). In this study, the term accuracy is used interchangeably with predictive accuracy. Furthermore, in this study, I mainly focus on the predictive methods for numerical data that are usually encountered in environmental sciences, with a brief discussion on categorical data. In this study, the following nine major components of spatial predictive modeling are identified and reviewed: (1) sampling design, sample quality control, and spatial reference systems; (2) selection of spatial predictive methods; (3) pre-selection of predictive variables; (4) exploratory analysis for variable selection; (5) parameter selection for relevant methods; (6) variable selection; (7) accuracy and error measures for predictive models (numerical vs. categorical); (8) model validation; and (9) spatial predictions, prediction uncertainty, and their visualization. In addition, reproducible examples using *spm* [20], an R package for machine learning, geostatistics, and their hybrid methods, are employed to demonstrate how to select and develop predictive models based on predictive accuracy for spatial predictive modeling; reproducible examples are also provided to generate and visualize spatial predictions in environmental sciences.

2. Sampling Design, Sample Quality Control, and Spatial Reference Systems

2.1. Sampling Design

Although samples are usually collected, stored, and ready to use for spatial predictive modeling, sometimes samples are not available and need to be collected. In the latter situation, a sampling design needs to be produced. In this study, I focus on sampling designs over space. To collect samples from a survey area for a certain survey purpose, a sampling design is an important step and must be created. A good sampling design ensures that data collected from a survey are capable of answering relevant research questions. Better designs, such as spatially stratified sampling designs, will also be as precise and efficient as possible [21,22]. Many methods were developed to generate sampling designs [23–26]. They typically fall into four main categories: (1) non-random sampling design; (2) unstratified random sampling design; (3) stratified random sampling design; and (4) stratified random sampling design with prior information.

The non-random sampling design can be ad hoc sampling based on expert knowledge, purely opportunistic when a certain type of environmental condition becomes available, or systematic sampling. This type of sampling design was applied to many surveys [27–29]. For spatial predictive modeling, this method is not recommended for future studies. However, an interesting comparison of non-random sampling designs was reported for spatial predictive modeling [26]; it may provide some useful clues for sampling designs (e.g., lattice plus close pairs) for spatial predictive modeling.

The unstratified random sampling design is that sampling locations are randomly selected. This can be (1) an unstratified equal probability design, or (2) an unstratified unequal probability design [30,31]. This type of sampling design is not recommended for spatial predictive modeling studies because (1) spatial information is available for sampling design and thus spatially stratified sampling design should be used as discussed below, and (2) it may even be over-performed by the non-random design (i.e., lattice design) [26].

The stratified random sampling design is often used when additional information is available. Such information can be spatial (or location) information, elevation, bathymetry, or geomorphological information. For spatial predictions, such information is important and should be considered when designing a survey for a region. A few recently developed randomized spatial sampling procedures were reviewed and compared using simple random sampling without replacement as a benchmark for comparison [22]. This study provided some empirical evidence for the improvement of sampling efficiency from using these designs and provided some guidance for choosing an appropriate spatial sampling design [22]. Furthermore, some R packages, such as *spsurvey* [31], *GSIF* [32], *spcosa* [33], *clhs* [34], and *BalancedSampling* [35], were developed for this type of sampling design. The stratified random sampling designs with spatial information (i.e., spatially stratified sampling design) are increasingly being used in practice [28,36,37].

The stratified random sampling design with prior information is a new development for sampling over a space. It incorporates the locations of legacy sites into new spatially balanced survey designs to ensure spatial balance among all sample locations [21]. It can be seen as a stratified unequal probability design. An R package, *MBHdesign*, was developed for this method [38].

2.2. Sample Quality Control

Sample quality is vitally important because samples provide the fundamental information for spatial predictive modeling. Many factors may affect sample quality and they are usually dataset-specific [39,40]. Consequently, relevant factors need to be identified for each dataset and then relevant data quality control (QC) criteria need to be developed to QC the dataset [39,40] prior to undertaking spatial predictive modeling. For example, in Geoscience Australia's Marine Samples Database (MARS; http://dbforms.ga.gov.au/pls/www/npm.mars.search), seabed sediment samples were initially quality controlled prior to and after entering the database according to various criteria [12,41]. However, the quality of the samples was still affected by many factors, including data credibility (e.g., non-dredge), data accuracy (e.g., non-positive bathymetry), completeness (e.g., no missing values), etc.; hence, data quality control approaches were developed to QC the samples of seabed mud content and sand content [12,41]. These approaches may provide examples about how to identify relevant factors and develop possible data QC criteria for a given dataset. In some instances, data noise may result from repeated measurements, and certain rules may need to be developed to clean such samples based on professional knowledge [42]. Moreover, exploratory analysis can be used to further detect abnormal samples, as detailed in Section 5.

2.3. Spatial Reference Systems

To generate spatial predictions (i.e., spatially continuous data) for a region using spatial predictive models, two types of georeferenced data are required: (1) point samples of response and predictive variables; and (2) grid data of predictive variables. Such georeferenced data are often stored according to various spatial reference systems [43]. The spatial reference system used to project or store the spatial information is often assumed to have certain effects on the performance of predictive models; thus, in practice, various spatial reference systems were used to minimize such effects [43]. When a study area is small and within one particular UTM zone, spatial data are often projected using either the UTM zone or an appropriate projection system; when the study area is spanning multiple UTM zones, the existing geographic coordinate system (i.e., WGS84) or another appropriate projection system can be used.

Although the spatial reference system by which the spatial information is stored is often considered as a potential source of error for spatial predictive modeling, a series of studies demonstrated that the effects of spatial reference systems on the performance of spatial predictive methods (i.e., inverse distance weighting and ordinary kriging) are negligible for areas at various latitudinal locations (up to 70 dd) and spatial scales (i.e., regional and continental) [43–45]. Therefore, it was recommended that spatial reference system selection and re-projection can be removed for spatial predictive modeling for areas with latitude less than 70 dd, and spatial data can be modeled in WGS84 or the spatial projection system already used for the data. Although new spatial reference systems (e.g., DGGS [46]) may be developed to remedy various limitations of existing ones, the above recommendation may still be applicable as discussed previously on why the effects of spatial reference systems on the predictive accuracy are negligible [43–45].

3. Selection of Spatial Predictive Methods

3.1. Spatial Predictive Methods

For spatial predictive modeling, there are many methods available [3]. Previously, over 20 spatial predictive methods were grouped into (1) non-geostatistical methods (e.g., inverse distance weighting (IDW)), (2) geostatistical methods (e.g., ordinary kriging (OK)), and (3) combined or hybrid methods [10]. Collectively, these methods are largely non-machine learning methods and a small portion of these methods, like regression tree (RT) and linear regression models (LM), use secondary information.

When sufficient secondary information is available, a number of other methods could be used. These methods include traditional statistical methods, machine learning methods, the hybrid methods of traditional statistical methods and geostatistical methods, and the hybrid methods of machine learning and geostatistical methods (Table 1). These methods were applied or compared in various spatial predictive modeling studies [12,41,47–55]. Of these methods, random forest (RF), hybrid method of RF and OK (RFOK), and hybrid method of RF and IDW (RFIDW) were among the most accurate methods in these applications. Generalized boosted regression modeling (GBM), hybrid method of GBM and OK (GBMOK), and hybrid method of GBM and IDW (GBMIDW) showed great potential based on our unpublished study. In the current study, these methods are presented in three main groups: (1) non-machine learning methods; (2) machine learning methods; and (3) the hybrid methods.

Table 1. Spatial predictive methods using predictive variables [6,12,41,48–55].

Non-Machine Learning Method and Hybrid Methods		Machine Learning Method and Hybrid Methods	
Non-machine learning method	Hybrid methods	Machine learning method	Hybrid methods
Generalized additive models		Cubist	Cubist and OK (cubistOK)
Generalized least squares trend estimation (GLS)	GLS and OK	Generalized boosted regression modeling (GBM)	GBM and IDS (GBMIDS)
Generalized linear models (GLM)	GLM and IDW (GLMIDW)		GBM and OK (GBMOK)
	GLM and OK (GLMOK)	General regression neural network (GRNN)	GRNN and IDS (GRNNIDS)
GLM with lasso or elastic net regularization			GRNN and OK (GRNNOK)
	Linear models and OK	Multivariate adaptive regression splines	
	RT and IDS (RTIDS)	Naïve Bayes	
	RT and OK (RTOK)	Random forest (RF)	RF and IDS (RKIDS)
			RF and OK (RKOK)
		Support vector machine (SVM)	SVM and OK (SVMOK)
			SVM and OK (SVMIDS)

3.2. Selecting Spatial Predictive Methods

Selection of appropriate spatial predictive methods for a response variable (or dependent variable, or primary variable in geostatistics) is critical. For data without predictive variables, geostatistical methods are the only methods that can be used. Method selection was discussed and guidelines were developed for using geostatistical methods in various studies [16,56–61]. A decision tree was developed for selecting the most appropriate method from a pool of 25 spatial predictive methods according to the availability and nature of data and the expected predictions, together with the features of each method [10]. However, it was argued that there was no simple answer regarding the choice of appropriate geostatistical methods, because the hallmark of a good geostatistical modeling work is customization of the approach to the dataset at hand [56]. This suggests that "no free lunch theorems" [15] are also applicable for spatial predictive modeling using geostatistical methods. Joint application of two spatial predictive methods might produce additional benefits such as the combined procedures in previous studies [10,62–64].

For data with predictive variables, there are many options available. It is often difficult to select an appropriate method because the performance of spatial predictive methods depends on many factors, including the assumptions and properties of each method, the nature and spatial structure of data for the response variable, sample size and distribution, the availability of predictive variables, availability of software, computational demands, and many other factors [10,11]. All of these factors need to be considered when making an appropriate selection.

Moreover, if more than one method can be applied, model comparison techniques such as cross-validation in combination with the measures of predictive error or accuracy can be used to select a method. This selection technique not only selects the most appropriate method but also the most accurate predictive model that can maximize the predictive accuracy [12,65,66]. This selection technique can be applied to methods irrespective of whether they use predictive variables.

4. Pre-Selection of Predictive Variables

Predictive variables are termed predictor variables, independent variables, predictors, and features. They are also called secondary variables/information in geostatistics. They are essential for spatial predictive methods that use predictive variables.

4.1. Principles for Pre-Selection of Predictive Variables and Limitations

Principles for pre-selecting predictive variables may change with disciplines. For environmental sciences, the main principle is that predictive variables need to be closely related to the variable to be predicted (i.e., the response variable) [67,68]; ideally, they should be causal variables, or variables directly caused by the response variable (e.g., optical reflectance of vegetation types, backscatter of seabed substrates). They are usually identified based on expert or professional knowledge. However, in many cases, it is hard to know what the causal variable(s) is (are) for a response variable. Proxy (or surrogate) variables are often used instead of causal variable for spatial predictive modeling. Again, they are usually identified based on expert or professional knowledge [69]. Certainly, predictive models can use causal variables, proxy variables, or both if causal variables are not all available.

When the accuracy of a resultant predictive model is unexpectedly poor, then we may need to consider that we may have missed some important predictive variables, for which we may have no knowledge or even awareness (e.g., hidden variables [70]). Further actions are required to expand the professional knowledge pool in order to identify such possible predictive variables.

For spatial predictive modeling, the selection of potential predictive variables is even more challenging. This is because the selection could be constrained by certain factors. For example, predictive variables need to be continuously available for a target region. Spatial resolution is also a critical issue as the resolutions of various predictive variables need to meet the desired resolution for the final predictions, although they can be rescaled. Sometimes, even though we know the possible causal predictive variables based on expert or professional knowledge, they may not meet these requirements and cannot be used for spatial predictive modeling. This is particularly true in marine environmental sciences.

4.2. Predictive Variables for Environmental Sciences

For terrestrial environmental modeling, many predictive variables are available. Many previous applications provided examples of variables used for terrestrial environmental modeling [13,42,49–51, 53,67,71,72].

In contrast, the information of predictive variables is often scarce for marine environmental modeling. In many cases, proxy variables are usually used for predictive modeling [69,73]. For example, to predict the spatial distribution of seabed sediments for Australian Exclusive Economic Zone (AEEZ) at a resolution of 0.01 or 0.0025 degrees, only a few predictive variables were available for the whole AEEZ [12,41,74] (Table 2). For spatial predictions over smaller areas, quite often more predictive variables became available at desired resolution such as for seabed sediment [4,75–77], seabed hardness [78–80], and sponge species richness [6,81] (Table 2). Bathymetry and backscatter were also used to predict seabed sediment at local scale [82]. Some derived information may be used as predictive variables. For example, in Table 2, predictive variables 5–13 were derived from bathymetry (bathy), while predictive variables 15–19 were derived from backscatter (bs). Some other variables were used for seabed grain size at small scale [48]. In addition, many variables were reviewed [69,83] and could be used for marine environmental modeling. Fuzzy geomorphic features were also used for spatial predictive modeling at local scales [77,84].

Table 2. A list of predictive variables used for some marine environmental variables.

No	Predictive Variables	Seabed Sediment/Grain Size	Seabed Hardness	Sponge Species Richness	Window/Kernel Size(s)
1	Longitude (long)	yes	yes	yes	
2	Latitude (lat)	yes	yes	yes	
3	Distance to coast (dist)	yes		yes	
4	Bathymetry (bathy)	yes	yes	yes	
5	Local Moran's I from bathymetry	yes	yes	yes	yes
6	Mean curvature	yes			yes
7	Planar curvature	yes	yes	yes	yes
8	Profile curvature	yes	yes	yes	yes
9	Relief	yes	yes	yes	yes
10	Rugosity (or surface, surface complexity)	yes	yes	yes	yes
11	Slope	yes	yes	yes	yes
12	Topographic or bathymetric position index (tpi or bpi)	yes	yes	yes	yes
13	Fuzzy morphometric features	yes			yes
14	Backscatter (bs) 10–36	yes	yes	yes	yes
15	Entropy from bs			yes	yes
16	Homogeneity from bs		yes	yes	yes
17	Local Moran's I from bs		yes	yes	yes
18	Prock from bs		yes		
19	Variance from bs		yes	yes	yes
20	Suspended particulate matter	yes			
21	Mean tidal current velocity	yes			
22	Peak orbital velocity of waves at seabed	yes			
23	Roughness from bathy *	yes			
24	Roughness from bs *	yes			
25	Sobel filter from bathy #	yes			

* The difference between the minimum and maximum of cell and its eight neighbors [48]. # A directional filter that emphasizes areas of large spatial frequency (edges) running horizontally (X) or vertically (Y) across the image [48].

5. Exploratory Analysis for Variable Pre-Selection

5.1. Non-Machine Learning Methods

Exploratory analysis is often used to detect the relationships between the response variable and predictive variables for non-machine learning methods, such as LM, generalized linear models (GLM), and kriging with an external drift (KED). By applying such analysis, people intend to find data nature and structure [85]. Key issues may include the identification of (1) outliers, (2) homogeneity in variance of response variable, (3) data distribution of response variables including normality, (4) collinearity (i.e., the correlation among predictive variables), (5) relationships or response curves of response variable to predictive variables, (6) how strong the relationships are between response variable and predictive variables, (7) possible interactions among predictive variables, (8) independence of a response variable, whether temporally, spatially, or both, and (9) source of random errors, which may lead to a mixed-effect model or additional predictive variable(s) [71]. On the basis of the above analyses, certain actions can be taken to deal with relevant samples or variables. For instance, some predictive variables can be removed if their correlations with the response variable are low. However, it would be wise to let the variable selection process determine which variables should be removed because some variables may be important predictive variables even with low correlations. Highly correlated predictive variables, usually determined based on correlation coefficient (r) or variance inflation factor (VIF), can also be eliminated to reduce the collinearity, although caution should be taken for this exercise [9,86]. Relevant issues about collinearity were also discussed [87]. Some predictive variables may need to be specified to their second or third orders if a non-linear relationship is detected. Moreover, some interactions may need to be considered and tested.

5.2. Machine Learning Methods and Hybrid Methods

For machine learning methods, exploratory analysis is useful for understanding data and interpreting modeling results [78]. However, some roles of exploratory analysis for non-machine learning methods are no longer needed for machine learning methods. This is because machine learning methods, like RF, are free of assumptions on the data distribution and can handle non-linear relationships and interactive effects [88,89]. They can also handle highly correlated predictive variables [6,47]. Furthermore, the use of highly correlated predictive variables is encouraged for RF because they may be able to make a meaningful contribution to improving predictive accuracy [6].

5.3. Hybrid Methods

For the hybrid methods, exploratory analysis is as useful as for the aforementioned methods. The residuals of a detrending method (e.g., GLM, RF) are assumed to be normal if kriging methods are applied. Thus, the residuals need to be analyzed to check this assumption [12,41].

6. Parameter Selection

6.1. Parameter Selection for Non-Machine Learning Methods

For non-machine learning methods, I mainly focus on two commonly used methods [11], IDW and kriging (e.g., OK). For IDW, it is really dependent on the selection of appropriate values for a power parameter and the number of nearest observations, which can be selected based on their resultant predictive accuracy [41,90]. The smoothness of the estimated surface increases as the value of power parameter increases [91]; however, manipulating the power parameter to smooth the predictions and to produce visually pleasant maps does not warrant the quality of the resultant predictions and is not recommended.

For kriging methods, a number of parameters, including window size, and isotropy and anisotropy of data, need to be considered, as well as the variogram model and its parameters. Data transformation needs to be considered when the data are skewed and anisotropic. Three methods of data transformation (i.e., logarithms, standardized rank order, and normal scores) can be employed to reduce the skewness [74, 92]. Some other methods, such as Box–Cox transformation [82], arcsine [88], square-root transformation [47], and double square root or square root and log [12], can be used to normalize the data. The selection of these transformation and normalization methods is largely data-dependent and careful examination should be taken. The selection can also be determined according to their effects on the predictive accuracy.

In addition, for anisotropy, non-stationary methods like KED should be used in cases with a general anisotropy or trend (i.e., drift) [75]. If different types of non-stationarity exist, application of different spatial predictive methods to each type may improve predictive accuracy [93].

The parameter selection for these methods can be determined according to the predictive accuracy of resultant predictive models. This is demonstrated in Section 11.

6.2. Parameter Selection for Machine Learning Methods

For machine learning methods, RF and GBM are considered. For RF, relevant parameters are mtry, ntree, and so on [94], while, for GBM, these include n.trees, learning.rate, interaction.depth, bag.fraction, and so on [95]. Some commonly used default parameter values can be used as they are quite often optimal [94,96], except the distribution parameter for GBM. The distribution parameter for GBM should be based on data type of the response variable; for example, Poisson should be used for count data. Relevant parameters can also be selected based on cross-validation [41,48,96].

6.3. Parameter Selection for Hybrid Methods

The parameter selections for the above non-machine learning methods and machine learning methods are equally applicable to relevant hybrid methods.

7. Variable Selection

For machine learning methods, variable selection is termed feature selection, while, for non-machine learning methods, it is often called model selection. However, model selection often leads to the most parsimonious fitted model rather than the most accurate predictive model [6]. In this study, we use the term "variable selection" for both non-machine learning and machine learning methods to identify and develop the most accurate spatial predictive model(s).

Variable selection is important for many predictive methods, although it is not required for all methods. For instance, classification and regression trees [97] and LIVES [98] are exempt from variable selection. However, as per all other methods, they assume that the predictive variables used are informative and not misleading because they treat each predictive variable as equally important. Thus, misleading predictive variables may considerably reduce predictive accuracy as discussed previously [98]. The variable selection procedure for machine learning methods and their hybrid methods is fundamentally different from the procedure for non-machine learning methods [47,79,99]. For geostatistical methods like IDW and OK, no variable selection is required, and it is really about the selection of appropriate values for relevant parameters, as discussed in Section 6. In this section, I focus on the following three methods: (1) GLM; (2) RF; and (3) GBM. This is because of their wide applications, robustness, or the recent developments in variable selection techniques for these methods.

For GLM, there are many methods available in R for variable selection [100–103]. These methods may include (1) *stepAIC* or *step*; (2) *dropterm*, *drop1*, or *add1*; (3) *anova*; (4) *regsubsets*; and (5) *bestglm* [100,102,103]. The application of these methods for spatial predictive modeling can be seen in recent studies [6,104]. Variables selected based on these methods may form the most parsimonious model, but the model may have low predictive accuracy or even be misleading [6,104], with the exception that *bestglm* is promising if cross-validation, instead of Akaike's information criterion (AIC) or Bayesian information criterion (BIC), is used for information criteria [104]. Alternatively, the variable selection for GLM can also be based on variables selected by other method such as RF [71]. It was found that traditional variable selection methods are unsuitable for identifying GLM predictive models, and joint application of RF and AIC can select accuracy-improved predictive models [6]. This highlights the importance of differentiating variable selection for predictive modeling [6] from variable selection for hypothesis testing [99] or inferential modeling [19]. The common mistakes associated with incorrectly distinguishing data analytic types were briefly summarized and discussed previously [19].

For RF, variable selection methods may include (1) variable importance (VI) [78], (2) averaged variable importance (AVI) [79], (3) Boruta [105], (4) knowledge informed AVI (KIAVI) [6,79], (5) recursive feature selection (RFE) [106], and (6) variable selection using RF (VSURF) [107]. Of these methods, KIAVI is recommended because it outperforms all other variable selection methods [6,79,104,108].

For GBM, variables can be selected in terms of the relative influence [95,108]. The recursive feature selection [106] can also be used for variable selection.

Two concepts were proposed for variable selection: important and unimportant predictive variables based on the predictive accuracy [6,79]. They were defined as follows:

1. Important variable based on the predictive accuracy (IVPA). This refers to the variable for which exclusion during the variable selection process would reduce the accuracy of a predictive model based on cross-validation. It may be more appropriate to call it predictive accuracy boosting variable (PABV).
2. Unimportant variable based on the predictive accuracy (UVPA). This refers to variables for which exclusion during the variable selection process would increase the accuracy of a predictive model based on cross-validation [6,79]. It may be more precise to call it predictive accuracy reducing variable (PARV).

Application of relevant variable selection methods and concepts can further improve the accuracy of predictive models [6,79]; this is demonstrated in Section 11. Although these concepts were developed based on RF and its hybridization with geostatistical methods, they can be equally applied to any other predictive methods.

8. Accuracy and Error Measures for Predictive Models

8.1. Relationship between Observed, Predicted, and True Values

Predictive accuracy is about the differences between observed and predicted values that are derived based on validation methods [18]. However, it is often questioned what the differences between the predicted values and true values are. Since the true values are mostly unknown, the observed values are used to validate predictive models. For an observed value, it may be again different from its corresponding true value. The difference between the true value and observed value is the error associated with the observed value. Let us refer to this error as an observational error that is the sum of random error associated with observed variable, sampling error, and measuring error. The sampling and measuring errors are the sum of errors resulting from various factors that may affect the accuracy of observation (i.e., measurement) and change with the variable observed. Let us take seabed sediment as an example; the factors may include sampling design, the position accuracy of survey vessel, equipment used for sample collection, field operation, sample storage, sample processing procedure and analysis in laboratory, data entry, etc. However, how much error can be attributed to each of these factors is unknown in most cases. Hence, we have to use observed and predicted values to assess the predictive accuracy in practice.

8.2. Error and Accuracy Measures of Predictive Models

Many error and accuracy measures were developed to assess the accuracy of predictive models for numerical data [65,109,110]. Some of these error and accuracy measures were assessed and their limitations were previously discussed [3,18]. Of these error and accuracy measures, VEcv (i.e., variance explained by a predictive model based on cross-validation) measures how accurate the predictive model is and was proven to be independent of unit, scale, data mean, and variance [18,111], while root mean squared error (RMSE) measures how wrong the predictive model and the resultant predictions can be. Therefore, VEcv and RMSE are recommended for numerical predictions. Legates and McCabe's E1 (E1) was recommended for numerical data as well [111]. The commonly used measure, r or r^2, is not recommended because it is an incorrect measure of predictive accuracy [111].

For categorical data, correct classification rate (CCR), kappa (kappa), sensitivity (sens), specificity (spec), and true skill statistic (TSS) are often recommended [112,113]. RMSE is also used for presence and absence data [114]. One commonly used measure, area under the curve (AUC) (or receiver operating characteristics (ROC)), is not recommended for reasons previously highlighted [112,115].

9. Model Validation

9.1. Model Validation Methods

The accuracy of predictive models is critical as it determines the quality of the resultant predictions. The accuracy is often assessed based on model validation methods that may include the following:

1. Hold-out validation;
2. K-fold cross-validation;
3. Leave-one-out cross-validation;
4. Leave-q-out cross-validation;
5. Bootstrapping cross-validation;
6. Using any new samples that are not used for model training.

In environmental sciences, the most commonly used validation methods are hold-out and leave-one-out [18]. However, of these validation methods, five- or 10-fold cross-validation was recommended [7,116].

9.2. Randomness Associated with Cross-Validation Methods

Although a five- or 10-fold cross-validation was recommended to evaluate the performance of spatial predictive models [116], the datasets are randomly generated for each fold of the cross-validation change when the process is repeated. Thus, the randomness associated with the cross-validation would produce predictive accuracy or error measures that change with each iteration of the cross-validation [47]. To reduce the influence of the randomness on predictive accuracy (i.e., to stabilize the resultant performance measures), the cross-validation needs to be repeated (e.g., 100 times) [47,74,78]. The choice of this iteration number is data-dependent and can be determined based on the method used in previous studies [47,78].

10. Spatial Predictions, Prediction Uncertainty, and Their Visualization

10.1. Spatial Predictions

In spatial predictive modeling, the goal is not only to develop the most accurate predictive model, but more importantly to generate spatial predictions. The spatial predictions are usually produced using the most accurate predictive model developed according to the above procedures. To make predictions, in addition to the spatial predictive model, we need relevant information of each model predictive variable to be available at each grid cell at a desired resolution. When all this information is prepared, the spatial predictions can then be generated. The predictions contain three columns, i.e., longitude, latitude, and predictions. Sometimes, uncertainty of the predictions can be produced.

10.2. Prediction Uncertainty

Prediction uncertainty in environmental modeling may refer to various aspects of the modeling process and is used to encompass many concepts [117–120]. It can result from various sources or factors as previously discussed [17,120,121]. In this study, the uncertainty which is produced by a predictive model is about spatial predictions. Prediction uncertainty is increasingly required for decision-making and many methods are used to produce such uncertainty. In this study, I focus on the uncertainty produced by some commonly used methods: OK, LM, and RF.

For OK, prediction variances can be produced [58]. However, it was shown that the variances produced are independent of the actual predicted values [122]. Thus, the resulting variances should not be used to measure uncertainty, although many studies used them for such purpose. Since they reflect the variations in spatial departures among samples, they can be used as good indicators where samples are sparse and, thus, may provide useful information for selecting future sampling locations.

For LM, prediction uncertainty (i.e., prediction intervals) produced are much wider than confidence intervals of a model fitted [101]. Such uncertainty, however, has little to do with the predictive accuracy of the model. For instance, it was found that the models developed according to goodness of fit could be misleading when they were used as predictive models [6]; hence, its prediction uncertainty could also be misleading, and further studies are recommended.

For RF, many types of uncertainty could be produced, which actually reflect the difference in sampling strategies [123–126]. For example, prediction uncertainty produced for RF in a previous study based on Monte Carlo resampling [127] was, in fact, measuring the variation in predictions among individual trees rather than by RF. A further example for RF is that an ensemble of equally probable realizations was generated and the differences amongst the realizations were used as a measure of uncertainty [128]. This type of uncertainty only measures the differences among the results of various runs of RF, that is, measuring the difference resulted from the randomness associated with each run. Hence, these values do not relate to predictive accuracy and do not measure prediction uncertainty.

In addition, for any of the spatial predictive methods above, predictive errors based on validation can be produced for a predictive model developed. However, this leads to only one error value, and all predictions would have the same uncertainty value if it is used as an uncertainty measurement [129].

It is apparent that the uncertainty values produced above are either not measuring prediction uncertainty, or they depend on various factors as discussed above. This consequently results in the need to question the uncertainty of uncertainty. In short, how to assess prediction uncertainty needs further study. Any uncertainty measures that can incorporate the information of predictive accuracy are worth further investigation and recommended for future studies.

10.3. Visualization

Spatial predictions can be visualized using various tools, most commonly ArcGIS and QGIS. The function, *spplot*, in R is often used to plot the distribution of spatially continuous predictions [59]. The R package, *raster*, can also be used for such purpose [130]. Joint application of R and Google Earth can be used to visualize the predictions. In this study, I demonstrate how to use the latter approach along with *spplot* to visualize the predictions as below.

11. Reproducible Examples for Spatial Predictive Modeling

In this section, reproducible examples using *spm*, an R package, are provided to demonstrate how to select and develop a predictive model according to the guidelines and recommendations provided in the previous sections for spatial predictive modeling in environmental sciences. The predictive model to be used was developed using RFOK [74], where data preparation, including pre-selection of predictive variables, relevant parameter selection, variable selection, and model validation, was detailed. Seabed gravel content samples in the Petrel sub-basin, northern Australia marine margin are used to demonstrate how to select relevant parameters, test the predictive accuracy, and generate and visualize spatial predictions. These examples for RFOK can be easily extended to other predictive methods including IDW, OK, RF, GBM, RFIDW, GBMOK, GBMIDW, RFOKRFIDW, and GBMOKGBMIDW by replacing *rfokcv* and its associated parameters with relevant functions and parameters for these methods in *spm*.

11.1. Accuracy of a Predictive Model for Seabed Gravel Content

In a previous predictive model [74], a spherical variogram model and a searching window size of 12 were used. The accuracy of this predictive model [74] can be shown using the *rfokcv* function in *spm* as shown below. To stabilize the accuracy derived, I repeat the cross-validation 100 times, which can be determined using the methods discussed in Section 9.2.

```
> library(spm)
> data(petrel)
> names(petrel)
[1] "long" "lat" "mud" "sand" "gravel" "bathy" "dist" "relief" "slope"
> set.seed(1234)
> n <- 100
> rfokvecv1 <- NULL
> for (i in 1:n) {
+   rfokcv1 <- rfokcv(petrel[, c(1,2)], petrel[, c(1,2, 6:9)], petrel[, 5], predacc = "VEcv")
+   rfokvecv1 [i] <- rfokcv1
+ }
> mean(rfokvecv1)
[1] 37.44799
```

It suggests that the predictive accuracy is 37.4% in terms of VEcv.

11.2. Parameter Selection

The *rfokcv* function in *spm* is used to demonstrate how to select the best parameters for a predictive model (by using above predictive model as an example), and to check if the parameters used are optimal.

```
> library(spm)
> data(petrel)
> nmax <- c(5:12); vgm.args <- c("Sph", "Mat", "Ste", "Log")
> rfokopt3 <- array(0, dim = c(length(nmax), length(vgm.args)))
> set.seed(1234)
> for (i in 1:length(nmax)) {
+   for (j in 1:length(vgm.args)) {
+   rfokcv1.1 <- NULL
+     for (k in 1:100) {
+     rfokcv1.1[k] <- rfokcv(petrel[, c(1, 2)], petrel[, c(1, 2, 6:9)], petrel[, 5], nmax = nmax[i],
+     vgm.args = vgm.args[j], predacc = "VEcv") }
+     rfokopt3[i, j] <- mean(rfokcv1.1) } }
> which (rfokopt3 == max(rfokopt3, na.rm = T), arr.ind = T)
   [1,]  6 4
> vgm.args[4]; nmax[6]
   [1] "Log"
   [1] 10
```

The results suggest that the model would achieve the best predictive accuracy if "Log" is used for variogram modeling, and the 10 nearest samples are used for nmax. Of course, a different range may be used to choose the best nmax, and other variogram models can also be tested if needed.

We can use *rfokcv* in *spm* to assess the accuracy of RFOK by using the parameters identified above.

```
> library(spm)
> data(petrel)
> set.seed(1234)
> n <- 100
> rfokvecv1 <- NULL
> for (i in 1:n) {
+   rfokcv1 <- rfokcv(petrel[, c(1, 2)], petrel[, c(1, 2, 6:9)], petrel[, 5], vgm.args = "Log",
+   nmax = 10,
+   predacc = "VEcv")
+   rfokvecv1 [i] <- rfokcv1
+ }
> mean(rfokvecv1)
[1] 38.30175
```

This finding suggests that the overall averaged accuracy of the RFOK predictive model for seabed gravel content in terms of VEcv is 38.3%, higher than that of the previous model. This demonstrates that the parameters used previously are not optimal and that parameter selection improved predictive accuracy.

11.3. Predictive Variable Selection

In this study, we use the predictive variables previously identified [74], where the predictive variables were selected based on VI. Since then, more advanced variable selection methods for RF, RFOK, and RFIDW, such as AVI, KIAVI, PABV, and PARV [6,79], were developed. Application of these model selection methods and concepts may further improve the predictive accuracy of the model above. It is apparent that latitude (lat) is a PARV, as shown in the previous study [74]; thus, the removal of lat is expected to improve the predictive accuracy. This can be demonstrated below.

```
> library(spm)
> set.seed(1234)
> rfokvecv1.1 <- NULL
> for (i in 1:n) {
>   rfokcv1 <- rfokcv(petrel[, c(1, 2)], petrel[, c(1, 6:9)], petrel[, 5], vgm.args = "Log",
+   nmax = 10,
+   predacc = "Vecv")
+   rfokvecv1.1 [i] <- rfokcv1
+ }
> mean(rfokvecv1.1, na.rm=T)
[1] 39.00298
```

A further improvement in predictive accuracy is achieved after applying PARV. This further demonstrates the role of variable selection, especially the importance of newly developed variable selection methods.

11.4. Generation of Spatial Predictions

The predictive model developed above can be used to generate spatial predictions. The function *rfokpred* in *spm* is used to produce the predictions.

```
> set.seed(1234)
> library(spm)
> data(petrel); data(petrel.grid)
```

```
> rfokpred1 <- rfokpred(petrel[, c(1, 2)], petrel[, c(1, 6:9)], petrel[, 5], petrel.grid[, c(1,
2)],   + petrel.grid, ntree = 500, nmax = 10, vgm.args = ("Log"))
> names(rfokpred1)
[1] "LON"  "LAT"  "Predictions"  "Variances"
```

The output dataset has four columns named longitude, latitude, predictions, and variances. Please note that the uncertainty information (i.e., variances) is produced for readers interested; however, be aware of the various limitations as discussed in Section 10 when using such information.

11.5. Visualisation of Spatial Predictions

Joint application of R and Google Earth can be used to visualize the predictions generated above.

```
> library(sp); library(plotKML)
> rfok1 <- rfokpred1
> gridded(rfok1) <- ~ longitude + latitude
> proj4string(rfok1) <- CRS("+proj=longlat +datum=WGS84")
> plotKML(rfok1, colour_scale = SAGA_pal[[1]], grid2poly = TRUE)
```

The resultant map is shown in Figure 1a. One of the advantages of using R and Google Earth is that it can place the prediction map into the context map by Google Earth, which provides additional information to final users. However, the labels of longitude and latitude are hard to place in Figure 1a. If these labels are required, *spplot* can be applied to the above gridded data as shown below (Figure 1b).

```
> par(font.axis=2, font.lab=2)
> spplot(s1, c("Predictions"), key.space=list(x=0.1,y=.95, corner=c(-1.2,2.8)),
+ col.regions = SAGA_pal[[1]], # this requires plotKML
+ scales=list(draw=T), colorkey = list(at = c(seq(0,80,5)), space="right",
+ labels = c("0%"," ","","","20%","","","","40%","","","","60%","","","","80%")),
+ at=c(seq(0,80, 5)))
```

With regard to the prediction map, it is obvious that there are artefacts (e.g., sharp vertical changes associated with longitude) in the predictions. These artefacts may disappear or be alleviated if more variables could be used; in other words, different predictive variables should be tested according to the recently development in variable selection [6,79], as discussed in Section 7.

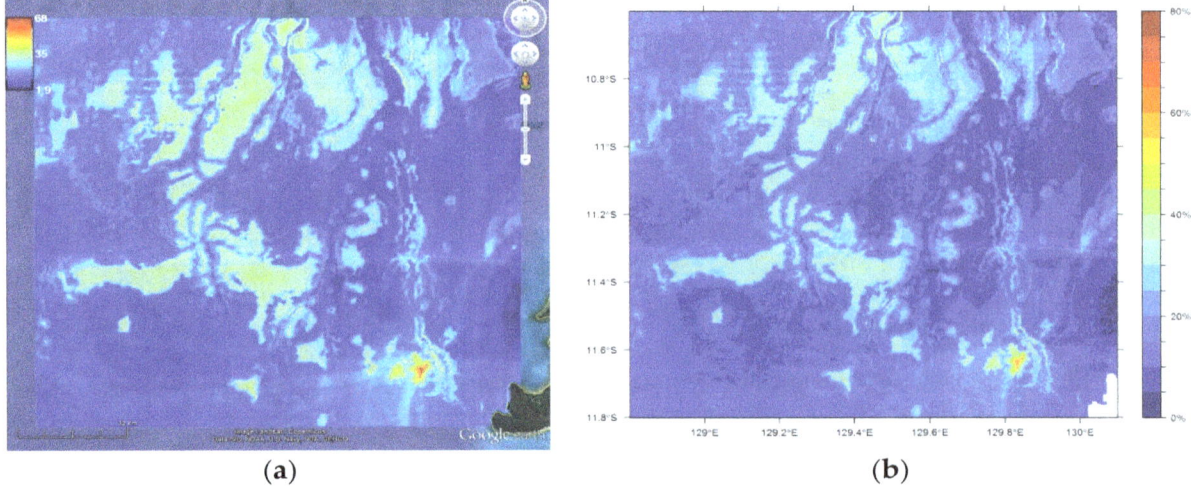

(a)   (b)

Figure 1. Predictions of seabed gravel in the Petrel sub-basin, northern Australian marine margin using a hybrid method of random forest and ordinary kriging (RFOK): (**a**) *plotKML* (left) and (**b**) *spplot* (right).

12. Summary

This study reviewed the modeling process for spatial predictive modeling in environmental sciences. The modeling process covers the following nine components:

1. Sampling design and data preparation;
2. Selection of predictive methods;
3. Pre-selection of predictive variables;
4. Exploratory analysis;
5. Parameter selection;
6. Variable selection;
7. Accuracy assessment;
8. Model validation;
9. Spatial predictions, prediction uncertainty, and their visualization.

Each of these components plays a significant role in model development. Incorrect or inappropriate implementation of any components may lead to less accurate or even misleading predictive model(s). To select the most accurate predictive model, all components and relevant requirements and factors for each component need to be considered and carefully implemented by following the guidelines, suggestions, and recommendations provided under relevant components in this study. Reproducible examples were provided to demonstrate how to select and identify the most accurate spatial predictive model using *spm*, and to generate and visualize spatial predictions in environmental sciences. For a predictive model, predictive accuracy is a key criterion for model selection and is critical for subsequent spatial predictions. This modeling process is not only important for spatial predictive modeling, but also provides valuable reference to other predictive modeling fields. Although this study attempts to cover relevant components, which may contribute to the improvement of predictive accuracy, as completely as possible, the spatial predictive modeling field is too broad to allow that to be done comprehensively in this study. This is because different disciplines have their own specific features and requirements. Therefore, further studies are needed to identify factors in relevant components or additional components that can further improve the accuracy of predictive models in various disciplines. This study would be expected to not only boost applications of appropriate spatial predictive modeling processes, but also provide spatial predictive modeling tools for various modeling components to improve the quality of spatial predictions.

Acknowledgments: I would like to thank Gareth Davies, Peter Tan, Trevor Dhu, Andrew Carroll, and Kim Picard for their valuable comments and suggestions. This study was supported by Geoscience Australia. This paper was published with the permission of the Chief Executive Officer, Geoscience Australia.

References

1. Marmion, M.; Luoto, M.; Heikkinen, R.K.; Thuiller, W. The performance of state-of-the-art modelling techniques depends on geographical distribution of species. *Ecol. Model.* **2009**, *220*, 3512–3520. [CrossRef]
2. Maier, H.R.; Kapelan, Z.; Kasprzyk, J.; Kollat, J.; Matott, L.S.; Cunha, M.C.; Dandy, G.C.; Gibbs, M.S.; Keedwell, E.; Marchi, A.; et al. Evolutionary algorithms and other metaheuristics in water resources: Current status, research challenges and future directions. *Environ. Model. Softw.* **2014**, *62*, 271–299. [CrossRef]
3. Li, J.; Heap, A. *A Review of Spatial Interpolation Methods for Environmental Scientists*; Record 2008/23; Geoscience Australia: Canberra, Australia, 2008; 137p.
4. Stephens, D.; Diesing, M. Towards quantitative spatial models of seabed sediment composition. *PLoS ONE* **2015**, *10*, e0142502. [CrossRef] [PubMed]

5. Sanabria, L.A.; Cechet, R.P.; Li, J. Mapping of australian fire weather potential: Observational and modelling studies. In Proceedings of the 20th International Congress on Modelling and Simulation (MODSIM2013), Adelaide, Australia, 1–6 December 2013; pp. 242–248.

6. Li, J.; Alvarez, B.; Siwabessy, J.; Tran, M.; Huang, Z.; Przeslawski, R.; Radke, L.; Howard, F.; Nichol, S. Application of random forest, generalised linear model and their hybrid methods with geostatistical techniques to count data: Predicting sponge species richness. *Environ. Model. Softw.* **2017**, *97*, 112–129. [CrossRef]

7. Hastie, T.; Tibshirani, R.; Friedman, J. *The Elements of Statistical Learning: Data Mining, Inference, and Prediction*, 2nd ed.; Springer: New York, NY, USA, 2009; p. 763.

8. Crawley, M.J. *The R Book*; John Wiley & Sons, Ltd.: Chichester, UK, 2007; p. 942.

9. Kuhn, M.; Johnson, K. *Applied Predictive Modeling*; Springer: New York, NY, USA, 2013.

10. Li, J.; Heap, A.D. Spatial interpolation methods applied in the environmental sciences: A review. *Environ. Model. Softw.* **2014**, *53*, 173–189. [CrossRef]

11. Li, J.; Heap, A. A review of comparative studies of spatial interpolation methods in environmental sciences: Performance and impact factors. *Ecol. Inform.* **2011**, *6*, 228–241. [CrossRef]

12. Li, J.; Potter, A.; Huang, Z.; Daniell, J.J.; Heap, A. *Predicting Seabed Mud Content across the Australian Margin: Comparison of Statistical and Mathematical Techniques Using a Simulation Experiment*; Record 2010/11; Geoscience Australia: Canberra, Australia, 2010; 146p.

13. Sanabria, L.A.; Qin, X.; Li, J.; Cechet, R.P.; Lucas, C. Spatial interpolation of mcarthur's forest fire danger index across australia: Observational study. *Environ. Model. Softw.* **2013**, *50*, 37–50. [CrossRef]

14. Tadić, J.M.; Ilić, V.; Biraud, S. Examination of geostatistical and machine-learning techniques as interpolaters in anisotropic atmospheric environments. *Atmos. Environ.* **2015**, *111*, 28–38. [CrossRef]

15. Wolpert, D.; Macready, W. No free lunch theorems for optimization. *IEEE Trans. Evol. Comput.* **1997**, *1*, 67–82. [CrossRef]

16. Burrough, P.A.; McDonnell, R.A. *Principles of Geographical Information Systems*; Oxford University Press: Oxford, UK, 1998; p. 333.

17. Jakeman, A.J.; Letcher, R.A.; Norton, J.P. Ten iterative steps in development and evaluation of environmental models. *Environ. Model. Softw.* **2006**, *21*, 602–614. [CrossRef]

18. Li, J. Assessing spatial predictive models in the environmental sciences: Accuracy measures, data variation and variance explained. *Environ. Model. Softw.* **2016**, *80*, 1–8. [CrossRef]

19. Leek, J.T.; Peng, R.D. What is the question? *Science* **2015**, *347*, 1314–1315. [CrossRef]

20. Li, J. spm: Spatial Predictive Modelling. R Package Version 1.1.0. Available online: https://CRAN.R-project.org/package=spm:2018 (accessed on 17 May 2019).

21. Foster, S.D.; Hosack, G.R.; Lawrence, E.; Przeslawski, R.; Hedge, P.; Caley, M.J.; Barrett, N.S.; Williams, A.; Li, J.; Lynch, T.; et al. Spatially balanced designs that incorporate legacy sites. *Methods Ecol. Evol.* **2017**, *8*, 1433–1442. [CrossRef]

22. Benedetti, R.; Piersimoni, F.; Postigione, P. Spatially balanced sampling: A review and a reappraisal. *Int. Stat. Rev.* **2017**, *85*, 439–454. [CrossRef]

23. Stevens, D.L.; Olsen, A.R. Spatially balanced sampling of natural resources. *J. Am. Stat. Assoc.* **2004**, *99*, 262–278. [CrossRef]

24. Benedetti, R.; Piersimoni, F. A spatially balanced design with probability function proportional to the within sample distance. *Biom. J.* **2017**, *59*, 1067–1084. [CrossRef]

25. Wang, J.-F.; Stein, A.; Gao, B.-B.; Ge, Y. A review of spatial sampling. *Spat. Stat.* **2012**, *2*, 1–14. [CrossRef]

26. Diggle, P.J.; Ribeiro, P.J., Jr. *Model-Based Geostatistics*; Springer: New York, NY, USA, 2010; p. 228.

27. Przeslawski, R.; Daniell, J.; Anderson, T.; Vaughn Barrie, J.; Heap, A.; Hughes, M.; Li, J.; Potter, A.; Radke, L.; Siwabessy, J.; et al. *Seabed Habitats and Hazards of the Joseph Bonaparte Gulf and Timor Sea, Northern Australia*; Record 2008/23; Geoscience Australia: Canberra, Australia, 2011; 69p.

28. Radke, L.C.; Li, J.; Douglas, G.; Przeslawski, R.; Nichol, S.; Siwabessy, J.; Huang, Z.; Trafford, J.; Watson, T.; Whiteway, T. Characterising sediments for a tropical sediment-starved shelf using cluster analysis of physical and geochemical variables. *Environ. Chem.* **2015**, *12*, 204–226. [CrossRef]

29. Radke, L.; Nicholas, T.; Thompson, P.; Li, J.; Raes, E.; Carey, M.; Atkinson, I.; Huang, Z.; Trafford, J.; Nichol, S. Baseline biogeochemical data from australia's continental margin links seabed sediments to water column characteristics. *Mar. Freshw. Res.* **2017**. [CrossRef]

30. Kincaid, T. GRTS Survey Designs for an Area Resource. 2019. Available online: https://cran.r-project.org/web/packages/spsurvey/vignettes/Area_Design.pdf (accessed on 17 May 2019).

31. Kincaid, T.M.; Olsen, A.R. spsurvey: Spatial Survey Design and Analysis. R Package Version 3.3. 2016. Available online: https://cran.r-project.org/web/packages/spsurvey/index.html (accessed on 17 May 2019).

32. Hengl, T. GSIF: Global Soil Information Facilities. R Package Version 0.4-1. 2014. Available online: https://cran.r-project.org/web/packages/GSIF/index.html (accessed on 17 May 2019).

33. Walvoort, D.J.J. Spatial Coverage Sampling and Random Sampling from Compact Geographical Strata. R Package Version 0.3-6. Available online: https://cran.r-project.org/web/packages/spcosa/index.html (accessed on 17 May 2019).

34. Roudier, P. CLHS: A R Package for Conditioned Latin Hypercube Sampling. 2011. Available online: https://cran.r-project.org/web/packages/clhs/index.html (accessed on 17 May 2019).

35. Grafström, A.; Lisic, J. Balancedsampling: Balanced and Saptially Balanced Sampling. R Package Version 1.5.4. 2018. Available online: https://cran.r-project.org/web/packages/BalancedSampling/index.html (accessed on 17 May 2019).

36. Radke, L.; Smit, N.; Li, J.; Nicholas, T.; Picard, K. *Outer Darwin Harbour Shallow Water Sediment Survey 2016: Ga0356—Post-Survey Report*; Record 2017/06; Geoscience Australia: Canberra, Australia, 2017. [CrossRef]

37. Siwabessy, P.J.W.; Smit, N.; Atkinson, I.; Dando, N.; Harries, S.; Howard, F.J.F.; Li, J.; Nicholas, W.A.; Picard, K.; Radke, L.C.; et al. *Bynoe Harbour Marine Survey 2016: Ga4452/sol6432—Post-Survey Report*; Record 2017/04; Geoscience Australia: Canberra, Australia, 2017.

38. Foster, S.D. MBHdesign: Spatial Designs for Ecological and Environmental Surveys. R Package Version 1.0.76. 2017. Available online: https://cran.r-project.org/web/packages/MBHdesign/index.html (accessed on 17 May 2019).

39. Cai, L.; Zhu, Y. The challenges of data quality and data quality assessment in the big data era. *Data Sci. J.* **2015**, *14*, 1–10.

40. Pipino, L.L.; Lee, Y.W.; Wang, R.Y. Data quality assessment. *Commun. ACM* **2002**, *45*, 211–218. [CrossRef]

41. Li, J.; Potter, A.; Huang, Z.; Heap, A. *Predicting Seabed sand Content across the Australian Margin Using Machine Learning and Geostatistical Methods*; Record 2012/48; Geoscience Australia: Canberra, Australia, 2012; 115p.

42. Li, J.; Hilbert, D.W.; Parker, T.; Williams, S. How do species respond to climate change along an elevation gradient? A case study of the grey-headed robin (*Heteromyias albispecularis*). *Glob. Chang. Biol.* **2009**, *15*, 255–267. [CrossRef]

43. Jiang, W.; Li, J. *The Effects of Spatial Reference Systems on the Predictive Accuracy of Spatial Interpolation Methods*; Record 2014/01; Geoscience Australia: Canberra, Australia, 2014; p. 33.

44. Jiang, W.; Li, J. Are Spatial Modelling Methods Sensitive to Spatial Reference Systems for Predicting Marine Environmental Variables. In Proceedings of the 20th International Congress on Modelling and Simulation, Adelaide, Australia, 1–6 December 2013; pp. 387–393.

45. Turner, A.J.; Li, J.; Jiang, W. Effects of spatial reference systems on the accuracy of spatial predictive modelling along a latitudinal gradient. In Proceedings of the 22nd International Congress on Modelling and Simulation, Hobart, Australia, 3–8 December 2017; pp. 106–112.

46. Purss, M. Topic 21: Discrete Global Grid Systems Abstract Specification, Open Geospatial Consortium [OGC 15-104r5]. 2017. Available online: https://www.google.com.au/url?sa=t&rct=j&q=&esrc=s&source=web&cd=4&cad=rja&uact=8&ved=2ahUKEwiHmPmnrqHiAhWFfisKHfTlB18QFjADegQIABAC&url=https%3A%2F%2Fportal.opengeospatial.org%2Ffiles%2F15-104r5&usg=AOvVaw3Ww2TasQntx17y99VlHwig (accessed on 17 May 2019).

47. Li, J. Predictive modelling using random forest and its hybrid methods with geostatistical techniques in marine environmental geosciences. In Proceedings of the Eleventh Australasian Data Mining Conference (AusDM 2013), Canberra, Australia, 13–15 November 2013; Volume 146.

48. Stephens, D.; Diesing, M. A comparison of supervised classification methods for the prediction of substrate type using multibeam acoustic and legacy grain-size data. *PLoS ONE* **2014**, *9*, e93950. [CrossRef] [PubMed]

49. Hengl, T.; Heuvelink, G.B.M.; Kempen, B.; Leenaars, J.G.B.; Walsh, M.G.; Shepherd, K.D.; Sila, A.; MacMillan, R.A.; de Jesus, J.M.; Tamene, L.; et al. Mapping soil properties of africa at 250 m resolution: Random forests significantly improve current predictions. *PLoS ONE* **2015**, *10*, e0125814. [CrossRef] [PubMed]

50. Zhang, X.; Liu, G.; Wang, H.; Li, X. Application of a hybrid interpolation method based on support vector machine in the precipitation spatial interpolation of basins. *Water* **2017**, *9*, 760. [CrossRef]

51. Seo, Y.; Kim, S.; Singh, V.P. Estimating spatial precipitation using regression kriging and artificial neural network residual kriging (rknnrk) hybrid approach. *Water Resour. Manag.* **2015**, *29*, 2189–2204. [CrossRef]

52. Demyanov, V.; Kanevsky, M.; Chernov, S.; Savelieva, E.; Timonin, V. Neural network residual kriging application for climatic data. *J. Geogr. Inf. Decis. Anal.* **1998**, *2*, 215–232.

53. Appelhans, T.; Mwangomo, E.; Hardy, D.R.; Hemp, A.; Nauss, T. Evaluating machine learning approaches for the interpolation of monthly air temperature at mt. Kilimanjaro, tanzania. *Spat. Stat.* **2015**, *14*, 91–113. [CrossRef]

54. Leathwick, J.R.; Elith, J.; Francis, M.P.; Hastie, T.; Taylor, P. Variation in demersal fish species richness in the oceans surrounding new zealand: An analysis using boosted regression trees. *Mar. Ecol. Prog. Ser.* **2006**, *321*, 267–281. [CrossRef]

55. Leathwick, J.R.; Elith, J.; Hastie, T. Comparative performance of generalised additive models and multivariate adaptive regression splines for statistical modelling of species distributions. *Ecol. Model.* **2006**, *199*, 188–196. [CrossRef]

56. Isaaks, E.H.; Srivastava, R.M. *Applied Geostatistics*; Oxford University Press: New York, NY, USA, 1989; p. 561.

57. Hengl, T. *A Practical Guide to Geostatistical Mapping of Environmental Variables*; Office for Official Publication of the European Communities: Luxembourg, 2007; p. 143.

58. Pebesma, E.J. Multivariable geostatistics in s: The gstat package. *Comput. Geosci.* **2004**, *30*, 683–691. [CrossRef]

59. Bivand, R.S.; Pebesma, E.J.; Gómez-Rubio, V. *Applied Spatial Data Analysis with R*; Springer: New York, NY, USA, 2008; p. 374.

60. Lark, R.M.; Ferguson, R.B. Mapping risk of soil nutrient deficiency or excess by disjunctive and indicator kriging. *Geoderma* **2004**, *118*, 39–53. [CrossRef]

61. Huang, H.; Chen, C. Optimal geostatistical model selection. *J. Am. Stat. Assoc.* **2007**, *102*, 1009–1024. [CrossRef]

62. Hernandez-Stefanoni, J.L.; Ponce-Hernandez, R. Mapping the spatial variability of plant diversity in a tropical forest: Comparison of spatial interpolation methods. *Environ. Monit. Assess.* **2006**, *117*, 307–334. [CrossRef] [PubMed]

63. Stein, A.; Hoogerwerf, M.; Bouma, J. Use of soil map delineations to improve (co-)kriging of point data on moisture deficits. *Geoderma* **1988**, *43*, 163–177. [CrossRef]

64. Voltz, M.; Webster, R. A comparison of kriging, cubic splines and classification for predicting soil properties from sample information. *J. Soil Sci.* **1990**, *41*, 473–490. [CrossRef]

65. Bennett, N.D.; Croke, B.F.W.; Guariso, G.; Guillaume, J.H.A.; Hamilton, S.H.; Jakeman, A.J.; Marsili-Libelli, S.; Newham, L.T.H.; Norton, J.P.; Perrin, C.; et al. Characterising performance of environmental models. *Environ. Model. Softw.* **2013**, *40*, 1–20. [CrossRef]

66. Gneiting, T.; Balabdaoui, F.; Raftery, A.E. Probabilistic forecasts, calibration and sharpness. *J. R. Stat. Soc. Ser. B* **2007**, *69*, 243–268. [CrossRef]

67. Austin, M. Species distribution models and ecological theory: A critical assessment and some possible new approaches. *Ecol. Model.* **2007**, *200*, 1–19. [CrossRef]

68. Elith, J.; Leathwick, J. Species distribution models: Ecological explanation and prediction across space and time. *Annu. Rev. Ecol. Evol. Syst.* **2009**, *40*, 677–697. [CrossRef]

69. McArthur, M.A.; Brooke, B.P.; Przeslawski, R.; Ryan, D.A.; Lucieer, V.L.; Nichol, S.; McCallum, A.W.; Mellin, C.; Cresswell, I.D.; Radke, L.C. On the use of abiotic surrogates to describe marine benthic biodiversity. *Estuar. Coast. Shelf Sci.* **2010**, *88*, 21–32. [CrossRef]

70. Huston, M.A. Hidden treatments in ecological experiments: Re-evaluating the ecosystem function of biodiversity. *Oecologia* **1997**, *110*, 449–460. [CrossRef]

71. Arthur, A.D.; Li, J.; Henry, S.; Cunningham, S.A. Influence of woody vegetation on pollinator densities in oilseed *brassica* fields in an australian temperate landscape. *Basic Appl. Ecol.* **2010**, *11*, 406–414. [CrossRef]

72. Elith, J.; Graham, C.H.; Anderson, R.P.; Dulik, M.; Ferrier, S.; Guisan, A.; Hijmans, R.J.; Huettmann, F.; Leathwick, J.R.; Lehmann, A.; et al. Novel methods improve prediction of species' distributions from occurrence data. *Ecography* **2006**, *29*, 129–151. [CrossRef]

73. Miller, K.; Puotinen, M.; Przeslawski, R.; Huang, Z.; Bouchet, P.; Radford, B.; Li, J.; Kool, J.; Picard, K.; Thums, M.; et al. *Ecosystem Understanding to Support Sustainable Use, Management and Monitoring of Marine Assets in the North and North-West Regions: Final Report for NESP d1 2016e*; Report to the National Environmental Science Program, Marine Biodiversity Hub; Australian Institute of Marine Science: Townsville, Australia, 2016; 146p. Available online:

https://www.nespmarine.edu.au/system/files/Miller%20et%20al%20Project%20D1%20Report%20summarising%20outputs%20from%20synthesis%20of%20datasets%20and%20predictive%20models%20for%20N%20and%20NW_Milestone%204_RPv3.pdf (accessed on 17 May 2019).

74. Li, J. Predicting the spatial distribution of seabed gravel content using random forest, spatial interpolation methods and their hybrid methods. In Proceedings of the International Congress on Modelling and Simulation (MODSIM) 2013, Adelaide, Austrialia, 1–6 December 2013; pp. 394–400.

75. Verfaillie, E.; Van Lancker, V.; Van Meirvenne, M. Multivariate geostatistics for the predictive modelling of the surficial sand distribution in shelf seas. *Cont. Shelf Res.* **2006**, *26*, 2454–2468. [CrossRef]

76. Verfaillie, E.; Du Four, I.; Van Meirvenne, M.; Van Lancker, V. Geostatistical modeling of sedimentological parameters using multi-scale terrain variables: Application along the belgian part of the north sea. *Int. J. Geogr. Inf. Sci.* **2008**. [CrossRef]

77. Huang, Z.; Nichol, S.; Siwabessy, P.J.W.; Daniell, J.; Brooke, B.P. Predictive modelling of seabed sediment parameters using multibeam acoustic data: A case study on the carnarvon shelf, western australia. *Int. J. Geogr. Inf. Sci.* **2012**, *26*, 283–307. [CrossRef]

78. Li, J.; Siwabessy, J.; Tran, M.; Huang, Z.; Heap, A. Predicting seabed hardness using random forest in R. In *Data Mining Applications with R*; Zhao, Y., Cen, Y., Eds.; Elsevier: Amsterdam, The Netherlands, 2014; pp. 299–329.

79. Li, J.; Tran, M.; Siwabessy, J. Selecting optimal random forest predictive models: A case study on predicting the spatial distribution of seabed hardness. *PLoS ONE* **2016**, *11*, e0149089. [CrossRef]

80. Siwabessy, P.J.W.; Daniell, J.; Li, J.; Huang, Z.; Heap, A.D.; Nichol, S.; Anderson, T.J.; Tran, M. *Methodologies for Seabed Substrate Characterisation Using Multibeam Bathymetry, Backscatter and Video Data: A Case Study from the Carbonate Banks of the Timor Sea, Northern Australia*; Record 2013/11; Geoscience Australia: Canberra, Australia, 2013; 82p.

81. Huang, Z.; Brooke, B.; Li, J. Performance of predictive models in marine benthic environments based on predictions of sponge distribution on the australian continental shelf. *Ecol. Inform.* **2011**, *6*, 205–216. [CrossRef]

82. Lark, R.M.; Marchant, B.P.; Dove, D.; Green, S.L.; Stewart, H.; Diesing, M. Combining observations with acoustic swath bathymetry and backscatter to map seabed sediment texture classes: The empirical best linear unbiased predi. *Sediment. Geol.* **2015**, *328*, 17–32. [CrossRef]

83. Diesing, M.; Mitchell, P.; Stephens, D. Image-based seabed classification: What can we learn from terrestrial remote sensing? *ICES J. Mar. Sci.* **2016**, fsw 118. [CrossRef]

84. Fisher, P.; Wood, J.; Cheng, T. Where is helvellyn? Fuzziness of multi-scale landscape morphometry. *Trans. Inst. Br. Geogr.* **2004**, *29*, 106–128. [CrossRef]

85. Zuur, A.; Leno, E.N.; Elphick, C.S. A protocol for data exploration to avoid common statistical problems. *Methods Ecol. Evol.* **2010**, *1*, 3–14. [CrossRef]

86. O'Brien, R.M. A caution regarding rules of thumb for variance inflation factors. *Qual. Quant.* **2007**, *41*, 673–690. [CrossRef]

87. Harrell, F.E., Jr. *Regression modelling strategies: with applications to linear models, logistic regression, and survival analysis*; Springer: New York, NY, USA, 1997.

88. Li, J.; Heap, A.D.; Potter, A.; Daniell, J. Application of machine learning methods to spatial interpolation of environmental variables. *Environ. Model. Softw.* **2011**, *26*, 1647–1659. [CrossRef]

89. Cutler, D.R.; Edwards, T.C.J.; Beard, K.H.; Cutler, A.; Hess, K.T.; Gibson, J.; Lawler, J.J. Random forests for classification in ecology. *Ecography* **2007**, *88*, 2783–2792. [CrossRef]

90. Collins, F.C.; Bolstad, P.V. A comparison of spatial interpolation techniques in temperature estimation. In Proceedings of the Third International Conference/Workshop on Integrating GIS and Environmental Modeling, Santa Fe, NM, USA, 21–25 January 1996.

91. Ripley, B.D. *Spatial Statistics*; John Wiley & Sons: New York, NY, USA, 1981; p. 252.

92. Wu, J.; Norvell, W.A.; Welch, R.M. Kriging on highly skewed data for dtpa-extractable soil zn with auxiliary information for ph and organic carbon. *Geoderma* **2006**, *134*, 187–199. [CrossRef]

93. Meul, M.; Van Meirvenne, M. Kriging soil texture under different types of nonstationarity. *Geoderma* **2003**, *112*, 217–233. [CrossRef]

94. Liaw, A.; Wiener, M. Classification and regression by randomforest. *R News* **2002**, *2*, 18–22.

95. Ridgeway, G. gbm: Generalized Boosted Regression Models. R Package Version 2.1.3. 2017. Available online: https://cran.r-project.org/web/packages/gbm/index.html (accessed on 17 May 2019).

96. Elith, J.; Leathwick, J.R.; Hastie, T. A working guide to boosted regression trees. *J. Anim. Ecol.* **2008**, *77*, 802–813. [CrossRef]

97. Breiman, L.; Friedman, J.H.; Olshen, R.A.; Stone, C.J. *Classification and Regression Trees*; Belmont: Wadsworth, OH, USA, 1984.

98. Li, J.; Hilbert, D.W. Lives: A new habitat modelling technique for predicting the distributions of species' occurrence using presence-only data based on limiting factor theory. *Biodivers. Conserv.* **2008**, *17*, 3079–3095. [CrossRef]

99. Johnson, J.B.; Omland, K.S. Model selection in ecology and evolution. *Trends Ecol. Evol.* **2004**, *19*, 101–108. [CrossRef]

100. Venables, W.N.; Ripley, B.D. *Modern Applied Statistics with S-Plus*, 4th ed.; Springer: New York, NY, USA, 2002; p. 495.

101. Chambers, J.M.; Hastie, T.J. *Statistical Models in S*; Wadsworth and Brooks/Cole Advanced Books and Software: Pacific Grove, CA, USA, 1992; p. 608.

102. Lumley, T.; Miller, A. leaps: Regression Subset Selection. R Package Version 3.0. 2009. Available online: https://cran.r-project.org/web/packages/leaps/index.html (accessed on 17 May 2019).

103. McLeod, A.I.; Xu, C. bestglm: Best Subset GLM. R Package Version 0.36. 2017. Available online: https://cran.r-project.org/web/packages/bestglm/index.html (accessed on 17 May 2019).

104. Li, J.; Alvarez, B.; Siwabessy, J.; Tran, M.; Huang, Z.; Przeslawski, R.; Radke, L.; Howard, F.; Nichol, S. Selecting predictors to form the most accurate predictive model for count data. In Proceedings of the International Congress on Modelling and Simulation (MODSIM) 2017, Hobart, Australia, 3–8 December 2017.

105. Kursa, M.B.; Rudnicki, W.R. Feature selection with the boruta package. *J. Stat. Softw.* **2010**, *36*, 1–13. [CrossRef]

106. Kuhn, M. caret: Classification and Regression Training. R Package Version 6.0-81. 2018. Available online: https://cran.r-project.org/web/packages/caret/index.html (accessed on 17 May 2019).

107. Genuer, R.; Poggi, J.M.; Tuleau-Malot, C. VSURF: Variable Selection Using Random Forests. R Package Version 1.0.2. 2015. Available online: https://cran.r-project.org/web/packages/VSURF/index.html (accessed on 17 May 2019).

108. Li, J.; Siwabessy, J.; Huang, Z.; Nichol, S. Developing an optimal spatial predictive model for seabed sand content using machine learning, geostatistics and their hybrid methods. *Geosciences* **2019**, *9*, 180. [CrossRef]

109. Han, J.; Kamber, M. *Data Mining: Concept and Techniques*, 2nd ed.; Elsevier: Amsterdam, The Netherlands, 2006; p. 770.

110. Moriasi, D.N.; Arnold, J.G.; Van Liew, M.W.; Bingner, R.L.; Harmel, R.D.; Veith, T.L. Model evaluation guidelines for systematic quantification of accuracy in watershed simulations. *Am. Soc. Agric. Biol. Eng.* **2007**, *50*, 885–900.

111. Li, J. Assessing the accuracy of predictive models for numerical data: Not r nor r^2, why not? Then what? *PLoS ONE* **2017**, *12*, e0183250. [CrossRef] [PubMed]

112. Allouche, O.; Tsoar, A.; Kadmon, R. Assessing the accuracy of species distribution models: Prevalence, kappa and true skill statistic (tss). *J. Appl. Ecol.* **2006**, *43*, 1223–1232. [CrossRef]

113. Fielding, A.H.; Bell, J.F. A review of methods for the assessment of prediction errors in conservation presence/absence models. *Environ. Conserv.* **1997**, *24*, 38–49. [CrossRef]

114. Thibaud, E.; Petitpierre, B.; Broennimann, O.; Davison, A.C.; Guisan, A. Measuring the relative effect of factors affecting species distribution model predictions. *Methods Ecol. Evol.* **2014**, *5*, 947–955. [CrossRef]

115. Lobo, J.M.; Jiménez-Valverde, A.; Real, R. Auc: A misleading measure of the performance of predictive distribution models. *Glob. Ecol. Biogeogr.* **2008**, *7*, 145–151. [CrossRef]

116. Kohavi, R. A study of cross-validation and bootstrap for accuracy estimation and model selection. In Proceedings of the International Joint Conference on Artificial Intelligence (IJCAI), Montreal, QC, Canada, 20–25 August 1995; pp. 1137–1143.

117. Refsgaard, J.C.; van der Sluijs, J.P.; Højberg, A.L.; Vanrolleghem, P.A. Uncertainty in the environmental modelling process - a framework and guidance. *Environ. Model. Softw.* **2007**, *22*, 1543–1556. [CrossRef]

118. Hayes, K.R. *Uncertainty and Uncertainty Analysis Methods*; CSIRO: Canberra, Australia, 2011; p. 131. Available online: https://publications.csiro.au/rpr/download?pid=csiro:EP102467&dsid=DS3 (accessed on 17 May 2019).

119. Barry, S.; Elith, J. Error and uncertainty in habitat models. *J. Appl. Ecol.* **2006**, *43*, 413–423. [CrossRef]

120. Oxley, T.; ApSimon, H. A conceptual framework for mapping uncertainty in integrated assessment. In Proceedings of the 19th International Congress on Modelling and Simulation, Perth, Australia, 12–16 December 2011.

121. Walker, W.E.; Harremoes, P.; Rotmans, J.; Van der Sluijs, J.P.; van Asselt, M.B.A.; Janssen, P.; Krayer von Krauss, M.P. Defining uncertainty: A conceptual basis for uncertainty management in model-based decision support. *Integr. Assess.* **2003**, *4*, 5–17. [CrossRef]

122. Goovaerts, P. *Geostatistics for Natural Resources Evaluation*; Oxford University Press: New York, NY, USA, 1997; p. 483.

123. Mentch, L.; Hooker, G. Quantifying uncertainty in random forests via confidence intervals and hypothesis tests. *J. Mach. Learn. Res.* **2016**, *17*, 1–41.

124. Slaets, J.I.F.; Piepho, H.-P.; Schmitter, P.; Hilger, T.; Cadisch, G. Quantifying uncertainty on sediment loads using bootstrap confidence intervals. *Hydrol. Earth Syst. Sci.* **2017**, *21*, 571–588. [CrossRef]

125. Wager, S.; Hastie, T.; Efron, B. Confidence intervals for random forests: The jackknife and the infinitesimal jackknife. *J. Mach. Learn. Res.* **2014**, *15*, 1625–1651. [PubMed]

126. Wright, M.N.; Ziegler, A. Ranger: A fast implementation of random forests for high dimensional data in c++ and r. *J. Stat. Softw.* **2017**, *77*, 1–17. [CrossRef]

127. Coulston, J.W.; Blinn, C.E.; Thomas, V.A.; Wynne, R.H. Approximating prediction uncertainty for random forest regression models. *Photogramm. Eng. Remote Sens.* **2016**, *82*, 189–197. [CrossRef]

128. Chen, J.; Li, M.-C.; Wang, W. Statistical uncertainty estimation using random forests and its application to drought forecast. *Math. Probl. Eng.* **2012**, *2012*, 915053. [CrossRef]

129. Bishop, T.F.A.; Minasny, B.; McBratney, A.B. Uncertainty analysis for soil-terrain models. *Int. J. Geogr. Inf. Sci.* **2006**, *20*, 117–134. [CrossRef]

130. Hijmans, R.J. raster: Geographic Data Analysis and Modeling. Available online: http://CRAN.R-project.org/package=raster (accessed on 17 May 2019).

11

Dual-Dense Convolution Network for Change Detection of High-Resolution Panchromatic Imagery

Wahyu Wiratama [1], **Jongseok Lee** [1], **Sang-Eun Park** [2] and **Donggyu Sim** [1,*]

[1] Department of Computer Engineering, Kwangwoon University, Seoul 139701, Korea; wiratama@kw.ac.kr (W.W.); suk2080@kw.ac.kr (J.L.)

[2] Department of Geoinformation Engineering, Sejong University, Seoul 143747, Korea; separk@sejong.ac.kr

* Correspondence: dgsim@kw.ac.kr

Abstract: This paper presents a robust change detection algorithm for high-resolution panchromatic imagery using a proposed dual-dense convolutional network (DCN). In this work, a joint structure of two deep convolutional networks with dense connectivity in convolution layers is designed in order to accomplish change detection for satellite images acquired at different times. The proposed network model detects pixel-wise temporal change based on local characteristics by incorporating information from neighboring pixels. Dense connection in convolution layers is designed to reuse preceding feature maps by connecting them to all subsequent layers. Dual networks are incorporated by measuring the dissimilarity of two temporal images. In the proposed algorithm for change detection, a contrastive loss function is used in a learning stage by running over multiple pairs of samples. According to our evaluation, we found that the proposed framework achieves better detection performance than conventional algorithms, in area under the curve (AUC) of 0.97, percentage correct classification (PCC) of 99%, and Kappa of 69, on average.

Keywords: change detection; convolutional network; deep learning; panchromatic; remote sensing

1. Introduction

Change detection is a challenging task in remote sensing for identifying changed areas between two images acquired at different times from the same geographical area. It has been widely used in both civil and military fields such as agricultural monitoring, urban planning, environment monitoring, and reconnaissance. In general, change detection is performed in three steps. First, a preprocessing step is commonly used to conduct registration of two images and to correct geometric and radiometric distortions. In the second step, a feature map is extracted, for example, a difference image is computed in order to generate change features with the assumption that two images are not perfectly registered for all of the pixels. Lastly, a classification or clustering algorithm is driven in order to distinguish changed pixels and unchanged pixels based on statistical characteristics.

For change detection, many manually designed features such as a difference image (DI) [1–7], local change vector [8], and texture vector [9–11] have been proposed. In further classification analysis, an unsupervised change detection was proposed based on fuzzy c-mean (FCM) clustering [12,13]. The optimization algorithm based on Markov random field (MRF) and genetic algorithm was employed so as to optimize the FCM. On the other hand, a supervised learning algorithm was presented based on an active learning and MRF in order to detect change areas [14]. In addition, a support vector machine (SVM) has widely been used to perform binary classification based on texture information and change vector analysis [9,15–17]. Since the classification process mainly depends on extracted features, the selection of handcrafted features for effective image representation is known to be crucial. In general, handcrafted features in change detection are sensitive due to geometric and radiometric distortions, as well as imperfect registration. All of those mentioned classification algorithms would be reasonably good for training data sets. However, those algorithms are not able to incorporate accurate and reliable statistical characteristics for a huge amount of data sets, and thus would not yield good detection performance for new data sets.

Recently, a deep convolution neural network (DCNN) was developed to produce a hierarchy feature-maps via learned filters, and it can automatically learn a complex feature space from a huge amount of image data. The DCNN can achieve superior performance compared to conventional classification algorithms with handcrafted features. Recently, several change detection methods using deep learning algorithms have been proposed [18–20]. A difference image is fed into the deep neural networks as input data [18]. In addition, the neighboring features on each pixel on the difference image are taken as inputs. The restricted Boltzmann machine (RBM) is used for pre-training and is then unrolled in order to create a deep neural network. On the other hand, the change detection is performed by combining a sparse autoencoder, convolutional neural network (CNN), and unsupervised clustering algorithm [19]. In addition, a log-ratio map was used and transformed by a sparse autoencoder into a suitable feature space. A change detection map is directly extracted from the two images using a pre-trained CNN [20]. A unique higher dimensional feature map is produced by the CNN through different convolutional layers. The change map is computed using pixel-wise Euclidean distance of hyper dimensional features. Another change detection algorithm has also been proposed that adopts a log-ratio difference [21]. It is used as a feature input for detecting changes between multi-temporal synthetic aperture radar (SAR) images. In addition, a deep neural network was developed by stacking RBMs to learn and recognize changed pixels and unchanged pixels. In addition, a combined algorithm with the deep belief networks (DBNs) and change analysis are presented to highlight changes [22]. The presented algorithm merges and vectorizes local pixel features into DBN inputs. Then, the DBN model is established in order to capture key information for discrimination and to suppress irrelevant variations. An unsupervised clustering algorithm is then used to classify changed and unchanged pixels. Another approach utilizes joint features for change detection [23]. This work proposed an efficient change rule with a reliable expression of difference information. It learns the reliable change rule by recording the change information for a long-term sequence of remote sensing data with long short-term memory (LSTM) model. As mentioned above, all of the deep learning-based change detection algorithms yield relatively good performance. However, most of them still rely on the difference image as a feature input of their networks, resulting in them being sensitive to noisy conditions caused by geometric, radiometric distortions, and different viewing angles. In order to solve these problems, an alternative approach for change detection was developed by measuring similarity. A Siamese convolutional network was proposed to detect changed areas for optical aerial images [24]. The Siamese convolutional network with shared weights learns to extract features directly from image pairs. This work uses shared weights that are dependent from those of two branch networks. The shared weights can reduce parameters to be optimized, resulting in faster convergence. However, this model is also less flexible, which leads to overfitting due to shared weights with some other neurons.

In order to overcome the problems described above, this paper proposes a dual-dense convolutional network for recognizing pixel-wise change based on dissimilarity analysis of neighborhood pixels on high resolution panchromatic (PAN) images. In this proposed algorithm, two fully convolutional neural networks are employed to measure dissimilarity of neighborhood pixels. Furthermore, dense connection in convolution layers is applied to reuse preceding feature maps by connecting them to all subsequent layers. It is proposed to enhance a feature-map representation. While the conventional change detection algorithm [24] and conventional Siamese network use shared weights, the proposed algorithm removes shared weights in order to obtain independent optimal weights for two points of input data. So, each network can independently learn for optimal weights, called the "dual-dense convolutional network (dual-DCN)". During its training, the dual-DCN is driven to learn more robust different representations to better distinguish different types of changes.

The proposed algorithm gives better performance compared to conventional methods in qualitative and quantitative evaluation. It yields AUC of 0.97, PCC of 99%, and Kappa of 69 on average.

The rest of this paper is organized into five sections: In Section 2, the conventional convolutional neural network and problem statements will be described. Section 3 presents the proposed algorithm in detail. Section 4 will present and analyze experiment results. Finally, we conclude it in the last section.

2. Convolutional Neural Network and Problem Statement

The convolutional neural networks (CNNs) are a category of neural networks which are very effective in image recognition, classification, and so on [25]. The CNN is one of the deep learning approaches that is composed of multiple convolutional and nonlinearity layers with optional pooling, followed by fully-connected layers, as shown in Figure 1.

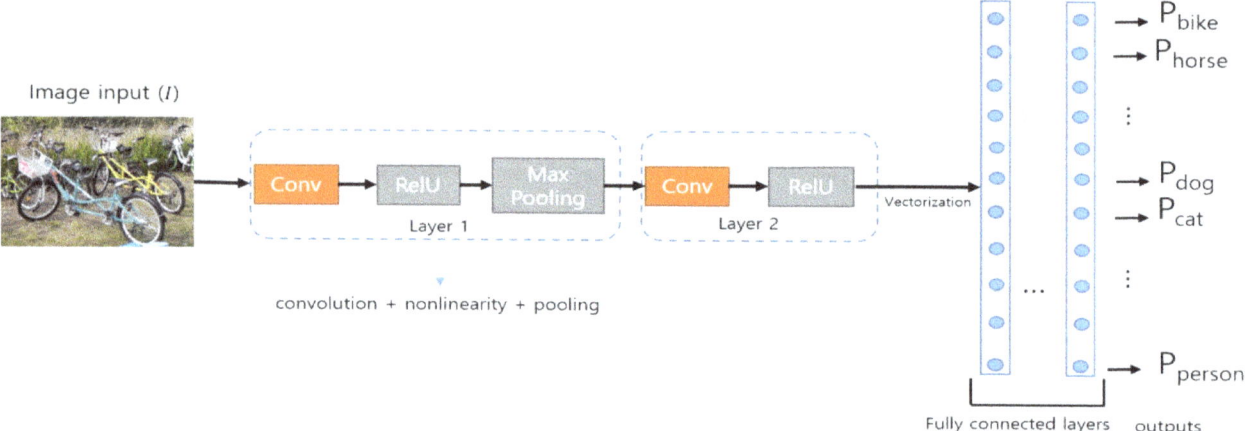

Figure 1. Traditional convolutional neural network (CNN) architecture.

Let I be an image ($m \times n \times c$) to be input, where m, n, and c are the height, width, and channel numbers of the image, respectively. In the convolutional layers, I is convolved by 2D k kernels and mapped by a nonlinearity function, called rectified linear units (ReLU), to build k feature-maps (F). The feature-map output of the lth layer is connected to the input of the ($l + 1$)th convolution and pooling layer. The final feature-maps are connected to a fully-connected layer. The last layer of fully-connected layer produces the class probability output (P_{class}). A cross-entropy function is then used to compute an objective loss. All of the weighting parameters of the network can be trained using the backpropagation algorithm.

Changes on remote-sensed images can be detected by analyzing two registered images over the same geographical area. For change detection, CNN could be employed to learn changed image characteristics and detect changed areas on remote-sensed imagery. However, the difference image (DI) or the feature fusion (FF) is widely used as an input feature of CNN, as shown in Figure 2. The DI is extracted by image subtraction or log ratio. Then, the FF is constructed by concatenating the two images. Note that these approaches are sensitive to noise as direct pixel-wise comparison features; thus, the traditional CNNs with DI or FF features could be weak to distorted data. In practice, distorted images and data are common in the remote sensing field. This distortion can be caused by not only radiometric but also geometric and viewing angle factors. In general, a geometric distortion is generated when satellites or aircrafts acquire images. In addition, image registration is required to align two images, even over the same geographical area, in a pre-processing stage. However, it is almost impossible to perfectly achieve distortion correction through automated methods. In addition, a viewing angle difference in acquisition is another challenging issue in registration and change detection. This problem cannot be resolved without precise 3-D building models, complicated algorithms, and manual intervention. For robust change detection, a robust and stable classification model is required that resolves all of the problems described above.

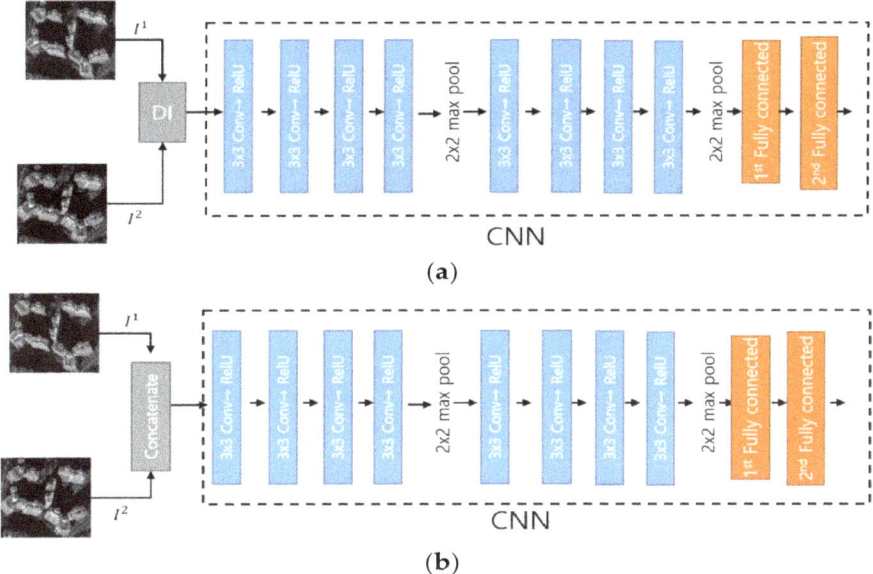

Figure 2. Conventional approaches. (**a**) Difference Image (DI) + CNN and (**b**) Feature Fusion (FF) + CNN.

3. Change Detection with the Proposed Dual-Dense Convolutional Neural Network

In general, generic change detection algorithms consist of two phases: pre-processing and change detection. Figure 3a depicts a general procedure of the conventional change detection system. The pre-processing stage performs radiometric correction, geometric rectification, and image registration. The registered images are then fed into a change detection algorithm in order to identify changed areas with feature vectors, for example, a difference image. In the general change detection systems, the radiometric correction and image registration stages are important and indispensable for better performance. The radiometric correction is performed in order to alleviate distortion for radiometric consistency. Then, the geometric correction is performed by aligning the global earth coordinates with the corresponding image points. Even though two images are compensated using multiple steps, they are still not perfectly registered, as they are independently processed with many error factors. Thus, an additional registration between two images is frequently required in order to reduce mis-alignment.

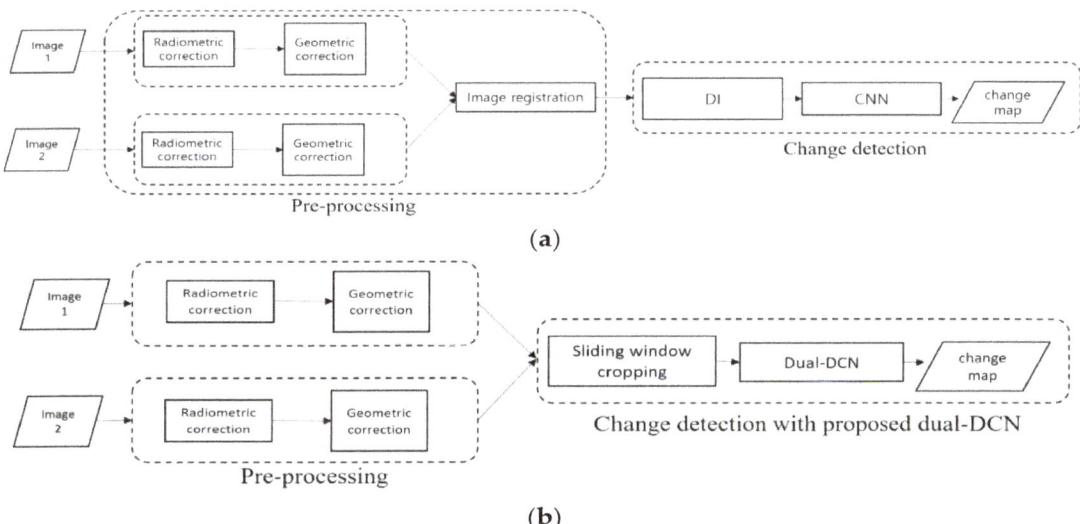

Figure 3. Change detection system schema. (**a**) Conventional change detection system and (**b**) the change detection system with the proposed dual-DCN.

In urban and mountainous areas in particular, most automatic image registration methods remain ineffective. It would degrade performance of change detection by direct pixel-wise comparison using the difference image. In order to resolve imperfect registration impacting on the change detection, a dual-DCN, as shown in Figure 3b is proposed by employing a dissimilarity distance in order to overcome the mis-alignment problem for better performance of change detection even without a perfect image registration. The proposed algorithm employs two deep convolution networks to keep all the information of the original data. The generic characteristics of the CNN handle some local distortion and alignment, thus, the proposed algorithm absorbs the misalignment problem. In addition, the dense connectivity in the convolution layer is introduced by reusing all preceding feature-maps to enhance the feature-map representation.

3.1. Pre-Processing for Change Detection

As mentioned previously, an atmospheric correction is required to remove scattering and absorption effects from the atmosphere to characterize the surface reflectance effects for a time-series image analysis. This work uses KOMPSAT-3 images with product level 1G. In these images, the radiometric correction has been done by converting the image pixel values (Digital Numbers/DNs) to surface reflectance values. It involves the conversion of DNs to a radiance value, and then to top-of-atmosphere (TOA) radiance. On the other hand, gain and offset values are provided by KOMPSAT-3 to derive the TOA reflectance values [26]. After the atmosphere errors are corrected, the geometric correction is performed in order to ensure that the pixels in the image are in their proper geometric positions on the Earth's surface. For our test images, geo-rectification and orthorectification are each conducted. For the geo-rectification, ground control points (GCPs) are identified in an unrectified image and correspond to their real coordinates to estimate the parameters (polynomial coefficients) of polynomial functions by the least square fitting. In addition, orthorectification can partly correct the image for image distortions caused by variations in the terrain topography in tandem with non-optimal satellite sensor viewing angle. Optical distortions are corrected, and terrain effects are corrected using coarse digital elevation model (DEM), namely shuttle radar topography mission DEM (SRTM DEM) for KOMPSAT-3 imagery [26].

In general change detection systems, an image registration is applied in order to ensure that two images become spatially aligned. Even though the correction of geometric distortion is performed, the spatial alignment of two images could contain a relatively large error of up to ± 6 pixels. In order to overcome this distortion, automatic image registration is widely used. However, it requires high computational load, and is furthermore not easy to obtain perfect registration. They impact the performance of change detection algorithms, resulting in the possibility that a great deal of false change areas could occur. The proposed dual-DCN is proposed so as to handle distortion problems and simplify image registration. The dissimilarity distance of local characteristics is measured in order to identify a change with the dense dual-DCN model.

3.2. Dual-Dense Convolutional Neural Network for Change Detection

In order to achieve accurate change detection without a perfect registration, this paper proposes a dual-dense convolutional network (dual-DCN) with two deep convolutional networks, as shown in Figure 4. This proposed network identifies the change areas by measuring the dissimilarity distance of two inputs at the last stage for use of all the information of the two input images. Two branch networks, N^1 and N^2, handle two input images acquired at different time instances, respectively. The proposed network is based on CNN, thus, it can robustly conduct a pixel-wise change detection by inspecting the neighboring pixels.

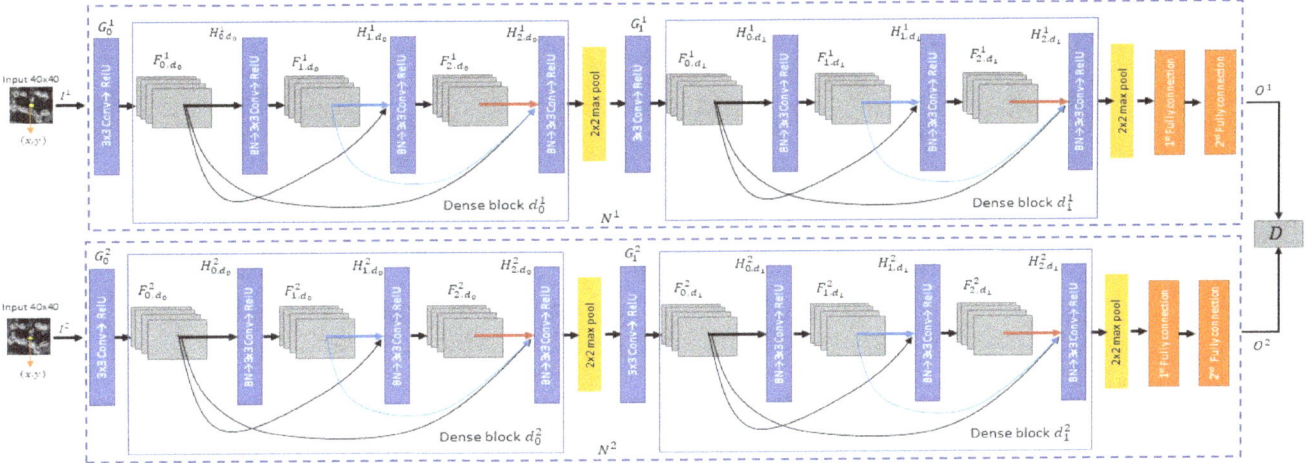

Figure 4. The proposed dual-DCN architecture for change detection.

A pair of images is cropped into two patches (40×40) by sliding in the raster scan order and two cropped patches (I^1 and I^2) are fed into the proposed network. The center pixel of the cropped patch is identified as changed or unchanged with the presence of a single dissimilar value between the cropped two patches. The Siamese network proposed by Reference [24] extracts features from an image pair. The pair of the convolutional networks is used to capture similarity characteristics by sharing the weights of the two network paths.

However, the shared weights of Siamese network reduce the parameters optimized during training for fast convergence. However, it is known to be frequently overfitted. Thus, the proposed network does not employ the shared weights to provide more flexible optimization than a restricted Siamese network. The parameters of each network branch of the proposed algorithm can be independently optimized in order to avoid early overfitted convergence. In addition, the proposed network employs dense connection [27] in the convolutional layers by reusing all the preceding feature-maps, in order to enhance representation capability of the feature-maps, as shown in Figure 4. The preceding feature maps are directly connected to all of the subsequent layers. The traditional CNN connects the output of the ($l - 1$)th layer as input to the lth [28]. In the proposed dual-DCN model, the lth layer receives all of the preceding feature-maps. The feature map of the lth layer at the rth dense block and the ith network can be computed by

$$F_{l,d_r}^i = H_{l-1,d_r}^i \left(\left[F_{0,d_r}^i, F_{1,d_r}^i, \ldots, F_{l-1,d_r}^i \right] \right), r = 0, 1; i = 1, 2 \tag{1}$$

where $\left[F_{0,d_r}^i, F_{1,d_r}^i, \ldots, F_{l-1,d_r}^i \right]$ indicates concatenation of the feature-maps of all of the previous layers, layer $0, \ldots,$ layer ($l - 1$). Each dense block is a group of convolution layers with the dense connectivity to avoid variant sizes of the feature maps. $H(\cdot)$ plays a role in batch normalization (BN) [29], 3×3 convolution, and ReLU. The BN is used to normalize parameters change of the preceding layers. The ReLU is used by thresholding at zero following 3×3 convolution. The convergence of the stochastic gradient descent algorithm can be accelerated. G including 3×3 convolution followed by ReLU is employed before a dense block in order to generate the feature-map F_0. In the proposed architecture, each dense block contains three $H(\cdot)$, including 64 feature maps of each layer. After a dense block is performed, a down-sampling operation is applied to produce variety scales with 2×2 maximum pooling. Furthermore, the feature maps at the last convolutional layer are vectorized and fed into the fully-connected layer consisting of 64 neurons and 0.5 drop-out. The probability output, O^i, at the last stage is computed by the sigmoid function. Furthermore, Euclidean distance (D) is employed in order to measure the dissimilarity between I^1 and I^2 computed by

$$D = O^1 - O^2{}_2 \tag{2}$$

When the value of D approaches 1, the center pixel of the 40×40 patch is set to a changed one, otherwise, it is set to unchanged.

3.3. Training of the Proposed Dual-DCN for Change Detection

Given a training set consisting of image pairs, the proposed network can be end-to-end trained by the backpropagation algorithm. For each image pair, let Y be a binary label of the ground truth in which $Y = 0$ if both inputs are similar, and $Y = 1$ if both inputs are dissimilar. The proposed dual-DCN is trained based on dissimilarity by computing the contrastive loss $L(D, Y)$ as an objective function [30]. This loss function employs a partial loss function for similar and dissimilar of a pair image. It produces a low value of D for unchanged pixels pair and high value for a pair of change pixels.

This proposed network is optimized using the stochastic gradient descent (SGD) optimizer. Each mini-batch arises from a single image pair that contains many changes and many absences of changes. The proposed algorithm randomly initializes all new layers by drawing weights from Glorot uniform [31]. The learning rate, decay rate, and momentum are set to 0.01, 1×10^{-6} and 0.9, respectively. The epoch number is set to 30.

4. Experimental Evaluation and Discussion

This paper uses a KOMPSAT-3 image dataset that was captured over South Korea. The KOMPSAT-3 image data set is provided by the Korea Aerospace Research Institute. Note that panchromatic band images which provide 0.7 m GSD are used for change detection. Figure 5 shows the example of an overlapped panchromatic images (1214×886) of the training dataset. These images were cropped by 40×40 sliding patch. The labels for the dataset were manually constructed for all of the center pixels of cropped patch pair.

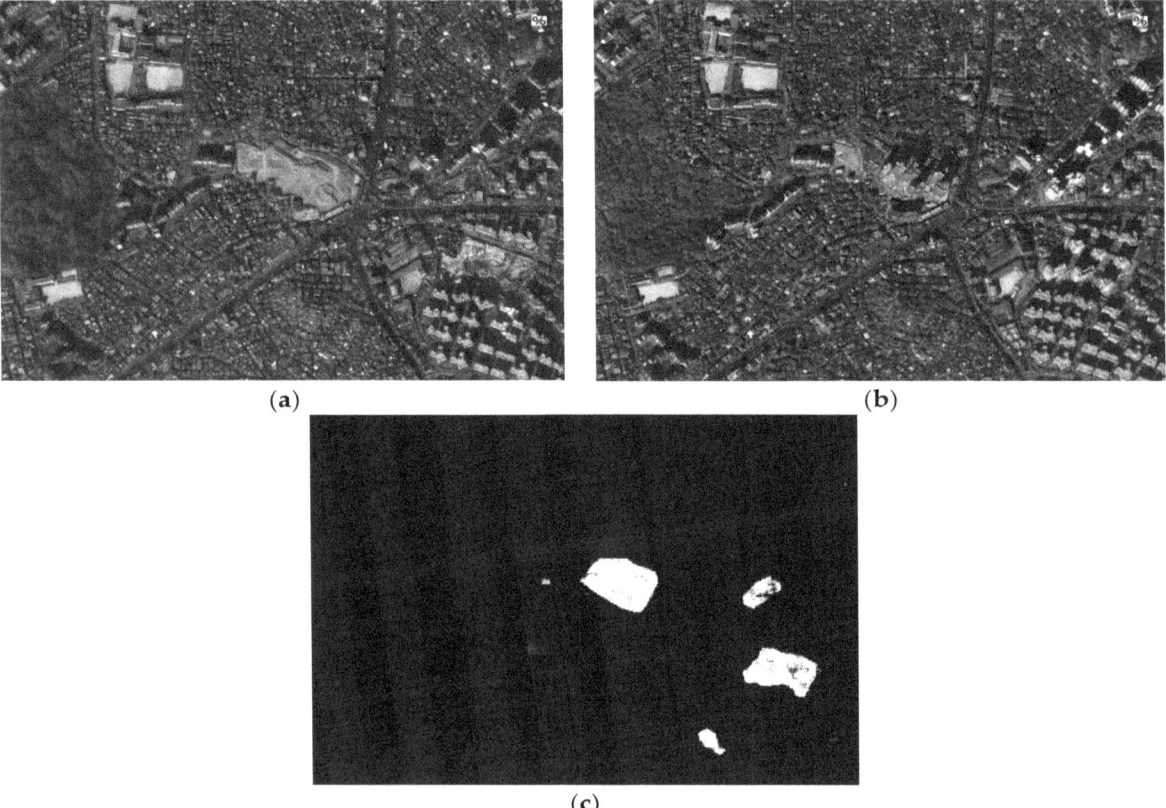

Figure 5. Seoul training data set: (**a**) image acquired in March 2014, (**b**) image acquired in December 2015, and (**c**) ground truth.

Figure 6 shows the two panchromatic images of $(29,368 \times 27,388)$ and $(29,188 \times 28,140)$ used in our experiments, which were acquired by KOMPSAT-3 on March 2014 and October 2015, respectively. These two images were acquired not only at different time instances, but also with different viewing angles. They have geometric misalignment of approximately ± 6 pixels for overlapped area.

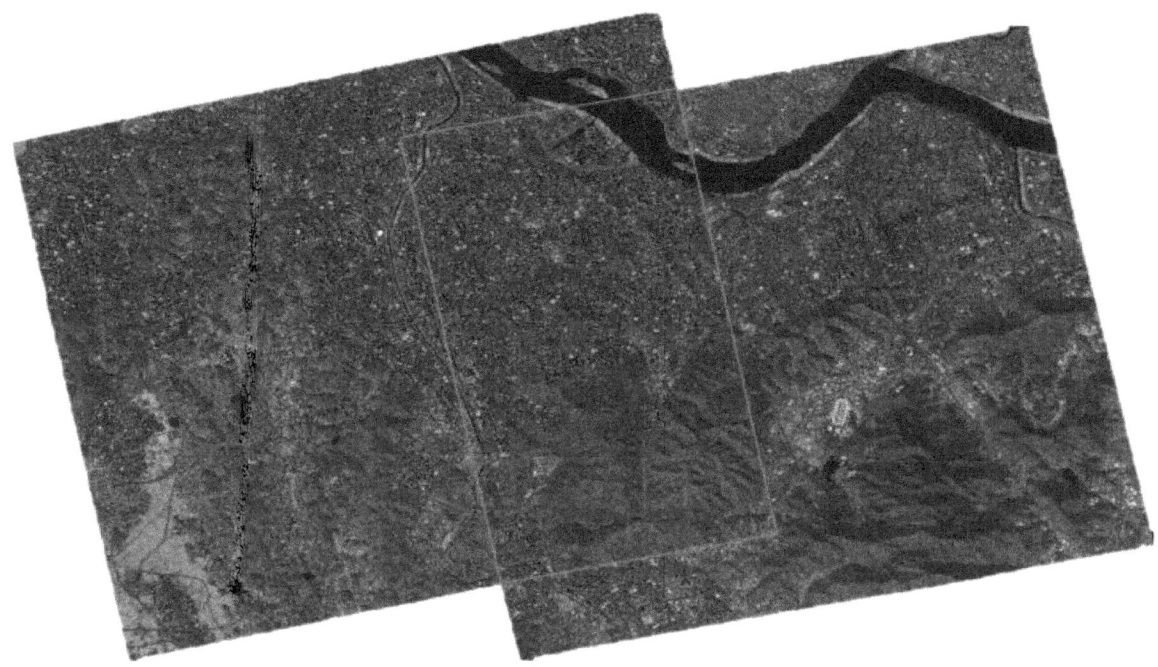

Figure 6. PAN images of Seoul area, overlapped area denoted by blue lines. (Left image: March 2014, right image: October 2015).

Figure 7 shows four selected urban areas from Figure 6, and they contain changed areas due to building construction. In Area 1, there are two types of construction changes, under construction changes and completed construction changes. Moreover, in certain areas, there are tall buildings, which could lead to false changes in change detection due to differing viewing angles. Rather than construction changes and tall buildings, we can find a forest area in Area 2. There are many tall buildings in Area 3, and accurate detection is not easy due to a large different viewing angle.

Area 4 is used to assess the influence of change detection due to differing seasons for a forest area. This case is challenging because the change due to the season should be disregarded for practical applications. Note that the labels for four areas were manually obtained, as shown in Figure 7.

In order to evaluate the change detection performance of the proposed algorithm and conventional algorithms, several metrics are used in this study, including receiver operating characteristic (ROC) curve, area under the curve (AUC), percentage correct classification (PCC), and Kappa coefficient [32]. For existing algorithms, DI + CNN, FF + CNN, and Siamese network were implemented. This CNN architecture includes 8 depth convolutional, 2 pooling, and 2 fully connected layers. For fair comparison, the same parameters of training parameters, the number feature maps, and training dataset were used in our evaluation.

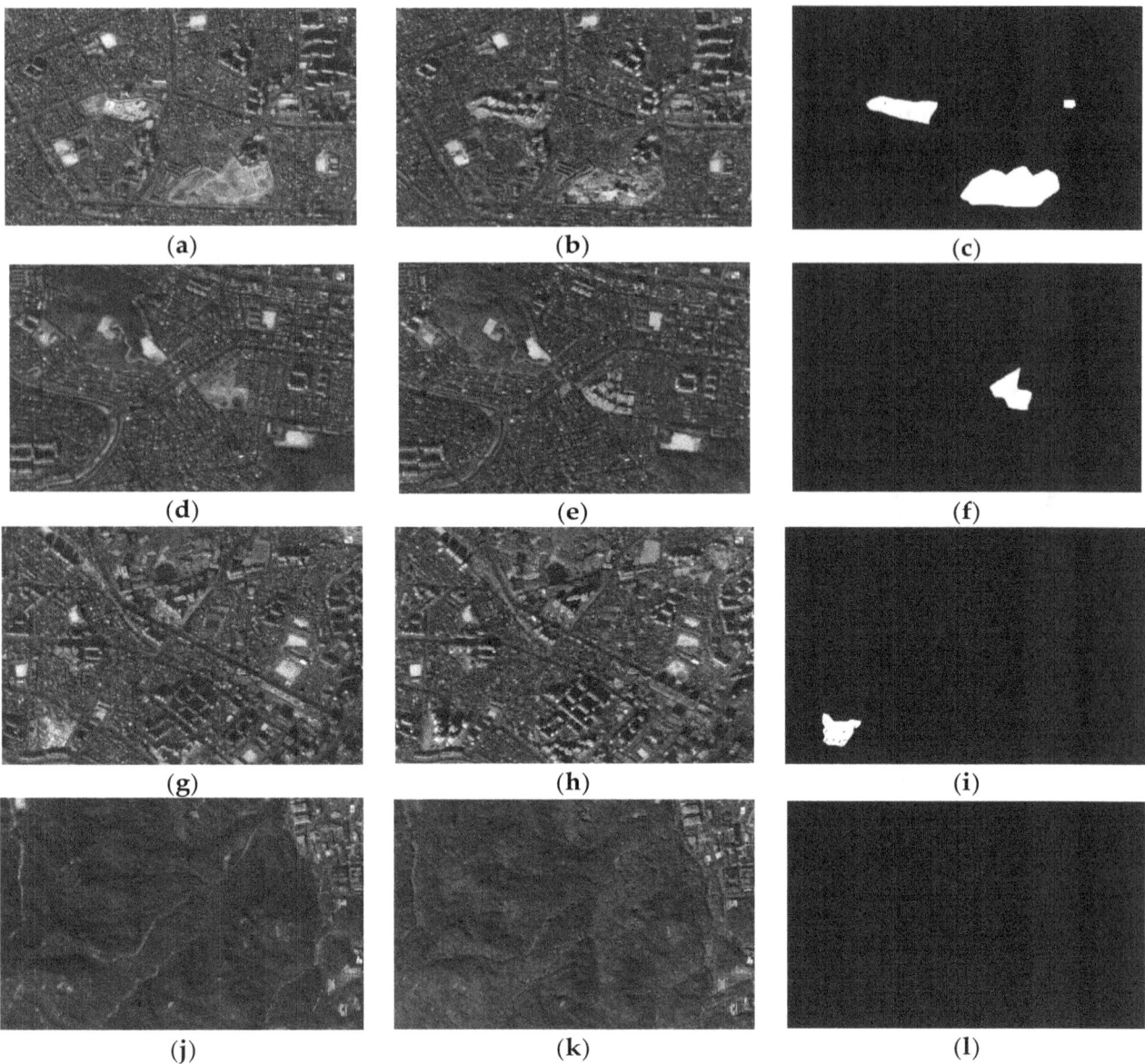

Figure 7. Four areas of Figure 6. (**a**) Input image for Area 1 (March 2014). (**b**) Input image for Area 1 (October 2015). (**c**) Ground truth for Area 1. (**d**) Input image for Area 2 (March 2014). (**e**) Input image for Area 2 (October 2015). (**f**) Ground truth for Area 2. (**g**) Input image for Area 3 (March 2014). (**h**) Image input for Area 3 (October 2015). (**i**) Ground truth for Area 3. (**j**) Input image for Area 4 (March 2014) (**k**) Input image for Area 4 (October 2015). (**l**) Ground truth for Area 4.

Figure 8 shows detection results for four areas with the exiting algorithms and the proposed algorithm. As shown in Figure 8, the proposed algorithm and FF + CNN generate better detection accuracy for Area 1. On the other hand, DI + CNN and Siamese network produce many false positives for the area. For urban surfaces, it is relatively difficult to handle the misalignment and the different viewing angle impacts because there exists tall buildings and complex constructions, resulting in the fact that false detections are likely to be performed. For Area 2, FF + CNN and DI + CNN yield more false positives, particularly in forest and urban areas. Moreover, Siamese net achieves a better detection result than other conventional algorithms. However, many false positives are still detected in certain areas. Overall, the proposed dual-DCN gives proper change detection performance, even in different viewing angle conditions. For Area 3, the proposed algorithm is still able to properly detect the changes.

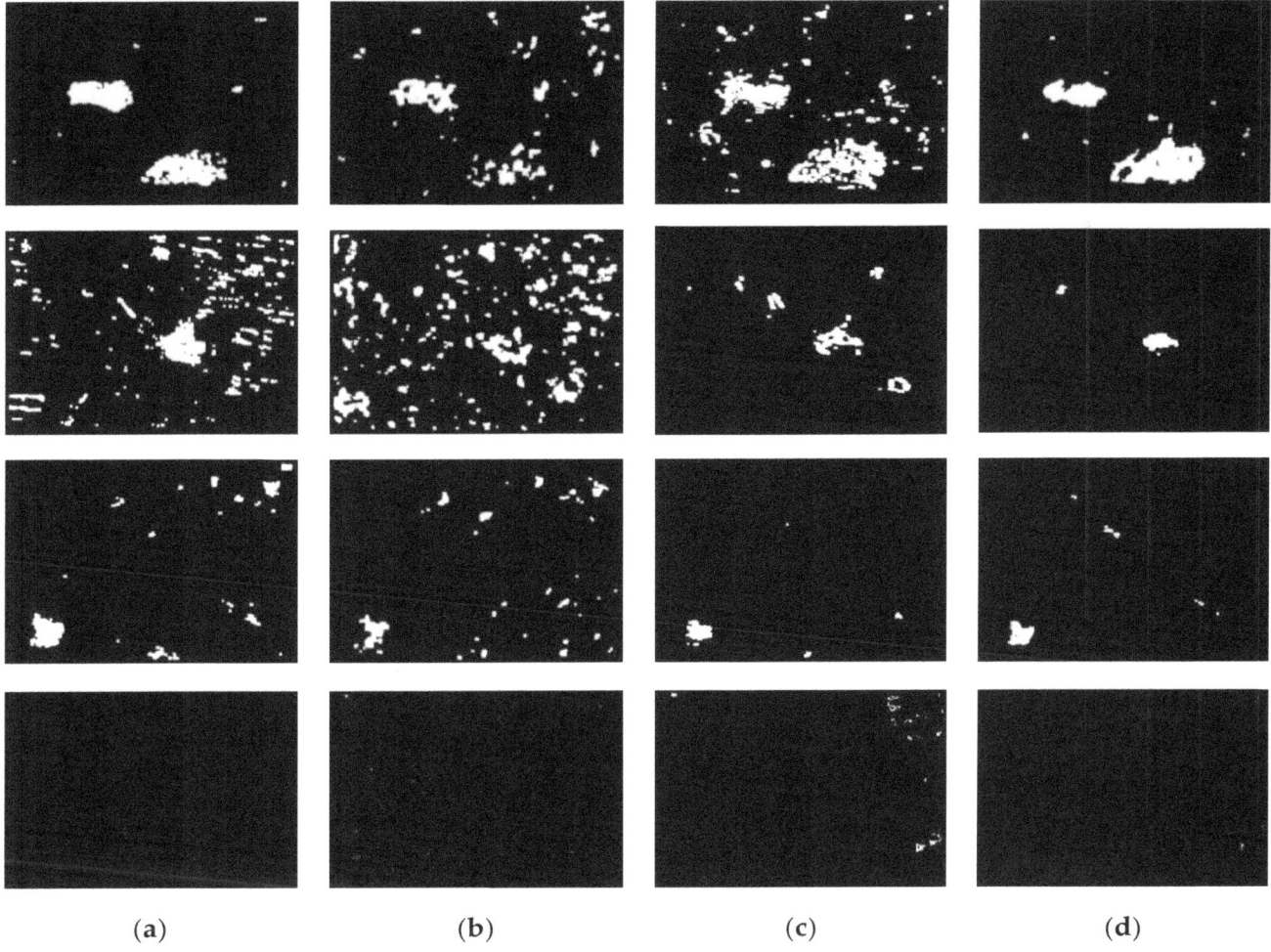

(a)   (b)   (c)   (d)

Figure 8. Detection results for four areas with the existing algorithms and proposed algorithm. (a) FF+ CNN, (b) DI + CNN, (c) Siamese network, and (d) the proposed dual-DCN.

The other algorithms result in more false positives. Note that input images for Area 4 were acquired in difference seasons for a forest area. For the test data, Siamese net produces some false positives. As shown in Figure 8, the proposed algorithm yields a better detection result with the proposed dual-DCN. The proposed algorithm can alleviate the impacts of distortions caused by imperfect geometric correction and different viewing angles. As mentioned previously, the proposed dual-DCN was designed to learn the dissimilarity of two local images in order to avoid false changes. That is why the false positive rate is relatively lower by the proposed algorithm. In contrast, DI + CNN and FF + CNN yield higher false positive rates, particularly for Areas 2 and 3. Moreover, the Siamese network produces higher false positives in Area 1, due to less optimized parameters. Figure 9 shows that the proposed algorithm can give better ROC than the conventional algorithms. According to ROC curves, the proposed dual-DCN shows better quantitative detection performance in AUC of 0.97, on average, as tabulated in Table 1. FF + CNN is slightly better in AUC than the proposed dual-DCN for Area 2, because it has better true positive for this case. However, the proposed algorithm has a lower false positive rate than FF + CNN. Table 1 summarizes the PCC and Kappa values of different methods for the three areas. As shown in Table 1, the proposed algorithm achieves higher PCC and Kappa values. We can say that the proposed dual dense convolutional network architecture has the ability to identify both changed and unchanged areas by disregarding irrelevant variations and false changes, even in cases of complicated urban surfaces, geometric distortion, and different viewing angles.

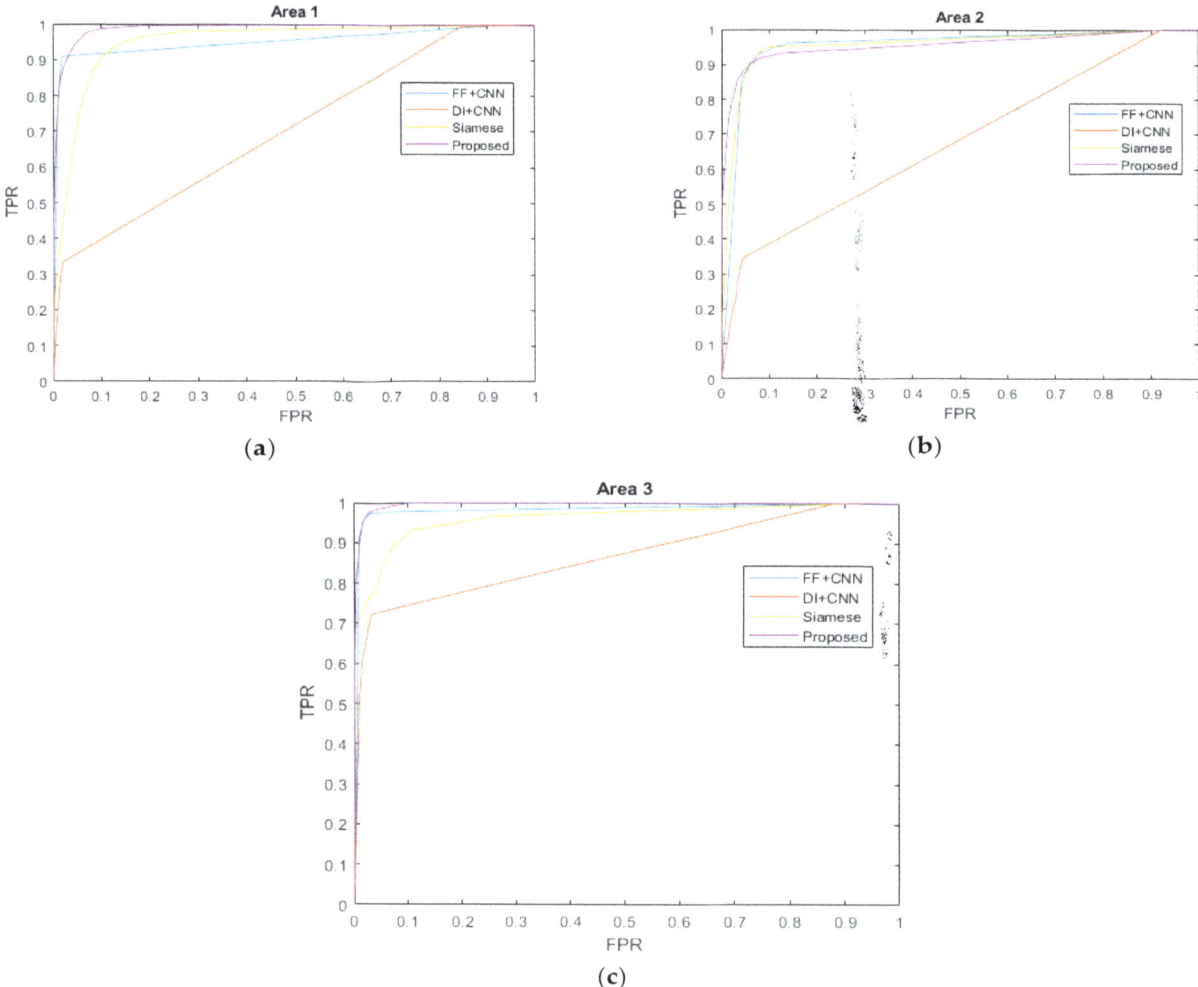

Figure 9. ROC for three areas. (**a**) ROC for Area 1, (**b**) ROC for area 2, and (**c**) ROC for Area 3.

Table 1. Quantitative assessments of the existing and proposed algorithms.

Metrics	Algorithms	Area 1	Area 2	Area 3	Avg
AUC	FF + CNN	0.95	0.95	0.98	0.96
	DI + CNN	0.70	0.68	0.88	0.75
	Siamese net	0.96	0.92	0.91	0.93
	The proposed	0.99	0.93	0.99	0.97
PCC (%)	FF + CNN	97	92	98	96
	DI + CNN	94	97	97	96
	Siamese net	96	98	99	98
	The proposed	98	99	99	99
Kappa	FF + CNN	78	19	47	48
	DI + CNN	30	32	28	30
	Siamese net	52	35	68	52
	The proposed	78	60	69	69

Regarding time complexity, the proposed DCN requires more computational complexity than the single architecture using FF + CNN and DI + CNN by a factor of approximately two with sequential machines. However, the proposed dual-DCN can work in parallel, thus, throughput can be enhanced with a parallel machine such as GPU. In addition, the proposed also takes about 20% more running time than the Siamese network because it includes additional preceding of feature maps.

5. Conclusions

In this paper, we presented a robust change detection algorithm for high-resolution panchromatic imagery. The proposed algorithm learns and analyzes the dissimilarity of two input images with the densely convolutional network by incorporating local information. We found that the proposed algorithm achieves higher detection accuracy, even with noisy conditions such as geometric distortion and different viewing angles in qualitative and quantitative analysis. Further work can be conducted to extend the framework for other modalities such as multi-spectrum images and SAR data.

Author Contributions: All authors contributed to the writing of the manuscript. W.W. and J.L. conceived and designed the experiments; W.W. performed the experiments and analyzed the data; S.-E.P. and D.S. supervised this study.

References

1. Coppin, P.R.; Bauer, M.E. Digital change detection in forest ecosystems with remote sensing imagery. *Remote Sens. Rev.* **1996**, *13*, 207–234. [CrossRef]

2. Bazi, Y.; Bruzzone, L.; Melgani, F. Automatic identification of the number and values of decision thresholds in the log-ratio image for change detection in SAR images. *IEEE Geosci. Remote Sens. Lett.* **2006**, *3*, 349–353. [CrossRef]

3. Singh, K.K.; Mehrotra, A.; Nigam, M.J.; Pal, K. Unsupervised change detection from remote sensing using hybrid genetic FCM. In Proceedings of the IEEE 2013 Students Conference on Engineering and Systems (SCES), Allahabad, India, 12–14 April 2013; pp. 1–5.

4. Bi, C.; Wang, H.; Bao, R. SAR image change detection using regularized dictionary learning and fuzzy clustering. In Proceedings of the 2014 IEEE 3rd International Conference on Cloud Computing and Intelligence Systems (CCIS), Shenzhen, China, 27–29 November 2014; pp. 327–330.

5. Gong, M.; Zhou, Z.; Ma, J. Change detection in synthetic aperture radar images based on image fusion and fuzzy clustering. *IEEE Trans. Image Process.* **2012**, *21*, 2141–2151. [CrossRef] [PubMed]

6. Gong, M.; Su, L.; Jia, M.; Chen, W. Fuzzy clustering with a modified MRF energy function for change detection in synthetic aperture radar images. *IEEE Trans. Fuzzy Syst.* **2014**, *22*, 98–109. [CrossRef]

7. Gong, M.; Zhao, J.; Liu, J.; Miao, Q.; Jiao, L. Change detection in synthetic aperture radar images based on deep neural networks. *IEEE Trans. Neural Netw. Learn. Syst.* **2016**, *27*, 125–138. [CrossRef] [PubMed]

8. Johnson, R.D.; Kasischke, E.S. Change vector analysis: A technique for the multispectral monitoring of land cover and condition. *Int. J. Remote Sens.* **1998**, *19*, 411–426. [CrossRef]

9. Gao, F.; Zhang, L.; Wang, J.; Mei, J. Change Detection in Remote Sensing Images of Damage Areas with Complex Terrain Using Texture Information and SVM. In Proceedings of the International Conference on Circuits and Systems (CAS 2015), Paris, France, 9–10 August 2015.

10. Guo, Z.; Du, S. Mining parameter information for building extraction and change detection with very high-resolution imagery and GIS data. *GISci. Remote Sens.* **2017**, *54*, 38–63. [CrossRef]

11. Huang, S.; Ramirez, C.; Kennedy, K.; Mallory, J.; Wang, J.; Chu, C. Updating land cover automatically based on change detection using satellite images: Case study of national forests in Southern California. *GISci. Remote Sens.* **2017**, *54*, 495–514. [CrossRef]

12. Hao, M.; Zhang, H.; Shi, W.; Deng, K. Unsupervised change detection using fuzzy c-means and MRF from remotely sensed images. *Remote Sens. Lett.* **2013**, *4*, 1185–1194. [CrossRef]

13. Hao, M.; Hua, Z.; Li, Z.; Chen, B. Unsupervised change detection using a novel fuzzy c-means clustering simultaneously incorporating local and global information. *Multimed. Tools Appl.* **2017**, *76*, 20081–20098. [CrossRef]

14. Yu, H.; Yang, W.; Hua, G.; Ru, H.; Huang, P. Change detection using high resolution remote sensing images based on active learning and Markov random fields. *Remote Sens.* **2017**, *9*, 1233. [CrossRef]

15. Habib, T.; Inglada, J.; Mercier, G.; Chanussot, J. Support vector reduction in SVM algorithm for abrupt change detection in remote sensing. *IEEE Geosci. Remote Sens. Lett.* **2009**, *6*, 606–610. [CrossRef]

16. Volpi, M.; Tuia, D.; Bovolo, F.; Kanevski, M.; Bruzzone, L. Supervised change detection in VHR images using contextual information and support vector machines. *Int. J. Appl. Earth Obs. Geoinf.* **2013**, *20*, 77–85. [CrossRef]

17. Bovolo, F.; Bruzzone, L.; Marconcini, M. A novel approach to unsupervised change detection based on a semisupervised SVM and a similarity measure. *IEEE Trans. Geosci. Remote Sens.* **2008**, *46*, 2070–2082. [CrossRef]

18. Zhao, J.; Gong, M.; Liu, J.; Jiao, L. Deep learning to classify difference image for image change detection. In Proceedings of the IEEE 2014 International Joint Conference on Neural Networks (IJCNN), Beijing, China, 61–1 July 2014; pp. 411–417.

19. Gong, M.; Yang, H.; Zhang, P. Feature learning and change feature classification based on deep learning for ternary change detection in SAR images. *ISPRS J. Photogramm. Remote Sens.* **2017**, *129*, 212–225. [CrossRef]

20. El Amin, A.M.; Liu, Q.; Wang, Y. Convolutional neural network features-based change detection in satellite images. In Proceedings of the First International Workshop on Pattern Recognition, Tokyo, Japan, 11–13 May 2016.

21. Liu, J.; Gong, M.; Zhao, J.; Li, H.; Jiao, L. Difference representation learning using stacked restricted Boltzmann machines for change detection in SAR images. *Soft Comput.* **2016**, *20*, 4645–4657. [CrossRef]

22. Zhang, H.; Gong, M.; Zhang, P.; Su, L.; Shi, J. Feature-level change detection using deep representation and feature change analysis for multispectral imagery. *IEEE Geosci. Remote Sens. Lett.* **2016**, *13*, 1666–1670. [CrossRef]

23. Lyu, H.; Lu, H.; Mou, L. Learning a transferable change rule from a recurrent neural network for land cover change detection. *Remote Sens.* **2016**, *8*, 506. [CrossRef]

24. Zhan, Y.; Fu, K.; Yan, M.; Sun, X.; Wang, H.; Qiu, X. Change Detection Based on Deep Siamese Convolutional Network for Optical Aerial Images. *IEEE Geosci. Remote Sens. Lett.* **2017**, *14*, 1845–1849. [CrossRef]

25. Yoo, H.-J. Deep convolution neural networks in computer vision. *IEIE Trans. Smart Process. Comput.* **2015**, *4*, 35–43. [CrossRef]

26. KOMPSAT-3 Product Specifications Version 2.0. Available online: http://www.si-imaging.com/resources/?pageid=2&uid=232&mod=document (accessed on 25 June 2018).

27. Huang, G.; Liu, Z.; Van Der Maaten, L.; Weinberger, K.Q. Densely Connected Convolutional Networks. In Proceedings of the 2017 IEEE Conference on Computer Vision and Pattern Recognition (CVPR), Honolulu, HI, USA, 21–26 July 2017; pp. 4700–4708.

28. Krizhevsky, A.; Sutskever, I.; Hinton, G.E. Imagenet classification with deep convolutional neural networks. *Adv. Neural Inf. Process. Syst.* **2012**, *25*, 1097–1105. [CrossRef]

29. Ioffe, S.; Szegedy, C. Batch normalization: Accelerating deep network training by reducing internal covariate shift. In Proceedings of the International Conference on Machine Learning, Lille, France, 6–11 July 2015.

30. Hadsell, R.; Chopra, S.; Le Cun, Y. Dimensionality reduction by learning an invariant mapping. In Proceedings of the 2006 IEEE Computer Society Conference on Computer Vision and Pattern Recognition, New York, NY, USA, 17–22 June 2006.

31. Glorot, X.; Bengio, Y. Understanding the difficulty of training deep feedforward neural networks. In Proceedings of the Thirteenth International Conference on Artificial Intelligence and Statistics, Sardinia, Italy, 13–15 May 2010; pp. 249–256.

32. Fitz, R.W.; Lees, B.G. Assesing the classification accuracy of multisource remote sensing data. *Remote Sens. Environ.* **1994**, *47*, 362–368.

12

Fusion Network for Change Detection of High-Resolution Panchromatic Imagery

Wahyu Wiratama and Donggyu Sim *

Department of Computer Engineering, Kwangwoon University, Seoul 139701, Korea; wiratama@kw.ac.kr
* Correspondence: dgsim@kw.ac.kr

Abstract: This paper proposes a fusion network for detecting changes between two high-resolution panchromatic images. The proposed fusion network consists of front- and back-end neural network architectures to generate dual outputs for change detection. Two networks for change detection were applied to handle image- and high-level changes of information, respectively. The fusion network employs single-path and dual-path networks to accomplish low-level and high-level differential detection, respectively. Based on two dual outputs, a two-stage decision algorithm was proposed to efficiently yield the final change detection results. The dual outputs were incorporated into the two-stage decision by operating logical operations. The proposed algorithm was designed to incorporate not only dual network outputs but also neighboring information. In this paper, a new fused loss function was presented to estimate the errors and optimize the proposed network during the learning stage. Based on our experimental evaluation, the proposed method yields a better detection performance than conventional neural network algorithms, with an average area under the curve of 0.9709, percentage correct classification of 99%, and Kappa of 75 for many test datasets.

Keywords: change detection; convolutional network; deep learning; panchromatic; remote sensing

---

1. Introduction

Change detection is a challenging task in remote sensing, used to identify areas of change between two images acquired at different times for the same geographical area. Such detection has been widely used in both civilian and military fields, including agricultural monitoring, urban planning, environment monitoring, and reconnaissance. In general, change detection involves a preprocessing step, feature extraction, and classification or clustering algorithm to distinguish changed and unchanged pixels. To obtain a good performance, the selected classification or clustering algorithm plays an important role in the field of change detection.

In prior studies, statistical approaches have been proposed to identify a change [1–3]. A corresponding maximal invariant statistic is obtained by analyzing a suitable group of transformations leaving problem invariant [2]. Then, a general problem of testing equality among M covariance metrices in the complex-valued Gaussian case is analyzed for synthetic aperture radar (SAR) change detection. A sample coherence magnitude as a change metric has been proposed by [3]. A new maximum-likelihood temporal change estimation and complex reflectance change detection is used for SAR coherent temporal change detection. Currently, a classification or clustering is becoming one approach to be used for change detection in remote sensed images by employing supervised or unsupervised learning, respectively. Feature selection and feature extraction are important aspects in this approach. Several detection algorithms using two images have been proposed with different features for different types of applications [3–19]. The methods used for change detection have mostly been designed to extract changed features such as in a difference image (DI) [3–9], a local change vector [10], or a texture vector [11–13]. A DI is a common feature used to represent a change in information through the subtraction of temporal images. Local change vectors have also been used by applying neighbor pixels to avoid a direct subtraction based on the log ratio. This method computes a mean value of the log ratio of temporal neighbor pixels. A texture vector [11–13] is employed to extract statistical characteristics. These changed features are then fed into a classification or clustering algorithm to determine changed/unchanged pixels. Some unsupervised change detection methods have been proposed based on the fuzzy c-mean (FCM) algorithm [14,16]. Such approaches are useful when labels in the training stages are unavailable. The learning algorithms in the aforementioned studies are based on observed data without any additional information, therefore, their application leads to overfitting for invariant changes. Furthermore, they do not yield a reasonably good performance in the change detection rates because they do not incorporate accurate information without supervision. Therefore, supervised change detection methods, such as a support vector machine (SVM) [11,16–18], have been proposed. The basic SVM can apply a binary classification to changed or unchanged pixels with texture information or using a change vector analysis. These algorithms are not perfect in terms of incorporating accurate and full statistical characteristics for large multi-dimensional data. Furthermore, they do not yield the best detection performance for new datasets.

Recently, a deep convolution neural network (DCNN) was developed to produce a hierarchy of feature-maps such as learned filters. The aforementioned DCNN can automatically learn a complex feature space from a huge amount of image data. A DCNN can achieve a superior performance compared to conventional classification algorithms. A restricted Boltzmann machine (RBM) [19], a convolutional neural network (CNN) [20–22], and deep belief networks (DBNs) [23] have been proposed for use in change detection. Such change detection algorithms based on deep learning yield a relatively good performance in terms of the detection accuracy. However, most can be categorized into front-end differential change detection using low-level features such as a difference image as a feature input of their networks, resulting in sensitivity to several deteriorated conditions caused by geometric/radiometric distortions, different viewing angles, and so on. This front-end differential change detection conducts an early feature extraction of two image inputs into a single-path network. In contrast, back-end differential detection methods by employing dual-path networks have been proposed for fusing higher-level features with a long short-term memory (LSTM) model [24] to avoid the use of low-level difference features such as a difference image. In addition, a Siamese convolutional network (SCN) [25–27] and dual-DCN (dDCN) [28] were also proposed to detect changed areas by measuring the similarity with high-level network features. These algorithms achieve a relatively good performance, although false negatives are still observed.

To reduce false positives and false negatives in change detection, a fusion network incorporating low- and high-level feature spaces in neural networks was proposed in this paper. For low-level differential features, the difference image is fed into the front-end differential DCN (FD-DCN). For a high-level differential feature, a back-end differential dDCN (BD-dDCN) is employed. In addition, a two-stage decision algorithm is incorporated for post-processing to enhance the detection rate during the inference stage. The intersection and union operations are employed to validate the change map. First, an intersection operation is used to avoid false positives. The second-stage decision operates a union by considering the local information of the first decision. This stage is developed to validate and repair the change map from the first decision. In addition, this study introduces a new loss function that combines a contrastive loss and weighted binary cross entropy loss function to optimize high- and low-level differential features, respectively. In our experiment, we found that the proposed algorithm can yield a better performance than existing algorithms by achieving an average area under the curve (AUC) of 0.9709, a percentage correct classification (PCC) of 99%, and a Kappa of 75 for several test datasets.

This work contributes three main key features as follows. (1) Unlike the mentioned existing works above, we propose a fusion network by combining a front- and back-end networks to perform the low- and high-level differential detection in one structure. (2) A combining loss function between contrastive loss and binary cross entropy loss is proposed to accomplish fusion of the proposed networks in training stage. (3) The two-stage decision as a post-processing is presented to validate and ensure the changes prediction at the inference stage to obtain better the final change map.

This paper is organized into five sections. In Section 2, related studies are briefly described. Section 3 presents the proposed algorithm in detail. Section 4 describes and analyzes the experiment results. Finally, we provide some concluding remarks regarding this research.

2. Deep Convolutional Network and Related Studies on Change Detection

Deep neural architectures with hundreds of hidden layers have been developed to learn high-level feature spaces. The recently developed convolutional neural network (CNN) is a deep learning architecture that has been shown to be effective in image recognition and classification [29]. The CNN architecture employs multiple convolutional layers, followed by an activation function, resulting in the development of feature maps. The rectified linear unit (ReLU) is widely used as the activation function in many CNN architectures. To progressively gather global spatial information, the feature maps are sub-sampled by the pooling layer. The final feature maps are connected to a fully connected layer to produce the class probability outputs (P_{class}), as shown in Figure 1. During the training stage, an objective loss such as cross-entropy is computed. All of the weighting parameters of the network are updated to reduce the cost function using the back-propagation algorithm.

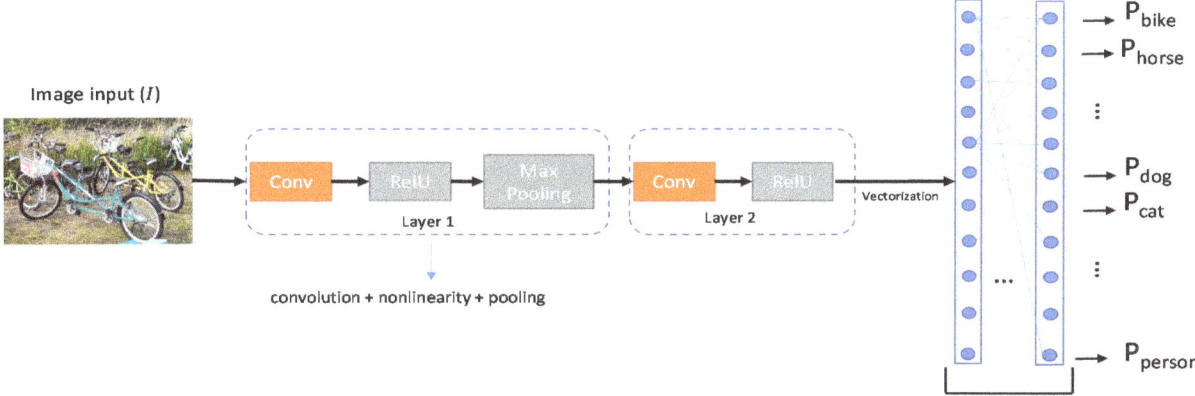

Figure 1. Convolutional neural network (CNN) architecture.

The related studies on change detection based on deep learning can be categorized into two categories based on the type of network that is used: A front-end differential network (FDN) and a back-end differential network (BDN). The front-end network uses low-level differential features such as a DI or joint feature (JF) as the feature input of the network, as shown in Figure 2a. In this case, a network with a single-path architecture receives the extracted DI as low-level differential features of the temporal images to identify changed pixels. Several studies based on an FDN have been proposed to improve the performance of the change detection rate. In addition, a deep neural network (DNN) is applied to detect objects from synthetic aperture radar (SAR) data [30]. The differential feature of temporal data is employed instead of a DI. This feature is used to solve the initial weight problem through pre-training using the restricted Boltzmann machine (RBM) algorithm. These pre-trained weights are then fed into the initial weights of the DNN during the training stage. In contrast, unsupervised change detection has been proposed by combining DBNs with a feature change analysis [23]. The feature maps of temporal input images are obtained using the DBN. The magnitude and direction of these feature maps are analyzed to distinguish the types of feature changes using an unsupervised fuzzy C-means algorithm. Other unsupervised systems have been proposed by combining a sparse autoencoder (SAE), unsupervised clustering, and a CNN to overcome the change detection problem without supervision [20]. First, a DI is computed using a log-ratio operator. The feature maps of the DI are extracted through the SAE and clustered into change classes as the labels for the training CNN. Next, some feature maps extracted by the SAE are taken as the training data for the CNN. In addition, an autoencoder and multi-layer perceptron (MLP) are combined to identify changed pixels [31]. Change detection using faster R-CNN has been proposed for high-resolution images [32]. This work detects changed areas with bounding boxes. The DI is extracted and then fed into faster R-CNN to detect changed locations. Each of these deep learning algorithms tackles the change detection problem using a front-end differential network. This network identifies changes by observing low-level feature such as the DI, which is sensitive to various distortions, including geometric and radiometric distortions, and different viewing angles. Another approach of FDN to detect the changes has been proposed by joining feature inputs (JF) [23]. Two temporal images are concatenated and they are fed into DBN to avoid a DI for change detection. However, by joining the features in the early network causes both low-level differential inputs to be dependently learned in the single network. It is for global change detection, resulting in more false positives.

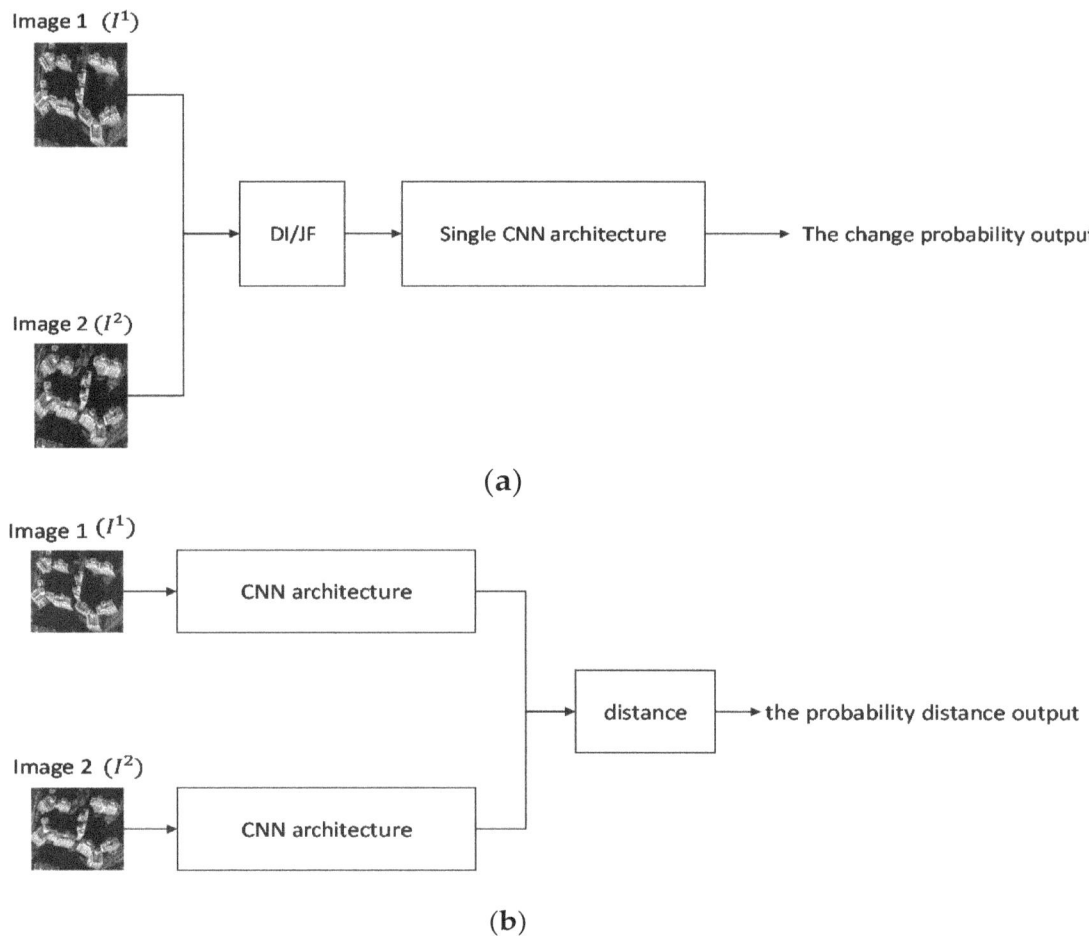

Figure 2. Front-end differential network (FDN) and back-end differential network (BDN) architectures: (a) Difference image (DI)/Joint features (JF) + single CNN, and (b) Dual-CNN.

Alternative algorithms for change detection were introduced by employing a high-level differential feature with a dual-path network, as shown in Figure 2b. Siamese CNN (SCNN) was proposed to detect changed areas for multimodal remote sensing data [27]. This architecture was employed to learn the different characteristics between multimodal remote sensing data. This approach learns the feature map of temporal images in each path network. The Euclidean distance was employed to measure the similarity at the back-end of the network. A similar method was developed based on an SCNN for optical aerial images [25]. A deep CNN was proposed by producing a change detection map directly from two images [33]. A change map was evaluated using the pixel-wise Euclidean distance from high-dimensional feature maps. Another method was proposed that incorporates a deep stacked denoising autoencoder (SDAE) and feature change analysis (FCA) for multi-spatial resolution change detection [34]. In the aforementioned study, denoising autoencoders were stacked to learn local and high-level features for unsupervised learning. Then, the inner relationship between the multi-resolution image pair was exploited by building a mapping neural network to identify any change representations. A dual-dense convolutional network was presented by incorporating information from neighbor pixels [28]. In the aforementioned study, a dense connection was used to enhance the features of the changed map information. All of the above-mentioned BDN architectures yield good performances by inspecting high-level differential features, which can reduce false positives. However, a BDN can achieve higher sensitivity and specificity through high-level differential features.

Although a high-level differential network can improve the sensitivity and specificity, the false negative rate is still too high for practical applications. The FDN architecture can achieve a relatively higher true-positive rate regardless of the number of false positives. In addition, the BDN architecture can reduce the false-positive rate by producing some false negatives caused by strict decision criteria in high-level differential features. In this work, an FDN and a BDN were fused to employ the advantages of both. A post-processing step was then employed during the inference stage to obtain the final decision for change detection.

3. Proposed Fusion Network for Change Detection with Panchromatic Imagery

In general, a change detection system involves a pre-processing step to reduce geometric and radiometric distortions for better results. A radiometric correction is applied to remove atmospheric effects for a time-series image analysis. In this study, a radiometric correction was applied by converting digital numbers (DNs) into a radiance value. Then, the top-of-atmosphere (TOA) reflectance values were computed using the gain and offset values provided by a satellite provider. In addition, to ensure that the pixels in the image were in their proper geometric position on the Earth's surface, a geometric correction was applied. The parameters (polynomial coefficients) of the polynomial functions were estimated using least square fitting with ground control points (GCPs) identified in an unrectified image and corresponding to their real coordinates. A digital elevation model (DEM), namely, shuttle radar topography mission DEM (SRTM DEM), was then used to correct the optical distortion and terrain effect. The corrected images were then incorporated into the proposed network to detect changes.

To achieve a change detection, the proposed network employs a fusion network by fusing the FDN and BDN architectures. Dual outputs were generated to solve low-level differential and high-level differential problems. For the training stage, a contrastive loss function and a weighted binary loss function were combined to optimize the proposed fusion network parameters. In addition, a pre-processing step was applied to validate and ensure false changes during the inference stage. Intersection and union operations were then applied from the dual outputs of the proposed network. According to the proposed change detection, the false-positive and false-negative rates could be reduced, resulting in high sensitivity and specificity for a proper change detection. Symbols used in the proposed method are tabulated in Table 1.

Table 1. Symbols used in the proposed fusion network for change detection.

Symbol	Description
I^1 and I^2	Cropped temporal input image in time 1 and 2, respectively
N^1 and N^2	Patch network 1 and 2 correspond to the back-end network
N^3	Patch network 3 correspond to the front-end network
F^i_{l,d_r}	Feature maps of the l-th layer at the r-th dense block and the i-th network
P^1 and P^2	Outputs of N^1 and N^2, respectively
D	Dissimilarity distance
O	Change detection probability output of N^3
H^i_{l-1,d_r}	Incorporation process of a batch normalization (BN), a 3×3 convolution, and ReLU of the $(l-1)$-th layer at the r-th dense block and the i-th network
$[F^i_{0,d_r}, F^i_{1,d_r}, \ldots, F^i_{l-1,d_r}]$	A concatenation of the feature-maps of all of previous layers, layer 0, $\ldots$, and layer $(l-1)$
L	Proposed loss function
E_c	Contrastive loss function
E_B	Weighted binary cross entropy loss function
Y	Ground truth
Ls and L_D	Partial loss function for a pair of similar and dissimilar pixels, respectively
m	Margin value
α	Weighted loss
W	Proposed weighted function
C and U	Changed and unchanged numbers of pixels, respectively
N	The number of full dataset
β_c and β_u	Penalization weights for false-negative and false-positive errors, respectively
M_1	Change map for first prediction
M_2	Change map for second prediction
Nb	Local information of M_1
T	Tested temporal images
m and n	Size of T
s	Size of I

3.1. Fusion Network for Change Detection

For a change detection, an FDN architecture is commonly used for identifying changed pixels. Such an architecture uses low-level differential features that are relatively sensitive to noises. It is caused by direct low-level features comparison, which misalignments of geometric error and a different angle view are very influential. This FDN assigns a DI or JF to a single path network. They conduct dependent learning of both low-level features together which lead to hard learning for invariant changes and above-mentioned noisy conditions. Thus, this approach would produce a global change detection, resulting in true positives and more false positives. In addition, BDN architectures are designed to avoid low-level differential features, thereby reducing the false-positive detection rate. These architectures apply strict identification for a high-level differential, which may cause some false negatives. Therefore, an FDN is suitable in terms of the true-positive detection rate, and a BDN is extremely reliable for overcoming false positives. To obtain a proper change detection, a fusion network architecture is proposed by fusing an FD-DCN and a BD-dDCN with a dense-connectivity of the convolution layers, as shown in Figure 3. There are three branch networks, N^1, N^2, and N^3, receiving two temporal images (I^1 and I^2) in which N^1 and N^2 correspond to the back-end network, and N^3 refers to the front-end network by concatenating these two inputs (I^1 and I^2). A dense convolutional connection was employed in the proposed fusion network to enhance the feature representation [35]. This dense architecture is very effective at covering invariant change representations by reusing all preceding feature maps of the network. The proposed network was designed using dual outputs, namely, the dissimilarity distance (D) and change probability (O) at the last layer, corresponding to the back-end and front-end networks, respectively. Let us assume that the feature maps of the l-th layer at the r-th dense block and the i-th network are computed as:

$$F^i_{l,d_r} = H^i_{l-1,d_r}([F^i_{0,d_r}, F^i_{1,d_r}, \ldots, F^i_{l-1,d_r}]), \; r = 0, 1; \; i = 1, 2, 3, \tag{1}$$

where $[F^i_{0,d_r}, F^i_{1,d_r}, \dots, F^i_{l-1,d_r}]$ indicates a concatenation of the feature-maps of all of previous layers, layer $0, \dots$, and layer $(l-1)$. In addition, $H(\cdot)$ incorporates a batch normalization (BN), a 3×3 convolution, and ReLU. A pair of temporal images were cropped into two patches (40×40) (I^1 and I^2) by sliding the window and fed into N^1 and N^2, respectively. The dissimilarity distance (D) was then computed based on the Euclidean distance, which is defined as follows:

$$D = \|P^1 - P^2\|_2 \tag{2}$$

where P^1 and P^2 are the outputs of N^1 and N^2 activated by sigmoid function, respectively. The proposed method applies a pixel-wise change detection by inspecting the neighboring pixels. The 40×40 patch images identify a change corresponding to the center pixel of the patch. Thus, when the value of D is close to 1, the center of I is assigned to a changed pixel. In addition, I^1 and I^2 were concatenated to be fed into N^3. The same dense convolution architecture was employed in this branch network to generate the change detection probability (O). The dual outputs (D and O) are a result of this fusion network. In addition, a post-processing step during the inference stage was proposed based on these outputs (D and O) to achieve a proper prediction.

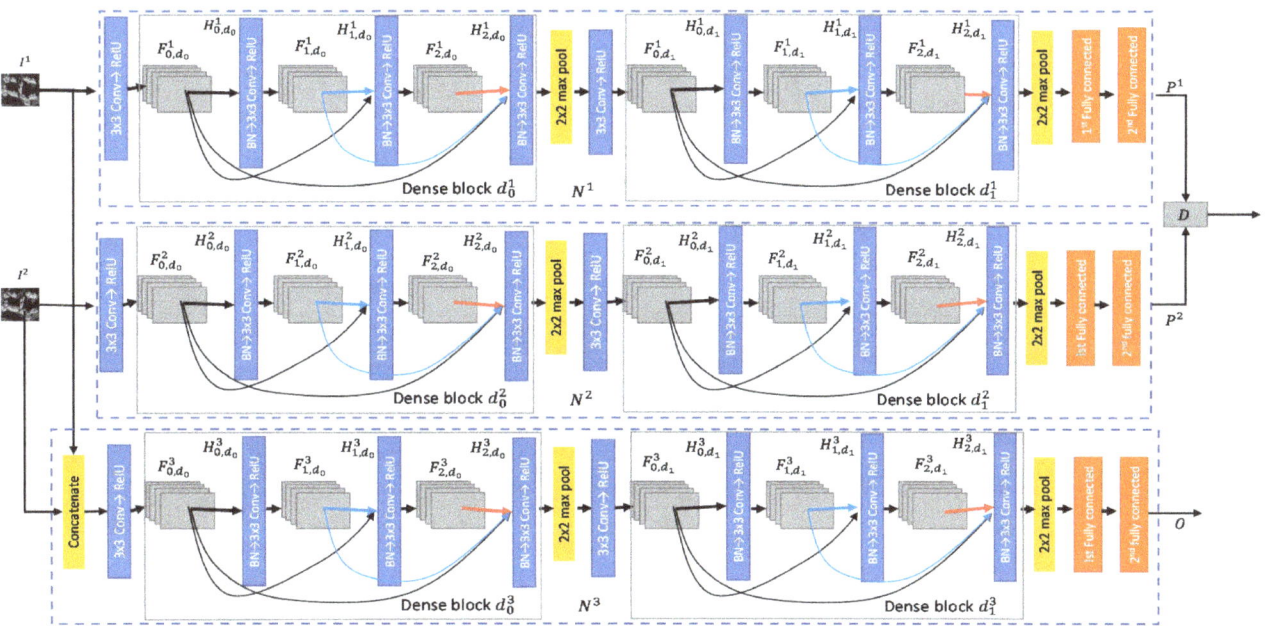

Figure 3. The proposed fusion network architecture for change detection.

3.2. Training of the Proposed Fusion Network for Change Detection

During the training stage, this paper introduced a loss function (L) by combining the contrastive loss (E_c) [36] and weighted binary cross entropy loss (E_B) as defined by:

$$L = \alpha E_c + (1 - \alpha) E_B \tag{3}$$

where α is a weight loss. Given a training set consisting of 40×40 image pairs and a binary label of the ground truth (Y), the proposed network was trained using backpropagation. Here, E_c was applied to optimize the parameters of N^1 and N^2, and is as computed as follows [36]:

$$E_c = \sum_i (1 - y_i) L_S(D_i) + (y_i) L_D(D_i) \tag{4}$$

where $y = 1$ is a changed pixel and $y = 0$ is an unchanged pixel. In addition, Ls is a partial loss function for a pair of similar pixels, and L_D is a partial loss function for a pair of dissimilar pixels, as defined by [36]:

$$Ls = \frac{1}{2}(D_i)^2 \tag{5}$$

$$L_D = \frac{1}{2}(max\{0, m - D_i\})^2 \tag{6}$$

The value of m is set to 1 as the margin value. In addition, E_B was used to optimize the parameters of N^3, as defined by:

$$E_B = -\sum_i W_i(y_i \log(O_i) + (1 - y_i) \log(1 - O_i)) \tag{7}$$

where W is the proposed weighted function used to penalize the false-positive and false-negative errors. Thus, W is computed by:

$$W_i = y_i \left(\beta_c \left(1 - \frac{C}{N} \right) \right) + (1 - y_i) \left(\beta_u \left(1 - \frac{U}{N} \right) \right) \tag{8}$$

where β_c and β_u are penalization weights for false-negative and false-positive errors, respectively. Moreover, C and U are the changed and unchanged numbers of pixels in the full dataset (N), respectively.

The proposed network was trained using a stochastic gradient descent (SGD) with the training parameters, including 0.001, 1×10^{-6}, and 0.9 as the learning rate, decay, and momentum, respectively. In addition, the epoch number was set to 30. The value of α was set to 0.7 to further penalize E_c. It was to prevent false positives, which are possible in a back-end network. The goal of prediction through the front-end was to obtain better true-positive rates regardless of the number of false positives. Thus, the false negatives were penalized ten times more than false positives, namely, $\beta_c = 10$ and $\beta_u = 1$.

3.3. Dual-Prediction Post-Processing for Change Detection

During the inference stage, post-processing was introduced using dual-prediction for change detection. In the counting rule, binary hypotheses can be passed to a fusion center, which then decides which one of the two hypotheses is true [37]. The proposed algorithm employed a hard-logical rule using an AND and OR operation with the same probability output thresholds to predict a changed pixel. This aimed to validate and ensure the change detection based on the proposed fusion network outputs (D and O). There were two steps to applying this post-processing. First, an intersection operation was employed to obtain a strict prediction and avoid false positives. Assume that ($m \times n$) images (T) will be tested using the proposed fusion network, resulting in an ($m \times n$) change map (M_1). This prediction was conducted by sliding in the raster scan order, as shown in Figure 4. The inputs (I^1 and I^2) with the central pixel position, x and y, were assigned to the proposed fusion network to generate the values of D and O. If D and O identified a changed pixel, then $M_1(x, y)$ was set to a value of 1; otherwise, it was set 0. This was performed for the entire image T.

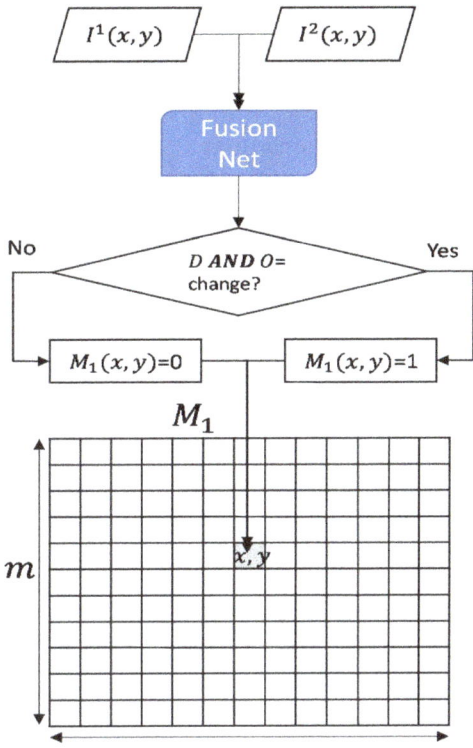

Figure 4. First prediction flowchart.

Then, the second prediction was performed to ensure the first prediction, as shown in Figure 5. Let us assume that $(m \times n)$ M_2 was a change map for the second prediction. Initially, a prediction noise was investigated by analyzing the local information from M_1 by computing Nb, as defined by:

$$Nb(x,y) = \sum_{i=x-\frac{q}{2}}^{x+\frac{q}{2}} \sum_{j=y-\frac{q}{2}}^{y+\frac{q}{2}} M_1(i,j). \tag{9}$$

where $Nb(x,y)$ computes the local information $M_1(x,y)$ using a $q \times q$ window. If the value of $Nb(x,y)$ is greater than the input size s (40) divided by 4, then the second prediction is applied, otherwise, $M_2(x,y)$ is assigned to 0. A union operation was operated from D and O for the second prediction. When it returned the changed pixel, $M_2(x,y)$ was assigned a value of 1, otherwise, it was assigned a value of 0. The final change map was obtained based on the result of M_2.

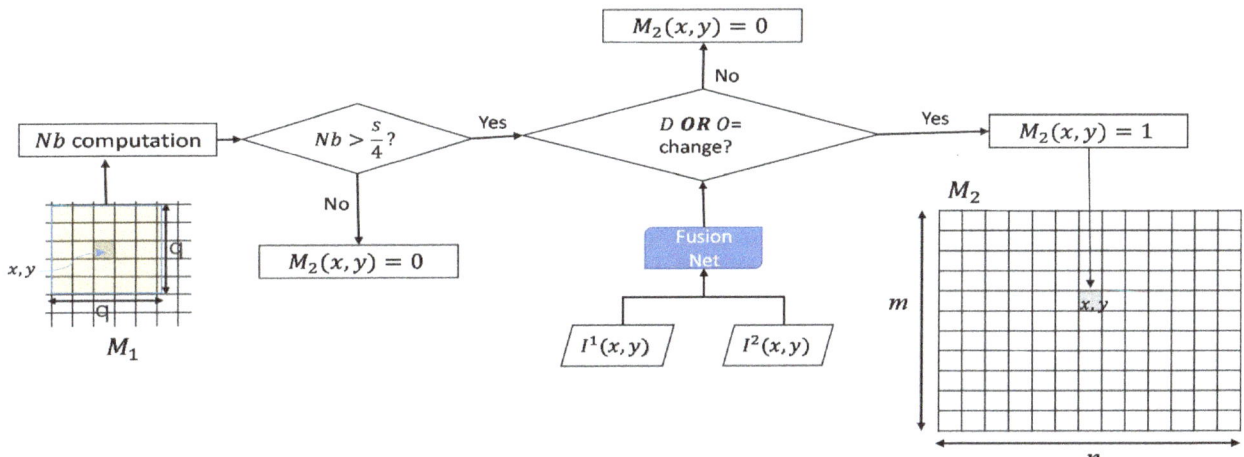

Figure 5. Second prediction flowchart.

4. Experimental Evaluation and Discussion

This study used a dataset of panchromatic imageries, which provided 0.7 GSD captured by the KOMPSAT-3 sensor. For the training dataset, this study used a scene of overlapped images (1214 × 886) over Seoul, South Korea, as shown in Figure 6. These images were cropped into a 40 × 40 sliding patch, and the center pixels of the cropped patch pair were labeled based on the ground truth.

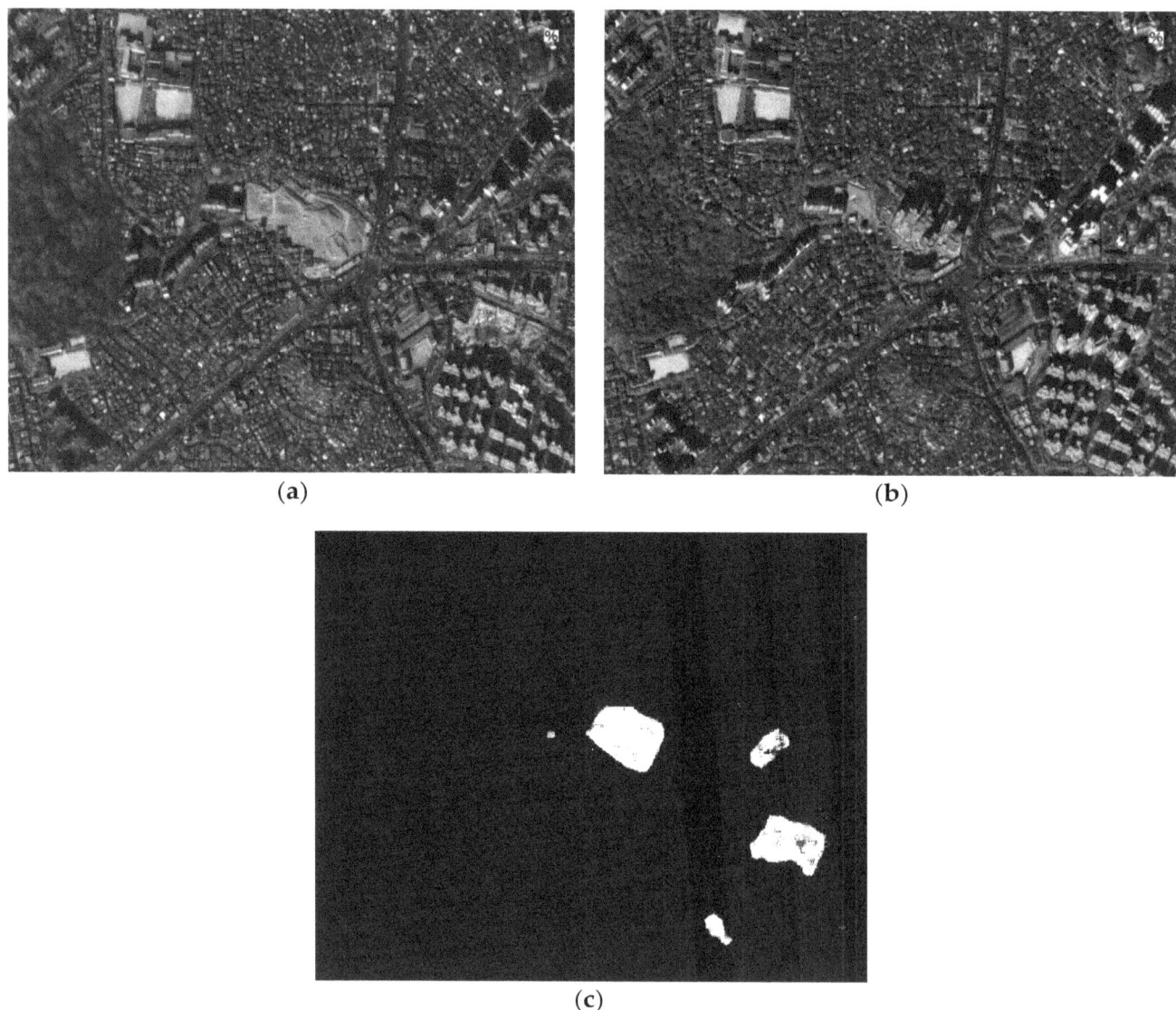

Figure 6. Training dataset: (**a**) Image acquired in March 2014, (**b**) image acquired in December 2015, and (**c**) the ground truth.

Figure 6 shows an area containing completed changes and changes under contraction. In addition, these images have many tall buildings, roads, houses, and forests to be trained for solving the misalignment and viewing angle problems. In our experiments, to assess the effectiveness of the proposed change detection system, three areas of the panchromatic datasets were used, namely, Areas 1, 2, and 3, as shown in Figure 7.

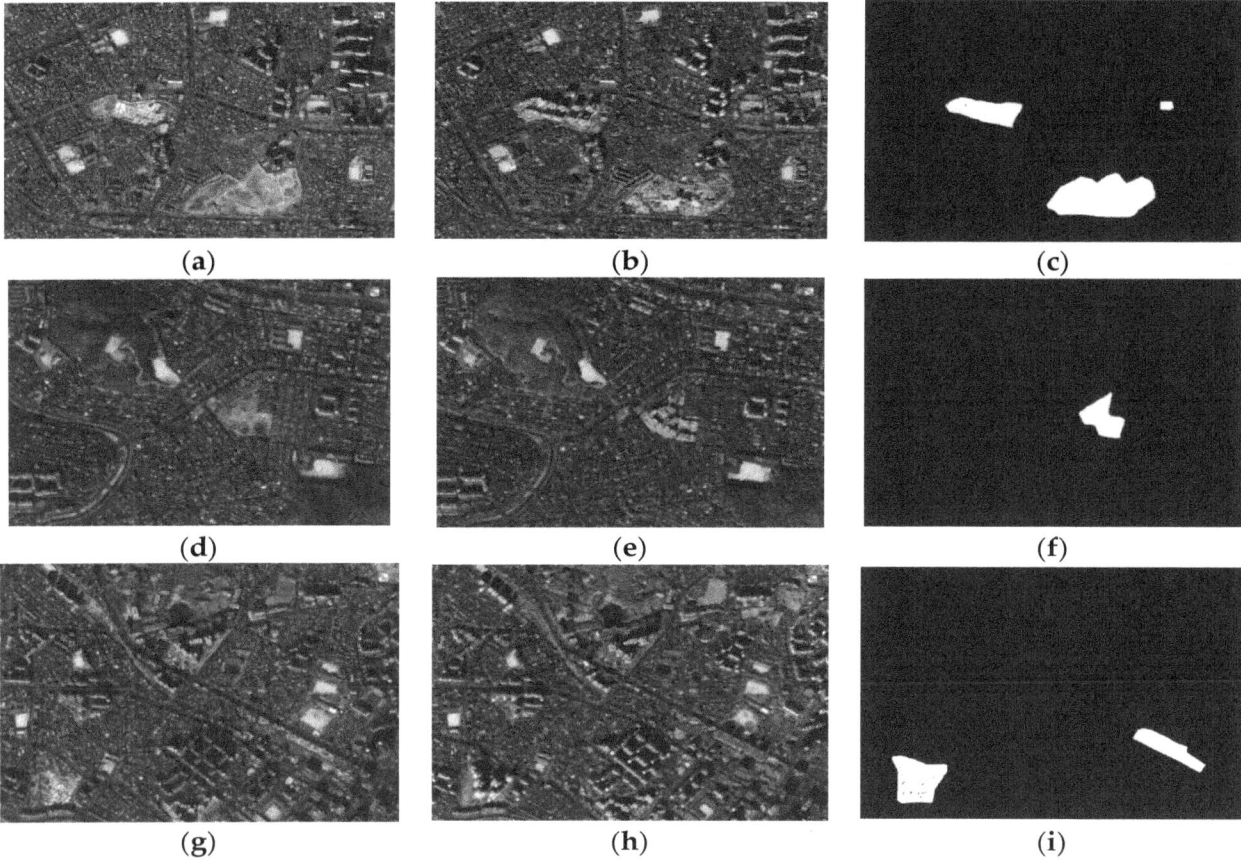

Figure 7. Experiment dataset: (**a**) Input image for Area 1 (March 2014), (**b**), input image for Area 1 (October 2015), (**c**) ground truth for Area 1, (**d**) input image for Area 2 (March 2014), (**e**) input image for Area 2 (October 2015), (**f**) ground truth for Area 2, (**g**) input image for Area 3, (March 2014), (**h**) image input for Area 3 (October 2015), and (**i**) ground truth for Area 3.

The images in Figure 7 were acquired in March 2014 and October 2015 over different areas of Seoul, South Korea. Each image pair had been radiometrically corrected and had a geometric misalignment of approximately ±6 pixels. In addition, it also had a different angle view, which cannot be resolved without precise 3D building models. Area 1 was located in a downtown part of Seoul, and contained areas changed through building construction. Moreover, the urban area had tall buildings and roads. These datasets included several factors of geometric distortion, misalignments, and different viewing angle effects, which could lead to many false changes. In addition, Area 2 represented a downtown area near a forest. These two images were acquired in different seasons. It was difficult to achieve robustness to seasonal changes for practical applications. Area 3 had many tall buildings, making it difficult to achieve an accurate detection rate owing to the different viewing angles.

In this study, the receiver operating characteristic (ROC) curve, AUC, PCC, and Kappa coefficient were used to quantitatively evaluate the performance of the proposed method. Moreover, to evaluate the effectiveness of the proposed method, it was compared with conventional algorithms having FDN and BD-dDCN architectures [28]. A DI and JF were incorporated into a single-path CNN architecture (DI + CNN and JF + CNN). These architectures included eight depth convolutional, two pooling, and two fully connected layers, which were the same as the proposed depth layers. In addition, Dual-DCN [28] was also compared to the proposed method.

Figure 8 shows an ROC curve, which indicates that the proposed method could achieve a better AUC compared to the existing algorithms. For Area 1, the proposed method yielded an AUC of 0.9904, which means that it could identify changes approximating the ground truth. It had a slightly higher dual-DCN of 0.9878. The FDN architectures provided an AUC lower than the proposed algorithm which JF + CNN and DI + CNN gave an AUC of approximately 0.9509 and 0.7060, respectively. Furthermore, the proposed method significantly outperformed the other algorithms with regard to the AUC for Areas 2 and 3 because it could properly detect the change events with the incorporation of low- and high-level differential features. Table 2 summarizes the PCC and Kappa values of the different methods applied for the three areas. The proposed method showed a higher PCC in Areas 1 and 3. The dual-DCN achieved a slightly higher PCC than the proposed method in Area 2. However, in terms of the Kappa value, the proposed fusion network outperformed all other existing algorithms. The proposed method achieved a Kappa value of 75.16 on average, which means that it yielded a good agreement in terms of the results.

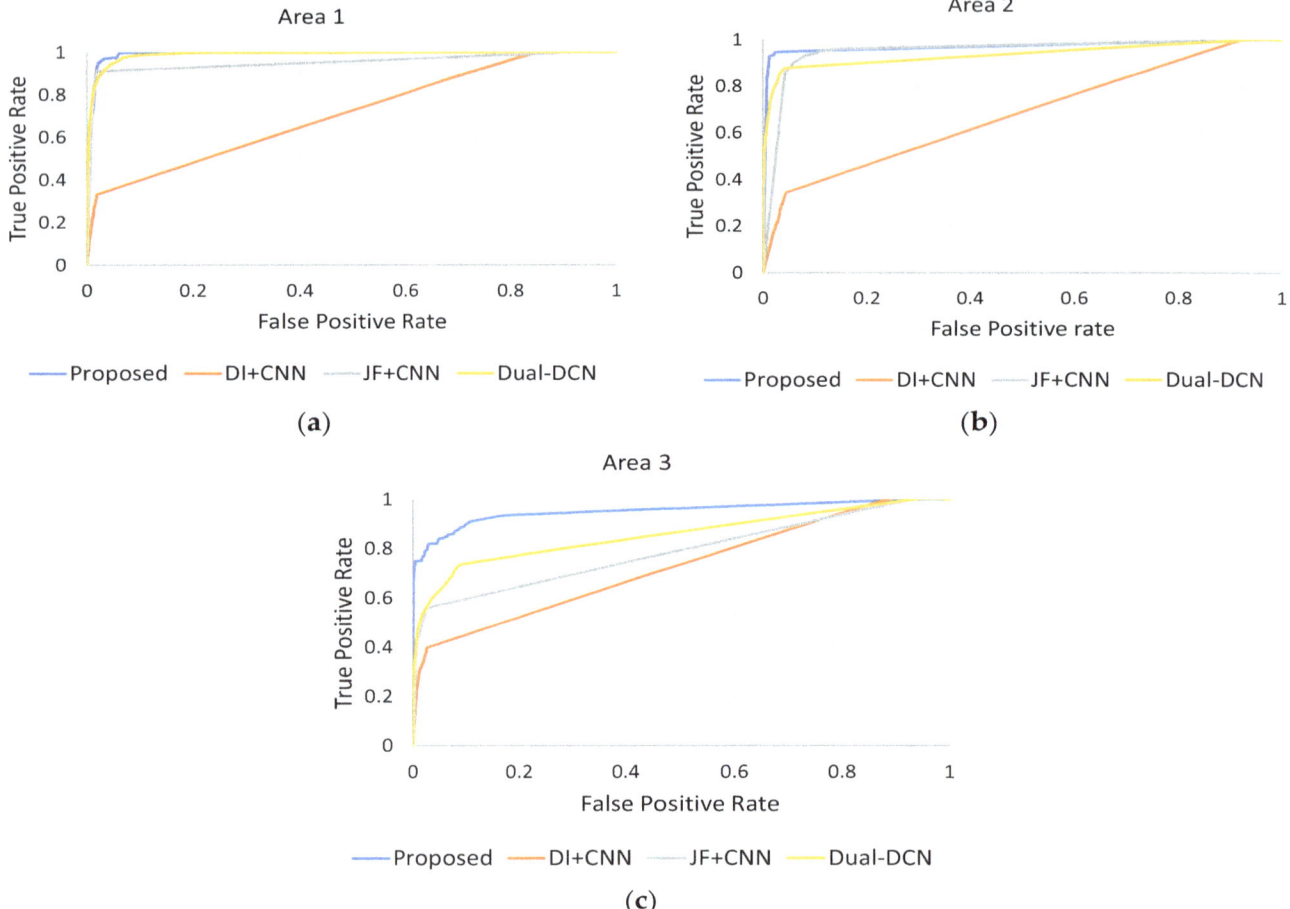

Figure 8. Receiver operating characteristic (ROC) for (**a**) Area 1, (**b**) Area 2, and (**c**) Area 3.

Table 2. Quantitative assessment of the existing and proposed algorithms.

Algorithm	Area 1			Area 2			Area 3		
	AUC	PCC	Kappa	AUC	PCC	Kappa	AUC	PCC	Kappa
DI + CNN	0.7060	0.9458	36.8938	0.6764	0.9571	11.8939	0.7213	0.9855	33.2651
JF + CNN	0.9509	0.9775	79.7190	0.9536	0.9570	29.7251	0.7847	0.9732	47.6066
Dual-DCN	0.9878	0.9774	78.4277	0.9546	0.9922	60.0070	0.8515	0.9751	50.7542
Proposed	0.9904	0.9782	80.7942	0.9707	0.9902	65.9929	0.9517	0.9892	78.6898

Figure 9 shows the change map results when applying the existing and proposed algorithms. Visually, the proposed method achieved a much better result than the existing algorithms. In Area 1, the proposed fusion network nearly approximated the ground truth. It could reduce the number of false positives while preserving the true positives. The proposed network produced a cleaner change map than the existing algorithms regarding false positives. Moreover, the proposed algorithm yielded reasonably good results for Areas 2 and 3. The proposed method significantly reduced the number of false positives and enhanced the true positives. This is caused by the proposed fusion network, which was designed and trained for low- and high-level differential problems. In addition, a post-processing step was employed to validate and repair the change map.

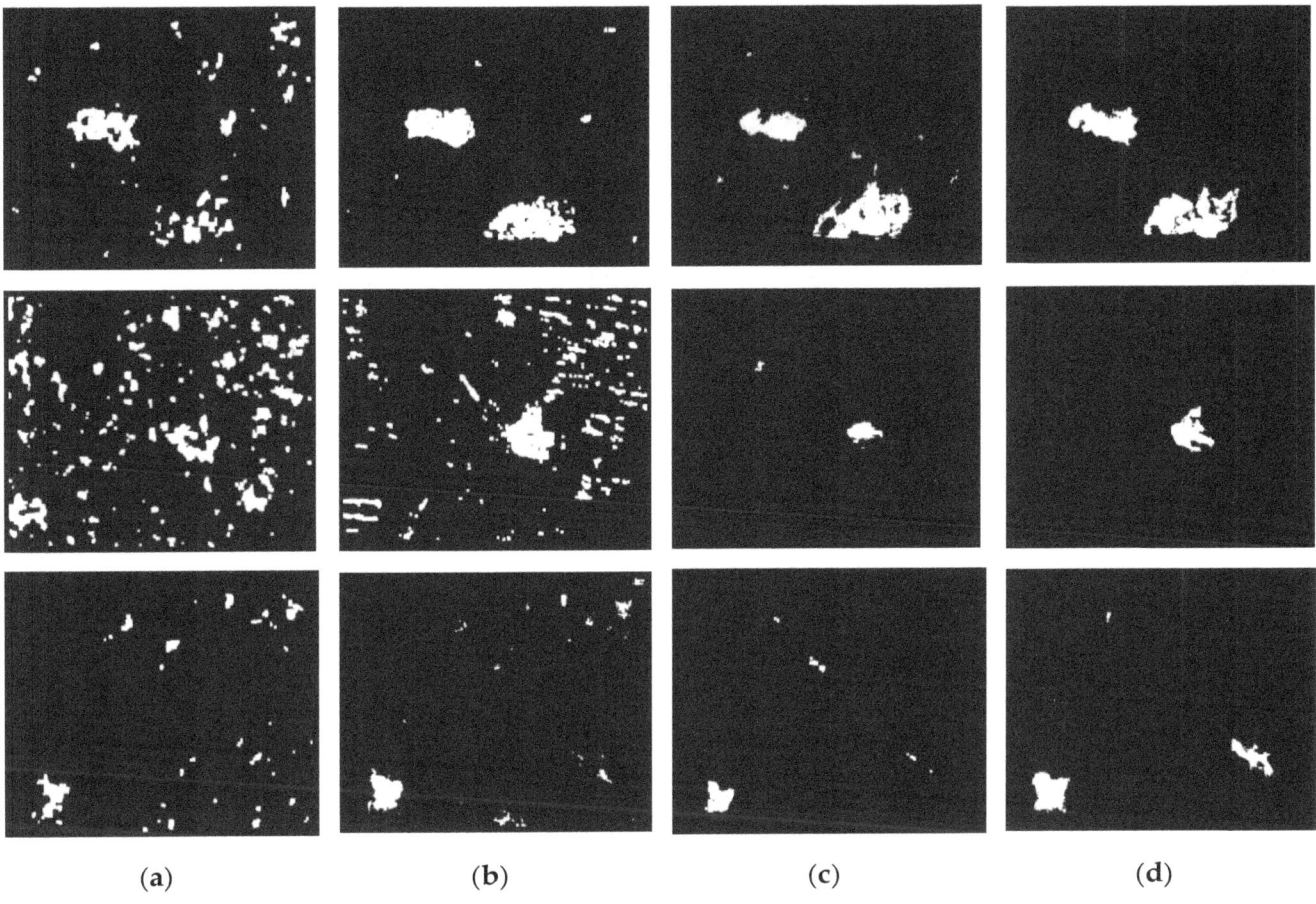

(a)   (b)   (c)   (d)

Figure 9. Detection results for three areas when using the existing and proposed algorithms: (**a**) DI + CNN, (**b**) JF + CNN, (**c**) dual-DCN, and (**d**) the proposed fusion network.

To evaluate the effectiveness of the proposed two-stage decision, the proposed algorithm also was compared to each individual network output (D and O) and the other decision method between two outputs of the proposed fusion network based on the mean operation. In addition, another single output fusion network (SOFN) architecture was designed same as the proposed fusion network architecture by fusing D and O outputs for more comparisons. This network was trained with the binary cross entropy loss function by the same training parameters. The objective and subjective evaluation are presented in Table 3 and Figure 10, respectively.

Table 3. Quantitative assessment of single output decision and proposed algorithms.

Network Outputs	Area 1			Area 2			Area 3		
	AUC	PCC	Kappa	AUC	PCC	Kappa	AUC	PCC	Kappa
D	0.9206	0.9655	65.3497	0.8154	0.9854	33.8273	0.8410	0.9794	55.1115
O	0.9357	0.9607	61.9808	0.8948	0.9879	50.0476	0.8667	0.9436	31.3879
Mean	0.9886	0.9781	78.4481	0.9588	0.9875	52.7712	0.9165	0.9803	59.5712
SOFN	0.9685	0.9661	65.7595	0.8903	0.9897	52.6660	0.8658	0.9820	61.2094
Proposed	0.9904	0.9782	80.7942	0.9707	0.9902	65.9929	0.9517	0.9892	78.6898

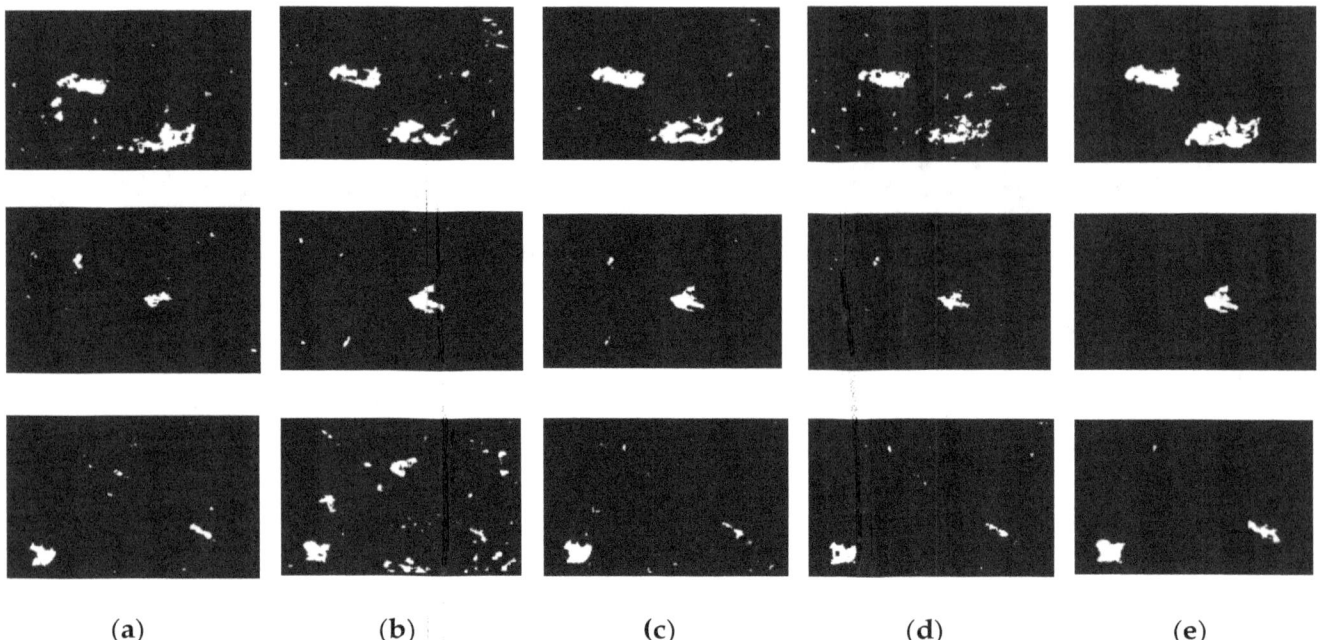

(a)   (b)   (c)   (d)   (e)

Figure 10. Detection results for three areas when using an individual network output and the proposed algorithms: (**a**) D, (**b**) O, (**c**) Mean, (**d**) SOFN, and (**e**) the proposed fusion network with a two-stage decision.

According to Table 3, the proposed two-stage decision shows better performance compared to individual outputs and mean operation. In term of AUC, PCC, and Kappa, the proposed gave significantly better results than that by individual outputs (D and O). Figure 10 shows that the output O produced more true positives regardless of the number of false positives. However, the output D can reduce the false-positive rate. This condition makes the proposed two-stage decision working as the goal that detection rates can be accelerated by the combining of two network outputs with a two-stage decision. In addition, the proposed algorithm still outperformed the mean operation between two network outputs for all areas. SOFN with the single output also gave worse results than the proposed one caused by no validation decision of post-processing for change detection. The proposed fusion network was employed with a two-stage decision to obtain a better prediction rate.

Regarding time complexity, the proposed fusion network consumed more computational complexity than the existing algorithm by a factor of approximately two over the dual-path network and three with the single-path network. It was due to the proposed architecture designed with more network paths. In addition, the proposed two-stage decision required an additional prediction process in the inference stage. Let us see that the general total time complexity of dense convolutional network [35] was $O(K^2)$ run-time complexity for a depth K network [38]. Dual-DCN [28] employed dual-path dense convolutional network with the depth of 6 that produced a run-time complexity of $O(2 \cdot 6^2)$. The proposed fusion network included three-path dense convolutional networks with the same depth by fusing back- and front-end differential network architectures, resulting in a run-time complexity of $O(3 \cdot 6^2)$. In the inference stage, a two-steps decision for the proposed made the run-time be $O(2 \cdot (3 \cdot 6^2))$ that gave it an expensive computational complexity while producing a better result.

5. Conclusions

This paper presented a robust fusion network for detecting changed/unchanged areas in high-resolution panchromatic images. The proposed method learns and identifies the changed/unchanged areas by combining front- and back-end neural network architectures. The dual outputs are efficiently incorporated for low- and high-level differential features with a modified loss function that combines the contrastive and weighted binary cross entropy losses.

In addition, a post-processing step was applied to enhance the sensitivity and specificity from false changes/unchanged detections based on the neighboring information. We found through qualitative and quantitative evaluations that the proposed algorithm can yield a higher sensitivity and specificity compared to the existing algorithms, even under noisy conditions such as geometric distortions and different viewing angles

For further work, the proposed algorithm can be extended for other modalities such as multi-spectrum images, Pan-sharpening, and SAR data. In addition, the proposed algorithm requires expensive time complexity caused by pixel-wise detection with a two-stage decision. To accelerate run-time complexity, block-wise prediction design would also be a focus of future work.

Author Contributions: All authors contributed to the writing of the manuscript. W.W. and D.S. conceived, designed the algorithm, and analyzed the data. W.W. developed it and conducted the experiments and analyzed the data. D.S. supervised the study.

References

1. Ciuonzo, D.; Salvo Rossi, P. DECHADE: Detecting slight changes with hard decisions in wireless sensor networks. *Int. J. Gen. Syst.* **2018**, *47*, 535–548. [CrossRef]

2. Ciuonzo, D.; Carotenuto, V.; de Maio, A. On multiple covariance equality testing with application to SAR change detection. *IEEE Trans. Signal Proc.* **2017**, *65*, 5078–5091. [CrossRef]

3. Wahl, D.E.; Yocky, D.A.; Jakowatz, C.V.; Simonson, K.M. A new maximum-likelihood change estimator for two-pass SAR coherent change detection. *IEEE Transon. Geosand. Remote Sens.* **2016**, *54*, 2460–2469. [CrossRef]

4. Coppin, P.R.; Bauer, M.E. Digital change detection in forest ecosystems with remote sensing imagery. *Remote Sens. Rev.* **1996**, *13*, 207–234. [CrossRef]

5. Bazi, Y.; Bruzzone, L.; Melgani, F. Automatic identification of the number and values of decision thresholds in the log-ratio image for change detection in SAR images. *IEEE Geosci. Remote Sens. Lett.* **2006**, *3*, 349–353. [CrossRef]

6. Singh, K.K.; Mehrotra, A.; Nigam, M.J.; Pal, K. Unsupervised change detection from remote sensing using hybrid genetic FCM. In Proceedings of the IEEE 2013 Students Conference on Engineering and Systems (SCES), Allahabad, India, 12–14 April 2013; pp. 1–5.

7. Bi, C.; Wang, H.; Bao, R. SAR image change detection using regularized dictionary learning and fuzzy clustering. In Proceedings of the 2014 IEEE 3rd International Conference on Cloud Computing and Intelligence Systems (CCIS), Shenzhen, China, 27–29 November 2014; pp. 327–330.

8. Gong, M.; Zhou, Z.; Ma, J. Change detection in synthetic aperture radar images based on image fusion and fuzzy clustering. *IEEE Trans. Image Process.* **2012**, *21*, 2141–2151. [CrossRef] [PubMed]

9. Gong, M.; Su, L.; Jia, M.; Chen, W. Fuzzy clustering with a modified MRF energy function for change detection in synthetic aperture radar images. *IEEE Trans. Fuzzy Syst.* **2014**, *22*, 98–109. [CrossRef]

10. Johnson, R.D.; Kasischke, E.S. Change vector analysis: A technique for the multispectral monitoring of land cover and condition. *Int. J. Remote Sens.* **1998**, *19*, 411–426. [CrossRef]

11. Gao, F.; Zhang, L.; Wang, J.; Mei, J. Change detection in remote sensing images of damage areas with complex terrain using texture information and SVM. In Proceedings of the International Conference on Circuits and Systems (CAS 2015), Paris, France, 9–10 August 2015.

12. Guo, Z.; Du, S. Mining parameter information for building extraction and change detection with very high-resolution imagery and GIS data. *GISci. Remote Sens.* **2017**, *54*, 38–63. [CrossRef]

13. Huang, S.; Ramirez, C.; Kennedy, K.; Mallory, J.; Wang, J.; Chu, C. Updating land cover automatically based on change detection using satellite images: Case study of national forests in Southern California. *GISci. Remote Sens.* **2017**, *54*, 495–514. [CrossRef]

14. Hao, M.; Zhang, H.; Shi, W.; Deng, K. Unsupervised change detection using fuzzy c-means and MRF from remotely sensed images. *Remote Sens. Lett.* **2013**, *4*, 1185–1194. [CrossRef]

15. Hao, M.; Hua, Z.; Li, Z.; Chen, B. Unsupervised change detection using a novel fuzzy c-means clustering simultaneously incorporating local and global information. *Multimed. Tools Appl.* **2017**, *76*, 20081–20098. [CrossRef]

16. Habib, T.; Inglada, J.; Mercier, G.; Chanussot, J. Support vector reduction in SVM algorithm for abrupt change detection in remote sensing. *IEEE Geosci. Remote Sens. Lett.* **2009**, *6*, 606–610. [CrossRef]

17. Volpi, M.; Tuia, D.; Bovolo, F.; Kanevski, M.; Bruzzone, L. Supervised change detection in VHR images using contextual information and support vector machines. *Int. J. Appl. Earth Obs. Geoinf.* **2013**, *20*, 77–85. [CrossRef]

18. Bovolo, F.; Bruzzone, L.; Marconcini, M. A novel approach to unsupervised change detection based on a semisupervised SVM and a similarity measure. *IEEE Trans. Geosci. Remote Sens.* **2008**, *46*, 2070–2082. [CrossRef]

19. Zhao, J.; Gong, M.; Liu, J.; Jiao, L. Deep learning to classify difference image for image change detection. In Proceedings of the IEEE 2014 International Joint Conference on Neural Networks (IJCNN), Beijing, China, 6–11 July 2014; pp. 411–417.

20. Gong, M.; Yang, H.; Zhang, P. Feature learning and change feature classification based on deep learning for ternary change detection in SAR images. *ISPRS J. Photogram. Remote Sens.* **2017**, *129*, 212–225. [CrossRef]

21. El Amin, A.M.; Liu, Q.; Wang, Y. Convolutional neural network features-based change detection in satellite images. In Proceedings of the First International Workshop on Pattern Recognition, Tokyo, Japan, 11–13 May 2016.

22. Liu, J.; Gong, M.; Zhao, J.; Li, H.; Jiao, L. Difference representation learning using stacked restricted Boltzmann machines for change detection in SAR images. *Soft Comput.* **2016**, *20*, 4645–4657. [CrossRef]

23. Zhang, H.; Gong, M.; Zhang, P.; Su, L.; Shi, J. Feature-level change detection using deep representation and feature change analysis for multispectral imagery. *IEEE Geosci. Remote Sens. Lett.* **2016**, *13*, 1666–1670. [CrossRef]

24. Lyu, H.; Lu, H.; Mou, L. Learning a transferable change rule from a recurrent neural network for land cover change detection. *Remote Sens.* **2016**, *8*, 506. [CrossRef]

25. Zhan, Y.; Fu, K.; Yan, M.; Sun, X.; Wang, H.; Qiu, X. Change detection based on deep siamese convolutional network for optical aerial images. *IEEE Geosci. Remote Sens. Lett.* **2017**, *14*, 1845–1849. [CrossRef]

26. Zhang, W.; Lu, X. The spectral-spatial joint learning for change detection in multispectral imagery. *Remote Sens.* **2019**, *11*, 240. [CrossRef]

27. Zhang, Z.; Vosselman, G.; Gerke, M.; Tuia, D.; Yang, M.Y. Change detection between multimodal remote sensing data using Siamese CNN. *arXiv*, 2018; arXiv:1807.09562.

28. Wiratama, W.; Lee, J.; Park, S.E.; Sim, D. Dual-dense convolution network for change detection of high-resolution panchromatic imagery. *Appl. Sci.* **2018**, *8*, 1785. [CrossRef]

29. Yoo, H.-J. Deep convolution neural networks in computer vision. *IEIE Trans. Smart Process. Comput.* **2015**, *4*, 35–43. [CrossRef]

30. Gong, M.; Jiaojiao, Z.; Jia, L.; Qiguang, M.; Jiao, L. Change detection in synthetic aperture radar images based on deep neural networks. *IEEE Trans. Neural Net. Learning Sys.* **2016**, *27*, 125–138. [CrossRef]

31. De, S.; Pirrone, D.; Bovolo, F.; Bruzzone, L.; Bhattacharya, A. A novel change detection framework based on deep learning for the analysis of multi-temporal polarimetric SAR images. In Proceedings of the 2017 IEEE International Geoscience and Remote Sensing Symposium (IGARSS), Fort Worth, TX, USA, 23–28 July 2017; pp. 5193–5196.

32. Wang, Q.; Zhang, X.; Chen, G.; Dai, F.; Gong, Y.; Zhu, K. Change detection based on Faster R-CNN for high-resolution remote sensing images. *Remote Sens. Lett* **2018**, *10*, 923–932. [CrossRef]

33. El Amin, A.M.; Liu, Q.; Wang, Y. Convolutional neural network features based change detection in satellite images. *Intern. Soc. Opt. Photonics.* **2016**, *10011*, 100110.

34. Zhang, P.; Gong, M.; Su, L.; Liu, J.; Li, Z. Change detection based on deep feature representation and mapping transformation for multi-spatial-resolution remote sensing images. *ISPRS J. Photo Remote Sens.* **2016**, *116*, 24–41. [CrossRef]

35. Huang, G.; Liu, Z.; van Der Maaten, L.; Weinberger, K.Q. Densely connected convolutional networks. In Proceedings of the IEEE Conference on Computer Vision and Pattern Recognition, Honolulu, HI, USA, 21–26 July 2017; pp. 4700–4708.

36. Hadsell, R.; Chopra, S.; Le Cun, Y. Dimensionality reduction by learning an invariant mapping. In Proceedings of the 2006 IEEE Computer Society Conference on Computer Vision and Pattern Recognition, New York, NY, USA, 17–22 June 2006.
37. Viswanathan, R.; Aalo, V. On counting rules in distributed detection. *IEEE Trans. Acous. Speech Signal Process.* **1989**, *37*, *772–775*. [CrossRef]
38. Hu, H.; Dey, D.; del Giorno, A.; Hebert, M.; Bagnell, J.A. Log-denseNet: How to sparsify a denseNet. *arXiv*, 2018; arXiv:1711.00002.

Permissions

All chapters in this book were first published in MDPI; hereby published with permission under the Creative Commons Attribution License or equivalent. Every chapter published in this book has been scrutinized by our experts. Their significance has been extensively debated. The topics covered herein carry significant findings which will fuel the growth of the discipline. They may even be implemented as practical applications or may be referred to as a beginning point for another development.

The contributors of this book come from diverse backgrounds, making this book a truly international effort. This book will bring forth new frontiers with its revolutionizing research information and detailed analysis of the nascent developments around the world.

We would like to thank all the contributing authors for lending their expertise to make the book truly unique. They have played a crucial role in the development of this book. Without their invaluable contributions this book wouldn't have been possible. They have made vital efforts to compile up to date information on the varied aspects of this subject to make this book a valuable addition to the collection of many professionals and students.

This book was conceptualized with the vision of imparting up-to-date information and advanced data in this field. To ensure the same, a matchless editorial board was set up. Every individual on the board went through rigorous rounds of assessment to prove their worth. After which they invested a large part of their time researching and compiling the most relevant data for our readers.

The editorial board has been involved in producing this book since its inception. They have spent rigorous hours researching and exploring the diverse topics which have resulted in the successful publishing of this book. They have passed on their knowledge of decades through this book. To expedite this challenging task, the publisher supported the team at every step. A small team of assistant editors was also appointed to further simplify the editing procedure and attain best results for the readers.

Apart from the editorial board, the designing team has also invested a significant amount of their time in understanding the subject and creating the most relevant covers. They scrutinized every image to scout for the most suitable representation of the subject and create an appropriate cover for the book.

The publishing team has been an ardent support to the editorial, designing and production team. Their endless efforts to recruit the best for this project, has resulted in the accomplishment of this book. They are a veteran in the field of academics and their pool of knowledge is as vast as their experience in printing. Their expertise and guidance has proved useful at every step. Their uncompromising quality standards have made this book an exceptional effort. Their encouragement from time to time has been an inspiration for everyone.

The publisher and the editorial board hope that this book will prove to be a valuable piece of knowledge for researchers, students, practitioners and scholars across the globe.

List of Contributors

Nan Wang and Jie Xue
Institute of Agricultural Remote Sensing and Information Technology Application, College of Environmental and Resource Sciences, Zhejiang University, Hangzhou 310058, China

Jie Peng
College of Plant Science, Tarim University, Alar 843300, China

Asim Biswas
School of Environmental Sciences, University of Guelph, Guelph, ON N1G2W1, Canada

Yong He
College of Biosystems Engineering and Food Science, Zhejiang University, Hangzhou 310058, China

Zhou Shi
Institute of Agricultural Remote Sensing and Information Technology Application, College of Environmental and Resource Sciences, Zhejiang University, Hangzhou 310058, China
Key Laboratory of Spectroscopy Sensing, Ministry of Agriculture, Hangzhou 310058, China

Piotr Janiec and Sébastien Gadal
Aix Marseille Univ, Université Côte d'Azur, Avignon Université, CNRS, ESPACE, UMR 7300, Avignon, 13545 Aix-en-Provence CEDEX 04, France
Department of Ecology and Geography, Institute of Natural Sciences, North-Eastern Federal University, 67000 Yakutsk, Russia

Alireza Arabameri
Department of Geomorphology, Tarbiat Modares University, Tehran 36581-17994, Iran

Biswajeet Pradhan
Centre for Advanced Modelling and Geospatial Information Systems (CAMGIS), Faculty of Engineering and IT, University of Technology Sydney, Ultimo, NSW 2007, Australia

Hamid Reza Pourghasemi
Department of Natural Resources and Environmental Engineering, College of Agriculture, Shiraz University, Shiraz 71441-65186, Iran

Khalil Rezaei
Faculty of Earth Sciences, Kharazmi University, Tehran 14911-15719, Iran

Norman Kerle
Department of Earth Systems Analysis (ESA), Faculty of Geo-Information Science and Earth Observation (ITC), University of Twente, 7522 Enschede, The Netherlands

Zenghui Sun and Jichang Han
Key Laboratory of Degraded and Unused Land Consolidation Engineering, Ministry of Land and Resources of China, Xi'an 710075, China
Shaanxi Provincial Land Engineering Construction Group Co. Ltd., Xi'an 710075, China

Wei Chen
Key Laboratory of Degraded and Unused Land Consolidation Engineering, Ministry of Land and Resources of China, Xi'an 710075, China
Shaanxi Provincial Land Engineering Construction Group Co. Ltd., Xi'an 710075, China
College of Geology & Environment, Xi'an University of Science and Technology, Xi'an 710054, Shaanxi, China

Yong Li, Guofeng Tong, Xiance Du, Xiang Yang, Jianjun Zhang and Lin Yang
College of Information Science and Engineering, Northeastern University, Shenyang 110819, China

Xuan Luan Truong
Faculty of Information Technology, Hanoi University of Mining and Geology, No.14 Vien Street, Bac Tu Liem, Hanoi 10000, Vietnam

Muneki Mitamura, Venkatesh Raghavan and Go Yonezawa
Graduate School for Creative Cities, Osaka City University, Osaka 558-8585, Japan

Thi Hang Do
Faculty of Information Technology, Hanoi University of Mining and Geology, No.14 Vien Street, Bac Tu Liem, Hanoi 10000, Vietnam
Graduate School for Creative Cities, Osaka City University, Osaka 558-8585, Japan

Yasuyuki Kono
Center for Southeast Asian Studies, Kyoto University, Kyoto 606-8502, Japan

Xuan Quang Truong
Faculty of Information Technology, Hanoi University of Natural Resources and Environment, No. 14 Phu Dien, Bac Tu Liem, Hanoi 10000, Vietnam

Dieu Tien Bui
Geographic Information System Group, Department of Business and IT, University College of Southeast Norway, Gulbringvegen 36, N-3800 Bø i Telemark, Norway

Saro Lee
Geological Research Division, Korea Institute of Geoscience and Mineral Resources (KIGAM), 124, Gwahak-ro, Yuseong-gu, Daejeon 34132, Korea
Department of Geophysical Exploration, Korea University of Science and Technology, 217 Gajeong-ro Yuseong-gu, Daejeon 305-350, Korea

Eric Ke Wang, Juntao Yu, Zuodong Liang and Xun Zhang
Harbin Institute of Technology, Shenzhen 518055, China

Yueping Li and Zhe Nie
School of Computer Engineering, Shenzhen Polytechnic, Shenzhen 518055, China

Siu Ming Yiu
Department of Computer Science, University of Hong Kong, Pokfulam Road, Hong Kong, China

Bang Liu and Di Niu
Electrical and Computer Engineering, University of Alberta, 9211-116 Street NW, Edmonton, AB T6G 1H9, Canada

Borislav Mavrin and Linglong Kong
Mathematical and Statistical Sciences, University of Alberta, 632 Central Academic Building, Edmonton, AB T6G 2G1, Canada

Geun-Ho Kwak and No-Wook Park
Department of Geoinformatic Engineering, Inha University, Incheon 22212, Korea

Jin Li
National Earth and Marine Observations Branch, Environmental Geoscience Division, Geoscience Australia, Canberra 2601, Australian Capital Territory, Australia

Jongseok Lee
Department of Computer Engineering, Kwangwoon University, Seoul 139701, Korea

Sang-Eun Park
Department of Geoinformation Engineering, Sejong University, Seoul 143747, Korea

Wahyu Wiratama and Donggyu Sim
Department of Computer Engineering, Kwangwoon University, Seoul 139701, Korea

Index

www.ingramcontent.com/pod-product-compliance
Lightning Source LLC
Chambersburg PA
CBHW080411190526
45161CB00003B/198